# Biomarkers of Hip Implant Function

# Biomarkers of Hip Implant Function

Edited by

**Ilona Świątkowska**
Institute of Orthopaedics and Musculoskeletal Science,
University College London, Stanmore, United Kingdom

ELSEVIER

ACADEMIC PRESS
An imprint of Elsevier

Academic Press is an imprint of Elsevier
125 London Wall, London EC2Y 5AS, United Kingdom
525 B Street, Suite 1650, San Diego, CA 92101, United States
50 Hampshire Street, 5th Floor, Cambridge, MA 02139, United States
The Boulevard, Langford Lane, Kidlington, Oxford OX5 1GB, United Kingdom



ISBN: 978-0-12-821596-8

For Information on all Academic Press publications
visit our website at https://www.elsevier.com/books-and-journals

*Publisher:* Stacy Masucci
*Acquisitions Editor:* Elizabeth A. Brown
*Editorial Project Manager:* Pat Gonzalez
*Production Project Manager:* Omer Mukthar
*Cover Designer:* Matthew Limbert

Typeset by MPS Limited, Chennai, India

# Contents

# List of contributors

**Obakanyin J. Akinfosile** Regenerative Medicine and Disability Laboratory, Department of Biomedical Sciences, University of Illinois College of Medicine Rockford, Rockford, IL, United States

**Ravindra V. Badhe** Regenerative Medicine and Disability Laboratory, Department of Biomedical Sciences, University of Illinois College of Medicine Rockford, Rockford, IL, United States

**Mark Barba** Orthoillinois, Department of Surgery, Rockford, IL, United States

**Andrew R. Beadling** Institute of Functional Surfaces (iFS), School of Mechanical Engineering, University of Leeds, Leeds, United Kingdom

**Reshid Berber** Nottingham University Hospital NHS Trust, Nottingham, United Kingdom

**Divya Bijukumar** Blazer Nanomedicine Laboratory, Department of Biomedical Sciences, University of Illinois College of Medicine, Rockford, IL, United States

**Benjamin Bloch** Nottingham University Hospitals, NHS Trust, Nottingham, United Kingdom

**Michael G. Bryant** Institute of Functional Surfaces (iFS), School of Mechanical Engineering, University of Leeds, Leeds, United Kingdom

**Alister J. Hart** Royal National Orthopaedic Hospital, Stanmore, United Kingdom

**Laura-Maria Horga** Institute of Orthopaedics and Musculoskeletal Science, University College London, Stanmore, United Kingdom

**Harry Hothi** Royal National Orthopaedic Hospital, Stanmore, United Kingdom

**Peter James** Nottingham University Hospitals, NHS Trust, Nottingham, United Kingdom

**Marc-Olivier Kiss** Department of Surgery, Hospital Maisonneuve-Rosemont, University of Montreal, Montreal, QC, Canada

**Andrew Manktelow** Nottingham University Hospitals, NHS Trust, Nottingham, United Kingdom

**Vincent Massé** Department of Surgery, Hospital Maisonneuve-Rosemont, University of Montreal, Montreal, QC, Canada

**Mathew T. Mathew** Regenerative Medicine and Disability Laboratory, Department of Biomedical Sciences, University of Illinois College of Medicine Rockford, Rockford, IL, United States

**Anne Neville** Institute of Functional Surfaces (iFS), School of Mechanical Engineering, University of Leeds, Leeds, United Kingdom

**Shiraz A. Sabah** Nuffield Department of Orthopaedics, Rheumatology, and Musculoskeletal Sciences, University of Oxford, Oxford, United Kingdom

**Angela Styhler** Surgery Department, Hôpital Maisonneuve-Rosemont, Montreal University, Montréal, QC, Canada

**Pascal-André Vendittoli** Surgery Department, Hôpital Maisonneuve-Rosemont, Montreal University, Montréal, QC, Canada

**Ilona Świątkowska** Institute of Orthopaedics and Musculoskeletal Science, University College London, Stanmore, United Kingdom

Part I

# Foundational information

Chapter 1

# Introduction to hip implants and biomarker testing

**Ilona Świątkowska[1], Shiraz A. Sabah[2], Laura-Maria Horga[1] and Alister J. Hart[3]**
*[1]Institute of Orthopaedics and Musculoskeletal Science, University College London, Stanmore, United Kingdom, [2]Nuffield Department of Orthopaedics, Rheumatology, and Musculoskeletal Sciences, University of Oxford, Oxford, United Kingdom, [3]Royal National Orthopaedic Hospital, Stanmore, United Kingdom*

## 1.1 Overview of the hip joint

The hip is a large synovial joint formed from a 'socket' on the pelvis (the acetabulum) and the 'ball' from the head of the thighbone (the femur). Its chief function is to allow the transmission of force from the lower extremity to the trunk, which is essential for upright standing and walking. The hip joint allows movement in three degrees of freedom: flexion/extension, abduction/adduction, and internal/external rotation. This is somewhat similar to the glenohumeral joint of the shoulder; however, to provide a stable platform for locomotion, the hip joint is deeper and more constrained than the shoulder joint.

Annually, the hip of an adult is subjected to 1–5 million steps and forces reaching up to eight times the body weight (Sedel and Raould, 2007). The joint is designed to withstand repetitive motions, with features to aid lubrication and minimise wear, which will be discussed in more detail in the following sections. Fig. 1.1 illustrates the basic structure of the hip joint.

### 1.1.1 Anatomy of the hip

The term 'acetabulum' comes from the Latin root *acetum* (vinegar) and the suffix *-abulum* (a small cup)—a nod to an item of Roman tableware. The structure is formed at the point of fusion of three bony components: the ilium, ischium, and pubis. The orientation of the acetabulum is described by the angle the face of the 'cup' makes to the transverse plane (anatomical inclination) and the coronal plane (anatomical anteversion). The normal

**Biomarkers of Hip Implant Function. DOI: https://doi.org/10.1016/B978-0-12-821596-8.00003-3**

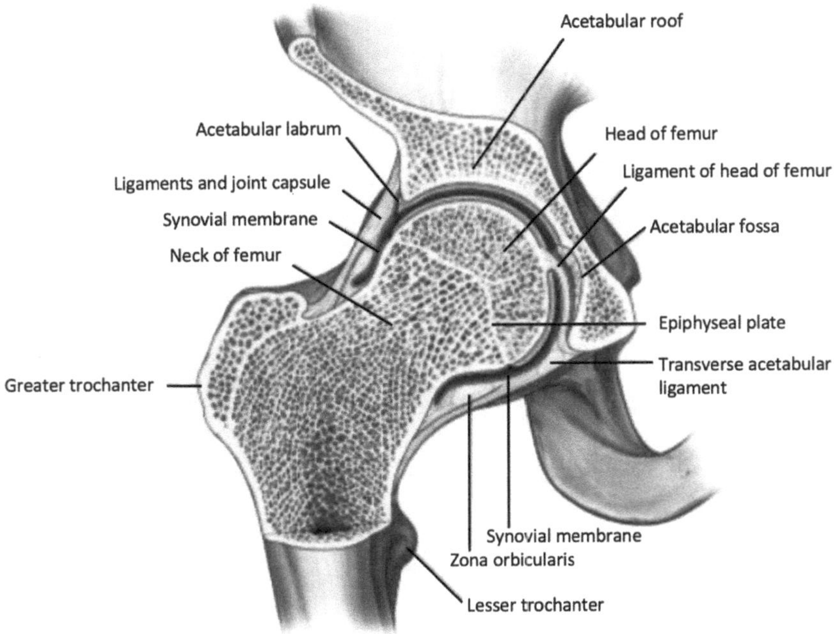


**FIGURE 1.1** Schematic representation of the hip joint. *Reprinted with permission from Gray's Atlas of Anatomy, 1st (ed.) (2007), Churchill Livingstone.*

anatomical inclination of the acetabulum is approximately 62°, with an anteversion of 19° (Merle et al., 2013).

Articulating with the acetabulum is the femoral head. In adults, the average diameter of the femoral head is 45 mm (49 mm in men and 43 mm in women) and the angle between the femoral neck and shaft ranges between 120° and 130° (Nakahara et al., 2011). The two bony prominences at the base of the femoral neck—the greater trochanter and the lesser trochanter—serve as important points for muscle attachment. The intertrochanteric line is a bony ridge connecting the two trochanters, marking the transition from femoral neck to shaft.

### *1.1.1.1 Cartilage*

In healthy individuals, the acetabulum and femoral head are covered with hyaline (meaning 'glass-like') cartilage—a thin layer of elastic connective tissue that cushions the ends of the bones, provides a smooth surface for articulation, and facilitates load transmission. Hyaline cartilage is composed of cells (chondrocytes), collagenous extracellular matrix, and ground substance, and it is devoid of blood vessels, lymphatics, or nerves. The relatively simple design allows it to withstand high compressive loads. However, the lack of a direct nutrient supply means that articular cartilage has limited capacity for self-repair after injury, and degenerative changes can permanently compromise its mechanical function (Fox et al., 2009).

The labrum (Latin for 'lip') is a specialised ring of cartilage that encircles the bony rim of the acetabulum. Its structure is similar to, but tougher than that of articular cartilage, with a higher number of collagen fibres, which form compact, parallel bundles. The purpose of the acetabular labrum is to increase the coverage of the femoral head, seal the joint, and regulate the intraarticular fluid pressure to maintain stability and flexibility.

#### *1.1.1.2 Joint capsule and synovial fluid*

The articulation is surrounded by a joint capsule—a dense layer of fibrous connective tissue that provides protection, stability, and vascular supply to the acetabulum and femoral head. The capsule is the thickest on the anterosuperior side where the most weight-bearing stresses occur and consists of two major types of fibres: circular (*zona orbicularis*) and longitudinal. The internal circular fibres form a collar around the femoral neck while the external longitudinal fibres spiral down the femoral neck and carry blood vessels.

Lining the inner surface of the capsule is a thin sheath of connective tissue known as the synovial membrane. The cells of the synovial membrane secrete a viscous fluid that lubricates the articulating surfaces, delivers nutrition to the surrounding structures, and removes metabolic waste from the joint. In healthy individuals, the biochemical composition of the synovial fluid is similar to that of the plasma. Conditions such as inflammation, infection, and trauma can affect the appearance, composition, and cellular content of the synovial fluid (Table 1.1). In patients with a painful joint effusion or suspicion of infection, a sample of synovial fluid may be collected for laboratory analysis to aid in the diagnosis and guide treatment.

#### *1.1.1.3 Ligaments*

The joint capsule is reinforced by strong ligaments, which are classified as intraarticular (inside the joint capsule) or extraarticular (outside the joint capsule). The only intraarticular ligament is *ligamentum teres*, also known as the ligament of the head of the femur. The structure extends from the acetabular fossa to *fovea capitis* and prevents further displacement following hip dislocation. It also encloses a branch of the obturator artery—a minor source of arterial supply to the femoral head.

The three extracapsular ligaments—the iliofemoral, pubofemoral, and ischiofemoral ligaments—become taut when the joint is extended and have an important role in stabilising the joint and preventing excessive range of movement. The Y-shaped iliofemoral ligament is the strongest ligament in the human body and functions to prevent hyperextension of the joint. The pubofemoral ligament guards against excess abduction and extension,

**TABLE 1.1** Laboratory evaluation and interpretation of synovial fluid characteristics.

| | Healthy | Non-inflammatory (degenerative joint disorder) | Inflammatory | Infectious | Haemorrhagic (trauma) |
|---|---|---|---|---|---|
| Appearance | Colourless to pale yellow; clear | Yellow; clear | Yellow to white; cloudy | Yellow to green; opaque | Red to brown (bloody) |
| Viscosity | High | Low | Low | Variable | Low |
| Total leukocyte count (cells/$mm^3$) | <200 | 200–2000 | 2000–50,000 | >50,000 | 50–10,000 |
| %PMN | <25 | <25 | >50 | >75 | >25 |
| Bacterial culture | Negative | Negative | Negative | Often positive | Negative |
| Protein content (g/dL) | 1–3 | <3 | >3 | >3 | >3 |
| Glucose level | Almost the same as in plasma | Lower than in plasma | Lower than in plasma | Lower than in plasma (up to 50%) | Almost the same as in plasma |

*%PMN*, polymorphonuclear cell percentage.

while the ischiofemoral ligament restricts internal hip rotation to impede excess extension.

#### 1.1.1.4 Muscles

The ligaments of the hip joint are overlaid by several large muscles, which can be broadly divided into flexors, extensors, abductors, adductors, and external and internal rotators.

The main hip flexors are the *iliopsoas*, *sartorius*, and *rectus femoris*. *Pectineus*, the adductors, and *glutei* are also active during flexion, but only via their accessory function. The extension is governed by *gluteus maximus* and the hamstring muscles. The action of gluteal muscles and *tensor fascia latae* allows abduction, while *adductores longus*, *brevis*, and *magnus*, assisted by the *gracilis*, *pectineus*, and *quadratus femoris*, enable adduction. External rotation is possible from the short external rotators (which include *piriformis*, *obturator internus*, and *externus*) and the *gemelli*. Internal rotation implicates *tensor fasciae latae* and *glutei medius* and *minimus*. A thick band of connective tissue known as the iliotibial tract runs along the femur and serves as an attachment site for several of the hip muscles.

#### 1.1.1.5 Blood supply and innervation

Blood is supplied to the hip joint primarily by the medial and lateral circumflex femoral arteries (branches of the deep femoral artery) and artery of *ligamentum teres* (a branch of the obturator artery). Innervation comes from the femoral, obturator, gluteal, and sciatic nerves. Since the same nerves innervate the knee joint, hip pain can be referred to the knee and vice versa (Lam and Amies, 2015).

### 1.1.2 Hip joint pathologies

As with many joints, the hip is susceptible to degenerative, inflammatory, vascular, and traumatic pathologies. Moreover, a variety of disorders of development may affect its anatomical structure and durability. The following paragraphs describe the most common hip joint pathologies in humans.

#### 1.1.2.1 Osteoarthritis

Osteoarthritis is a degenerative joint disease characterised by loss of articular cartilage, formation of bone spurs (osteophytes), periarticular cysts, and sclerotic bone, which manifests clinically by pain, stiffness, and an altered gait. Osteoarthritis is a complex, multifactorial disorder with many risk factors, including advanced age, obesity, joint overuse, and family history. Diagnosis is based on a physical examination and imaging studies. First-line treatments include lifestyle changes, physical therapy, and medication; however, in severe cases, surgical reconstruction of the hip is necessary. Osteoarthritis is the leading reason for hip replacement: out of the 1,191,253 primary hip replacements recorded in the United Kingdom's National Joint Registry (NJR) between 2003 and 2019, 88.4% were performed for osteoarthritis as the sole indication (National Joint Registry, 2020). Fig. 1.2 presents the radiographic features of an osteoarthritic hip.

### *1.1.2.2 Rheumatoid arthritis*

Rheumatoid arthritis is an autoimmune disorder characterised by the production of antibodies that attack healthy joint tissue, leading to inflammation of the joint. The aberrant immune response is triggered by a combination of genetic susceptibility and environmental and lifestyle risk factors, such as cigarette smoking. Disease-modifying antirheumatic drugs (DMARDs) and biologics have revolutionised the treatment of rheumatoid arthritis. However, many patients still develop end-stage joint disease and elect to undergo total hip replacement (THR) to alleviate joint pain and improve function.

### *1.1.2.3 Hip dysplasia*

Adult hip dysplasia is a general term that refers to the abnormal development of the hip joint. In most cases, it is due to developmental dysplasia of the hip (DDH), a type of intrauterine packaging disorder. Risk factors for DDH include female sex, breech presentation, and positive family history. Many countries now have national screening programmes to detect cases at birth,

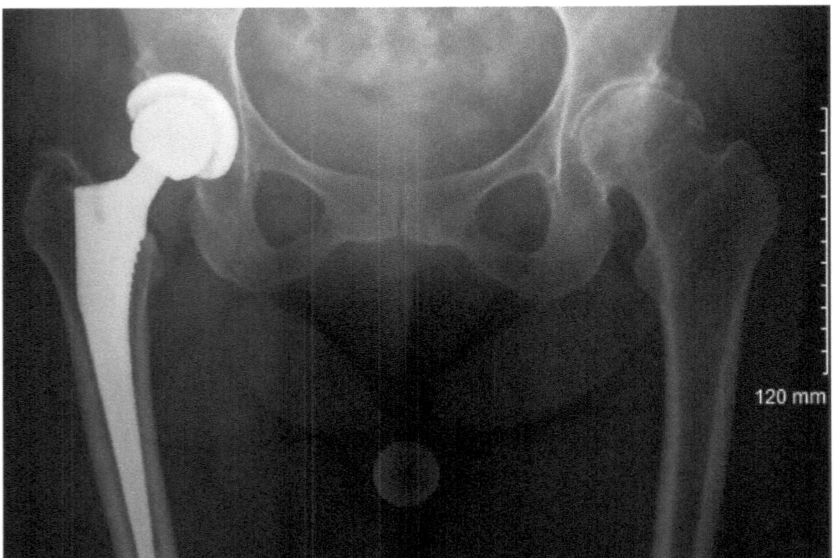


**FIGURE 1.2** An osteoarthritic hip showing loss of joint space, subchondral sclerosis, cysts, and osteophytes. A total hip replacement is present on the opposite side.

with the majority of infants treated using non-surgical modalities, such as the Pavlik harness. The pathology of the condition commonly involves an acetabulum that is too shallow to provide adequate coverage of the femoral head, which may result in accelerated osteoarthritis.

#### *1.1.2.4 Avascular necrosis*

Avascular necrosis is a condition where the blood supply to the femoral head is disrupted, resulting in the death of bone tissue. The underlying nature of the disease is poorly understood, but risk factors such as trauma, high-dose corticosteroid use, alcoholism, pregnancy, and hereditary coagulation disorders (e.g., thrombophilia) have been implicated in its development. Avascular necrosis is often asymptomatic in the early stages, but as the disease progresses, symptoms similar to those of osteoarthritis may develop. A variety of treatments have been used for the early stages of avascular necrosis (such as core decompression and bone grafting), but patients with the end-stage disease may require a hip replacement.

#### *1.1.2.5 Femoroacetabular impingement*

Femoroacetabular impingement occurs when there is a mismatch between the femoral head and the acetabular socket, resulting in abnormal contact between the two structures. The condition may lead to damage to the labrum and articular cartilage. Diagnosis is made through physical examination, assessment of the patient's medical history, and pelvic radiography. Non-operative treatment is usually employed in the first line. Patients without significant degenerative changes may benefit from procedures to reshape the femoral head (femoroplasty) or acetabular cup (acetabuloplasty). However, for end-stage disease, hip replacement may be indicated.

#### *1.1.2.6 Traumatic injuries*

Femoral neck fractures are common in elderly patients and may result from minor trauma if the bone has been weakened by osteoporosis or other pathology (such as cancer). In younger patients, these injuries are comparatively rare but may be the result of high-energy trauma. Femoral fractures can interrupt the blood supply to the femoral head, causing avascular necrosis. The goal of surgical management is to restore function to the joint, which usually involves fracture fixation or replacement of the joint. In the United Kingdom, femoral neck fracture is the second most common indication for primary hip replacement, accounting for 3.4% of primary hip replacements recorded between 2003 and 2019 (National Joint Registry, 2020).

## 1.2 Overview of hip implants

Hip arthroplasty is a surgical reconstruction of the hip joint using artificial material. The term encompasses a wide range of procedures, including THR, in which the femoral head and the acetabulum are replaced; hip resurfacing (HR), whereby the patient's femur is simply reshaped and capped (akin to crowning, rather than removing, a tooth); and hip hemiarthroplasty, in which the femoral head is replaced and the patient's acetabulum is conserved. Hip arthroplasty is a highly successful and cost-effective intervention. Based on data from 24 European joint replacement registries, over 3.1 million primary procedures have been performed in Europe since 1975 (Lübbeke et al., 2018). In the United States, approximately 2.5 million people are living with a hip replacement (Kremers et al., 2015). Estimations based on current trends in the United Kingdom indicate a significant increase in primary hip arthroplasty, with an average cost of up to £7000 per procedure (Culliford et al., 2015; Jenkins et al., 2013). Outside Europe, the United States report a 50% expansion of indications for primary hip replacement in the young population, while the Australian healthcare system expects a rise of 208% between 2013 and 2030, and an overall associated cost exceeding $AUD 5.32 billion in 2030 (Ackerman et al., 2019; Kurtz et al., 2009).

In general, hip implants consist of two basic parts: the ball (femoral head) and socket (acetabular cup). In THR, a long metal stem is inserted into the femur, which connects with the femoral head via a tapered neck. In some designs, the head and stem components form one piece while others are modular, featuring interchangeable heads and femoral necks.

### 1.2.1 Implant biomaterials

The components of a hip implant incorporate several different materials. Femoral heads are typically made of metal or ceramic, acetabular cups are made of metal, polyethylene, or ceramic, and all modern femoral stems and necks are metallic.

#### *1.2.1.1 Metals*

Cobalt-based, iron-based, and titanium-based alloys are all popular in hip implant manufacture because they are hard, tough, resistant to fracture and corrosion, and well tolerated by the body. The most important medical alloys are defined by the American Society for Testing Materials (ASTM) or International Organisation for Standardization (ISO).

Cobalt-based alloys contain 26%–30% chromium, which facilitates the formation of a thin, protective layer of chromium oxide on the surface of the metal that helps to prevent corrosion. The addition of nickel at 3%–5% and molybdenum at 4%–6% serves to increase plasticity and improve the material's technological properties (Matusiewicz, 2014). The most widely used cobalt-based alloy is cobalt-28 chromium-6 molybdenum (CoCrMo), that is, ASTM F75.

Surgical and orthopaedic stainless steel (ASTM F55 and ASTM F138, respectively) are iron-based alloys, with 17%–20% chromium, 13%–15% nickel, and 2%–3% molybdenum. Stainless steel is strong and relatively cheap to produce, but its wear characteristics and corrosion resistance are inferior to those of other orthopaedic alloys. The marked allergenic potential, stemming from the relatively high nickel content, has further restricted its use in hip implant manufacture (Matusiewicz, 2014).

Titanium is softer and less stiff than cobalt and stainless steel, which allows it to more closely mimic the Young's modulus of bone. Commercially pure titanium (ASTM F67) has found application in porous coatings that promote bone growth onto the surface of implants while titanium alloyed with 4% vanadium and 6% aluminium (Ti-6Al-4V; ASTM F136) is well suited to use in non-moving parts of implants, whose fixation relies on bone integration, for example, femoral stems, acetabular cups, and screws. Concerns over the toxic potential of vanadium and aluminium ion release from Ti-6Al-4V have led to the development of alternative versions of the alloy, such as titanium-6 aluminium-7 niobium (Ti-6Al-7Nb; ASTM F1295), titanium-13 niobium-13 zirconium (Ti-13Nb-13Zr; ASTM F1713), and titanium-12 molybdenum-6 zirconium-2 iron (TMZF; ASTM F1813). The latter alloy was once a popular femoral stem material but is no longer used commercially due to high reported failure rates when combined with CoCrMo alloy femoral heads (Morlock et al., 2018).

Despite excellent mechanical properties, metals are susceptible to wear and corrosion once implanted. Excessive release of metal particles and ions can lead to adverse local tissue reactions (ALTR) in the periprosthetic area, such as damage to the soft tissue or formation of cystic or solid masses in the vicinity of the implant.

#### *1.2.1.2 Plastic polymers*

Ultra-high-molecular-weight polyethylene (UHMWPE) is a semicrystalline polymer with a long history of use in orthopaedic applications, most notably in acetabular liners for THR implants. The material has a low coefficient of friction, is biocompatible, and inexpensive to produce. However, when in contact with harder surfaces, UHMWPE releases micrometre-sized particles, which can lead to bone resorption around the implant (periprosthetic osteolysis), aseptic loosening (loss of implant fixation in the absence of infection), and early mechanical failure. To reduce the prevalence of these adverse effects, much effort has been made to increase the degree of crosslinking within the UHMWPE. First-generation highly crosslinked UHMWPE (HXLPE) liners, clinically introduced in the 1990s, were gamma irradiated and then thermally processed (annealed or remelted) to improve their resistance to free radicals created during irradiation. Neither process yielded perfect results: annealing failed to eliminate all the free radicals, while remelting resulted in a material with undetectable free radicals but reduced crystallinity and increased susceptibility to fatigue cracking (Kurtz et al., 2011). To try to address these drawbacks, the next generation of HXLPE liners aimed to achieve oxidative resistance while maintaining the high wear resistance of the first-generation material and the mechanical strength of conventional polyethylene; the two approaches used were sequential irradiation and annealing, and vitamin E doping (vitamin E acts as a free-radical scavenger) (D'Antonio et al., 2012; Oral and Muratoglu, 2011). Despite initial concerns, first-generation HXLPE demonstrates excellent radiographic results and longevity, even in young and active patients (Lim et al., 2019). Second-generation HXLPE delivered promising short- to mid-term results, but long-term follow-up will be required to ascertain whether these designs have a clinical advantage over first-generation liners (Langlois and Hamadouche, 2020).

#### *1.2.1.3 Ceramics*

Owing to increased hardness, lower wear rate, better wettability, and higher biological inertness compared with metal alloys, ceramics are the new preferred bearing surface, particularly in younger patients. Ceramic materials with an established place in hip implant manufacture are alumina ($Al_2O_3$), zirconia (95% $ZrO_2$, 5% $Y_2O_3$), and zirconia-toughened alumina (ZTA). The latter is a composite consisting of an alumina matrix reinforced with 17% zirconia and trace amounts of strontium aluminate and chromium oxide. Despite excellent wear properties, ceramic is a brittle material that can chip or shatter following errors in surgical technique (i.e., if the acetabular liner is malseated). Ceramic fragments are abrasive and difficult to completely remove from the joint, which means that they will accelerate the wear of the new metal-bearing surface, leading to high release of toxic metal ions and subsequent complications. Older ceramics were prone to fracture due to large grain size and presence of impurities, but advances in ceramic manufacturing technology have helped to increase the strength and toughness of the material, so fracture is very rare with modern ceramics. Alternative bearing materials that are currently investigated include silicon nitride coatings for CoCrMo femoral heads and ceramicised metal alloys, such as Oxinium (zirconium metal alloy whose outer surface had been transformed into a layer of zirconia). These new materials promise improved fracture resistance while preserving the ultra-low wear rates and biocompatibility, but data on long-term outcomes are as yet limited (Carli et al., 2020; Piconi et al., 2017).

### 1.2.2 Bone fixation

The stem and cup components of a hip implant can be fixed to the bone with or without bone cement. Cemented designs use fast-drying polymethylmethacrylate (PMMA) to 'grout' the implant and bone together, while cementless hips rely on bone growth onto a roughly textured or porous surface of the prosthesis—a process referred to as osseointegration. In cementless THR, the implant is pushed into a slightly undersized bone cavity to encourage pressure bonding, that is, 'press-fit', which can be reinforced with metal screws for the acetabular component. While the cementless approach may offer a better long-term bond between the implant and bone, its success is dependent on the quality of the bone. For this reason, patients with osteoporosis or osteopenia may not be suitable candidates for press-fit fixation.

THR implant constructs in which the stem is cemented and the acetabulum is uncemented are referred to as 'hybrid' replacements, while those in which the situation is reversed are called 'reverse hybrid' replacements.

### 1.2.3 Implant classification

Hip arthroplasties are classified in terms of the combination of the femoral head and acetabular cup materials; by convention, the bearing material of the femoral head is listed before that of the acetabular socket. The five possible combinations of bearing surfaces are metal-on-polyethylene (MoP), ceramic-on-polyethylene (CoP), ceramic-on-ceramic (CoC), ceramic-on-metal (CoM), and metal-on-metal (MoM). For MoM hips, the femoral component can be either a resurfacing or modular with a large ($\geq$36 mm diameter) or small (<36 mm diameter) head attached to a traditional femoral stem. Typical hip implant designs are depicted in Fig. 1.3.

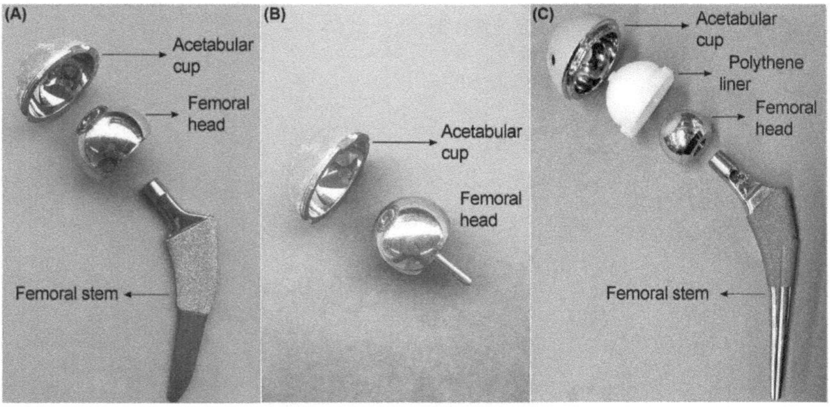


**FIGURE 1.3** Typical hip implant designs. (A) Large-head metal-on-metal total hip replacement. (B) Metal-on-metal hip resurfacing. (C) Metal-on-polyethylene total hip replacement. *Images by Eltit, Wang, and Wang (Eltit et al., 2019: Eltit, F., Wang, Q., Wang, R., 2019. Mechanisms of adverse local tissue reactions to hip implants. Front. Bioeng. Biotechnol. 7, 1–17/CC-BY-4.0).*

Despite vast advancements in prosthetic design, no bearing combination is ideal as each of the commonly used biomaterials has drawbacks that limit its longevity in vivo (Table 1.2). The choice of bearing surface is influenced by factors such as implant cost, age and activity level of the patient, and complications during surgery.

**TABLE 1.2 Advantages and limitations of the main bearing couples.**

| Bearing type | Material combination | Advantages | Limitations |
|---|---|---|---|
| Hard-on-soft | MoP (CoCr on HXLPE) | Longest track record<br>Lowest cost | Polyethylene debris linked to osteolysis-mediated aseptic loosening |
| | CoP (ZTA on HXLPE) | Lower wear rate than MoP | Polyethylene debris linked to osteolysis-mediated aseptic loosening<br>More expensive than MoP |
| Hard-on-hard | MoM (CoCr on CoCr) | Lower volumetric wear rate and a potentially longer lifespan than MoP<br>Use of larger head sizes possible, leading to a lower risk of dislocation than MoP | Metal debris linked to systemic (rare) and local toxicity symptoms<br>The highest risk of hypersensitivity (allergic) reactions |
| | CoC (ZTA on ZTA) | The best wear properties and longest lifespan<br>Most biologically inert | Can produce squeaking sounds on movement<br>Most expensive<br>Potential for liner fracture if malseated |
| | CoM[a] (ZTA on CoCr) | Lower wear rate than MoM<br>Lower risk of squeaking than CoC<br>Allows use of larger ceramic heads to reduce the risk of ceramic fracture | Elevated blood metal ion levels<br>Limited clinical data to support favourable in vitro results |

*CoC/M/P*, ceramic-on-ceramic/metal/polyethylene; *CoCr*, cobalt-chromium alloy; *HXPLE*, highly crosslinked polyethylene; *MoM/P*, metal-on-metal/polyethylene; *ZTA*, zirconia-toughened alumina.
[a]*CoM bearings are no longer used clinically.*
Source: Adapted from Swiatkowska, I., 2019. Toxicity of Metal Debris From Hip Implants (Ph.D. thesis). University College London, London, UK/CC-BY-4.0.

### 1.2.4 Evolution of a total hip replacement

The development of a THR began in the 1930s with Phillip Wiles' MoM implant, in which the articulating components were both made of stainless steel. A similar design manufactured of cast CoCrMo alloy (Vitallium) was popularised in the 1950s and implanted in high numbers by surgeons such as George McKee and John Watson-Farrar. First-generation MoM devices, such as the McKee–Farrar prosthesis, achieved good long-term survival, but the high rate of wear and aseptic loosening prompted their temporary abandonment in favour of the low-friction MoP arthroplasty pioneered by Sir John Charnley (Charnley, 1961). Charnley's joint, comprising a stainless steel stem, small-diameter (22 mm) femoral head, and acetabular liner made of UHMWPE, was remarkably successful. Reports of excellent patient function after as long as 25 years encouraged their increasing use in younger and more active patients (Caton and Prudhon, 2011). These designs came with their problems, however, and during the 1990s, rising revision rates were reported. It became clear that wear of the UHMWPE bearing produced micrometre-sized particles that stimulated periprosthetic osteolysis and led to aseptic loosening.

In an attempt to reduce wear rates and the extent of adverse reactions to UHMWPE debris, metal-bearing surfaces returned in the late 1990s with the introduction of second-generation MoM implants. Osteolysis was rarely seen with these bearing surfaces, as the nanometre-sized metal debris was too small to stimulate a macrophage response. However, raised circulating blood metal ion levels and ALTR were noted, and in the very rare cases where ion levels were extremely elevated, systemic toxicity symptoms were also observed. Because of these concerns and reports of high failure rates, the use of MoM-bearing surfaces has declined and currently constitutes less than 1% of all hip replacement procedures.

The use of ceramic implants has been increasing since their introduction in the 1970s (Boutin, 1972). The tissue response to ceramic wear debris was found to be much less damaging than that to metal or UHMWPE particles; however, the brittleness of the material resulted in high fracture rates in the early CoC designs. Susceptibility to fracture was overcome through research into more fracture-resistant ceramics, such as ZTA (Massin et al., 2014). CoP bearings, which combine the advantages of the smooth, hard ceramic surface with the softer, less rigid polyethylene surface, have steadily increased in popularity to become one of the most popular bearing surface types.

Nowadays, MoP and CoP implants make up the majority of THR undertaken in the United Kingdom. MoM implants continue to fail at higher than expected rates and their use is now rare (Fig. 1.4).

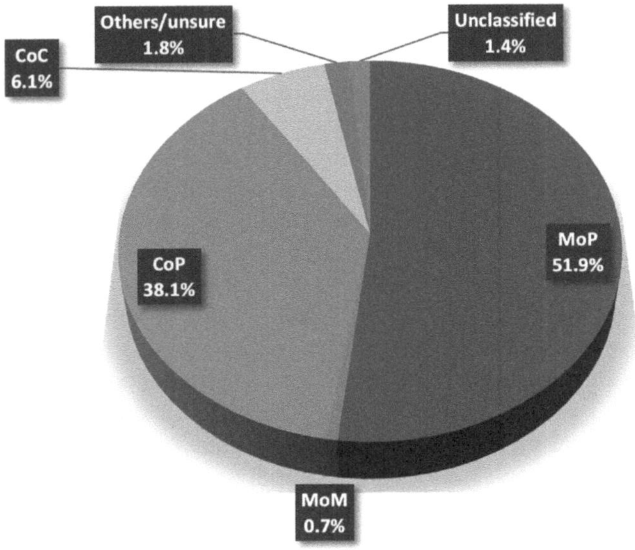


**FIGURE 1.4** Percentage of primary hip replacements in 2019 by bearing type. *CoC/P*, ceramic-on-ceramic/polyethylene; *MoM/P*, metal-on-metal/polyethylene. *Data taken from the National Joint Registry for England, Wales, Northern Ireland, and Isle of Man: 17th Annual Report, 2020.*

### *1.2.4.1 Implant modularity*

Early THR implants were of a one-piece (monobloc) design, which offered a reduced risk of breakage but permitted little intraoperative freedom. In the 1960s, the first modular implants became available, where the femoral head was attached to the stem by a Morse-type taper (head/stem modularity) (Krishnan et al., 2013). The introduction of modularity into the traditional design of a THR allowed surgical practices to tailor the operation to individual patients without large increases in the component inventory needing to be carried. Modularity can also be advantageous in cases where further surgery is required for the hip joint, as some components can be retained while others are exchanged.

More recently, additional forms of implant modularity, namely dual-modular femoral necks, have been introduced. Neck/stem modularity allows for intraoperative customisation of leg length, femoral offset, and version, to better reconstruct the natural biomechanics of the hip and enhance implant function (Jacobs et al., 2014). Despite the perceived benefits, the neck/stem junction is particularly vulnerable to metal ion release, often leading to ALTR, component dissociation, and mechanical failure of the implant (Banerjee et al., 2015; De Martino et al., 2015; Ko et al., 2016). The condition, referred to as 'trunnionosis', is estimated to account for approximately 2% of the revision THR burden. As a result of these complications, several authors have recommended against using femoral stems with exchangeable necks (Gill et al., 2012; Pivec et al., 2014).

Dual-mobility hip replacements include an additional bearing surface (a mobile polyethylene layer) between the femoral head and the acetabular shell. These designs combine Charnley's low-friction principle with a large head-to-neck ratio to maximise stability, and are mainly indicated to reduce the incidence of dislocation of the joint replacement. Their disadvantages include increased component wear at the additional interface and the introduction of the possibility of intraprosthetic dislocation (Cuthbert et al., 2019).

### 1.2.5 The rise and fall of hip resurfacing

First attempts at HR include the Smith–Peterson cup arthroplasty in the 1930s and Charnley's polytetrafluoroethylene (PTFE) shells in the 1950s. Although these implants produced good short-term results, the high wear rate ultimately led to severe osteolysis and implant failure. Resurfacing re-emerged in the 1990s in the form of the McMinn prosthesis (McMinn et al., 1996)—later called the Birmingham Hip Resurfacing (BHR). The reason for the rise in the use of MoM HR was to satisfy a desire to preserve bone stock in the femur, achieve near-human replication of bony anatomy, eradicate polyethylene from the bearing surface, and allow the use of the largest possible diameter bearings to reduce the incidence of joint dislocation (Hart et al., 2015). The BHR had highly promising initial success rates in the hands of the designing surgeon (Daniel et al., 2004), prompting several manufacturers to release competing devices, such as DePuy's ASR and Stryker/Corin Cormet system. These implants were marketed as long-lasting and better suited for young and active patients, with the advantage of an easy conversion to a traditional THR should the device require revision, but most were subsequently withdrawn due to design differences (Ball et al., 2007). The use of MoM HR peaked in the late 2000s before issues surrounding the effects of metal wear debris became apparent, leading to a shift toward conventional stemmed MoP THR. In 2019, HR accounted for only 0.6% of all primary hip replacement procedures performed in the United Kingdom (National Joint Registry, 2020).

Despite a considerable fall in popularity, HR devices continue to be implanted in limited numbers, mostly in active young male patients with good femoral bone, for whom they perform well. There are several patient groups, including women, for whom metal resurfacing is contraindicated despite good bone stock. The use of ceramic materials in resurfacing is intended to increase the population of patients who can receive this device (de Villiers et al., 2020).

### 1.2.6 Implant degradation

Cobalt-, iron-, and titanium-based alloys are considered biocompatible thanks to the spontaneous formation of an oxide-rich passivation layer on their surface. This protective film forms a boundary at the interface between the biological medium and the implant, limiting corrosion and increasing the alloy's mechanical stability. Once implanted, friction induces selective fractures of the surface layer, which can lead to material release from where the protective film had been compromised. Implant degradation products are complex and include metal and polyethylene wear particles, metal—protein complexes, inorganic salts, and free metal ions. These can remain in the joint space and cause local toxicity, or enter the blood or lymph and affect distant organs and systems (Fig. 1.5). In cases of high-wearing prostheses, the soft tissue surrounding the implant can take on a dark-grey discoloration, which is commonly referred to as 'metallosis' (Choi et al., 2018; Pisanu et al., 2018).

Implant debris can be generated at the articulation or any other implant interface, that is, head/neck, neck/stem, stem/cement, bone/implant, or bone/cement junction. The amount and physicochemical properties of metal debris released at the different interfaces are dependent on factors such as implant

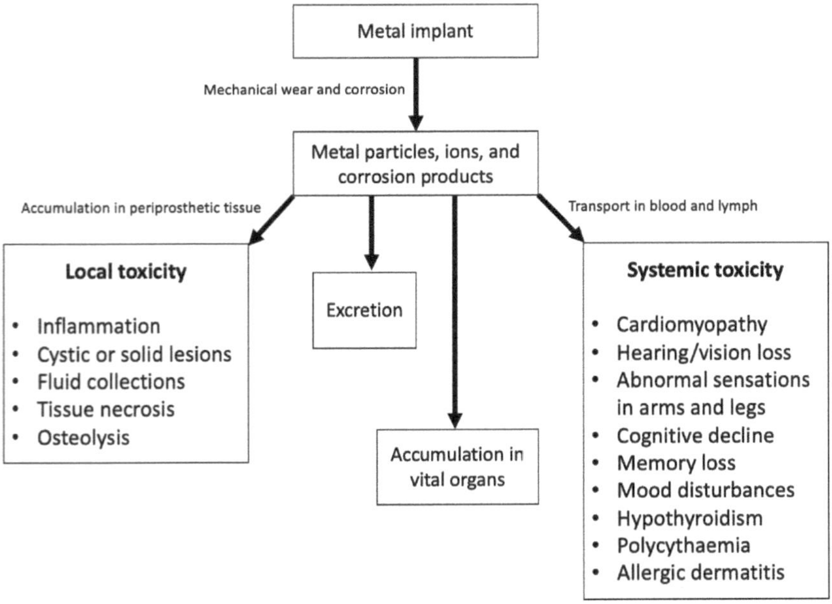


**FIGURE 1.5** Summary of the proposed local and systemic adverse effects of hip implant-derived metal debris. Systemic toxicity is chiefly associated with cobalt, except for allergic reactions, which are primarily due to nickel. *Adapted from Swiatkowska, I., 2019. Toxicity of Metal Debris From Hip Implants (Ph.D thesis). University College London, London, UK/CC-BY-4.0.*

type, implant positioning, the exact degradation mechanisms at play, and host characteristics (Bijukumar et al., 2018; Cobelli et al., 2011). Particles generated through different wear processes can display different elemental compositions, crystal structures, sizes, and shapes, which may result in differing bioactivity (Hallab and Jacobs, 2009; Xia et al., 2017). Particulate wear debris generated by MoM implants has an average particle size range of 30–100 nm while polyethylene wear particles are generally 500 nm in size (Catelas and Wimmer, 2011; Jacobs et al., 2008). In vitro studies have indicated that nanoparticles (≤100 nm) are often more harmful than coarser particles of the same composition because they enter cells more easily and have a larger surface area for ion release (Shi et al., 2013). The particle shape is also important: elongated particles (fibres) can elicit a more potent inflammatory reaction than round particles.

The principles governing the degradation of metal hip implants and toxicity of implant-derived metal debris are discussed in detail in Chapters 2 and 3, respectively.

### 1.2.7 Implant performance

Once a hip replacement has been implanted there is a ‘race to failure’ between the component and its host. This is due to processes such as implant degradation, adverse reactions to metal debris, osteolysis, periprosthetic infection, fracture, and dislocation. A recent systematic review and meta-analysis of case series and national implant registry reports estimated that 75% of THRs will last 15–20 years and 58% will last 25 years in patients with osteoarthritis (Evans et al., 2019). The failure of a hip replacement is influenced by a multitude of surgeon-, implant-, and patient-specific factors, many of which are interrelated. For example, the wear rate of a component may be influenced by the orientation in which it was implanted, the materials making up the bearing surfaces, and the activity level of the patient.

#### *1.2.7.1 Surgeon factors*

Surgeon factors that may impact clinical outcomes include the surgical approach to the joint, surgeon experience and caseload, and component positioning. Accurate positioning of the acetabular and femoral components has been shown to improve the longevity of some devices. The orientation of the hip replacement components is described by the inclination and anteversion of the acetabular component, and the offset and version of the femoral component. These factors influence the range of motion of the joint, determine whether hip impingement occurs, and may affect the risk of dislocation and implant wear (Dargel et al., 2014; Hart et al., 2011; Mellon et al., 2015; Widmer and Zurfluh, 2004). Higher cup inclination angles are associated with edge-loading (contact between the femoral head and the edge or rim of

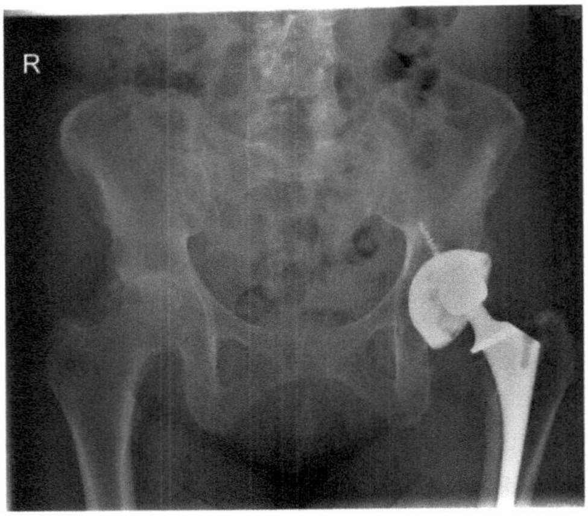


**FIGURE 1.6** Antero-posterior pelvic plain radiograph showing a left hip replacement with a worn polyethylene liner caused by edge-loading as a result of a high ($>40^\circ$) cup inclination angle.

the acetabular cup), which disrupts lubrication and increases bearing wear (Fig. 1.6). For polyethylene, this can be high enough to cause mechanical instability and wearing through of a liner to the titanium cup underneath (in a cementless hip).

### *1.2.7.2 Implant factors*

Implant factors include the design of the components, the materials used for the bearing surfaces, and the manufacturing process. Examples of constructs or components with poor performance have been large-diameter MoM bearings on a titanium stem, modular-neck femoral stems with CoCr/Ti taper junctions, cemented titanium stems, early-generation ceramics, and acetabular liners made of non-crosslinked UHMWPE.

### *1.2.7.3 Patient factors*

Patient factors, including age, sex, body mass index, nutritional status, presence of comorbidities, level of activity, and genetic make-up, are the most complex to investigate because of the high diversity of genotypes and multiple medical, environmental, and lifestyle influences. Several lines of evidence suggest that the risk of revision changes according to the age of the patient at primary procedure, with younger individuals having a higher lifetime risk of revision (Evans et al., 2019). The lifetime revision risk for men having a THR in their 50s is approximately 35% compared with 5% when the procedure is performed when they are in their 70s (Bayliss et al., 2017). Moreover, it is known that some patients are particularly sensitive to polyethylene or CoCr wear debris, leading to an increased risk of severe osteolysis, soft-tissue reactions, and/or systemic toxicity. The specific factors underlying this sensitivity have not been completely elucidated.

## 1.3 Introduction to biomarkers

Biological markers, or biomarkers, are quantifiable biological features that can be used to assess the health or disease state of an individual, monitor response to a medicine or toxic agent, or estimate an individual's risk of developing a certain condition or adverse effect (Califf, 2018). Although the exact definition is still the subject of certain controversy, generally, the term 'biomarker' has been extended to 'any useful characteristic that can be objectively measured and evaluated as an indicator of normal biological processes, pathogenic processes, or pharmacologic responses to a therapeutic intervention' (FDA-NIH Biomarker Working Group, 2016). Qualitative biomarkers, such as genetic mutations, can be either present or not, while quantitative biomarkers, such as blood glucose concentration, are recorded on a continuous scale and have preselected cut-off values that define abnormal results.

The most fundamental classification of biomarkers is based on which steps in the relationship between external exposure to a substance and its clinically relevant health effects they can clarify (Fig. 1.7). Biomarkers of exposure measure the internal dose of an exogenous chemical, its metabolites, or products of its interaction with target molecules or cells. These biomarkers are often used to monitor occupational, accidental, or intentional

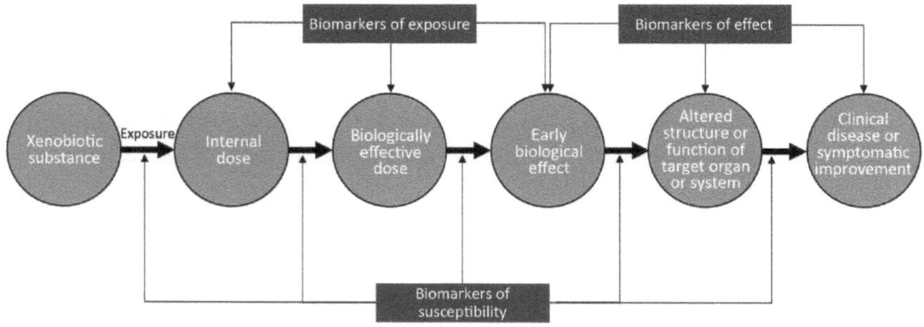


**FIGURE 1.7** Simplified diagram of the exposure–outcome continuum showing how biomarkers can be used to monitor or inform the process. Xenobiotic substances can be hazardous, resulting in negative biological effects and health impairment, or therapeutic, leading to positive biological effects and health improvement. The internal dose is a fraction of the external dose that is systemically available as a function of absorption, distribution, metabolism, internal storage, and excretion. The biologically effective dose is a fraction of the internal dose that reaches the target tissue and interacts with the target molecule(s). Early biological effects of such an interaction may include functional alterations in the target tissue. Increasing the duration and/or intensity of exposure may be accompanied by subclinical/clinical manifestations of disease (hazardous exposure) or symptomatic improvement and resolution of disease (therapeutic exposure). *Modified from Chen, X., Huang, S., Kerr, D., 2011. Biomarkers in clinical medicine. IARC Sci. Publ. 163, 303–322.*

exposure to toxic agents, such as heavy metals, pesticides, or tobacco. Biomarkers of effect reflect the biochemical, structural, physiological, or behavioural alterations that occur as a result of exposure to a xenobiotic substance, be it a hazardous chemical or medicinal compound. These are sometimes referred to as biomarkers of disease because they can be used for early detection of pathology, prognostication, and monitoring of treatment response. Finally, biomarkers of susceptibility assess the biological differences that may render some individuals particularly sensitive to a xenobiotic substance or infectious agent, for example, genetic variations or changes caused by environmental or lifestyle factors. Once identified, these individuals can be monitored for the development of adverse health effects using biomarkers of effect (WHO, 1993).

Depending on their characteristics, biomarkers can be further subdivided into four classes: (1) anthropomorphic (markers of the body shape or form), (2) physiological (measurements of body processes), (3) imaging (measurements of structural and metabolic features of the body derived from in vivo imaging studies), and (4) molecular (measurements of biochemical, cellular, histologic, proteomic, and genomic features of an individual derived from in vitro laboratory analyses of biological samples). The latter is the largest of the four categories and includes levels of various biomolecules, metabolites, xenobiotics, phenotypic characteristics of cells, and genetic and epigenetic traits. In practice, the full spectrum of biomarkers encompasses everything from measurements of body temperature and blood pressure, through assessment of basic body chemistry, to complex laboratory tests and sophisticated imaging studies. The clinical applications of biomarkers are similarly diverse and include screening, diagnosis, prognostication, susceptibility/risk assessment, disease monitoring, and evaluation of drug efficacy and safety (Table 1.3). Some biomarkers have just one known role while others can be applied in several contexts or inform the diagnosis of more than one disease. For example, C-reactive protein (CRP) is a non-specific marker of inflammation whose levels are elevated in a variety of conditions, including trauma, rheumatoid arthritis, infection, and cancer. Serum CRP is a common screening biomarker for periprosthetic infection following hip replacement surgery.

The paradigm shift toward precision medicine and personalised treatment of disease relies on the continued identification of new biomarkers that reflect the individual's health status, and their successful integration into clinical practice. The process of biomarker development comprises four main steps: (1) biomarker discovery, (2) assay optimisation, (3) evaluation of clinical validity in retrospective studies and prospective clinical trials, and (4) assessment of how the biomarker performs in routine clinical practice (Marchiò et al., 2011).

**TABLE 1.3 Examples of biomarkers and their clinical applications.**

| Context of use | Biomarker | Characteristic | Target condition or state |
|---|---|---|---|
| To screen individuals for subclinical disease | Faecal calprotectin | Molecular | Inflammatory bowel disease |
| | Serum PSA | Molecular | Prostate cancer |
| | Serum CRP | Molecular | Infection |
| To assess a healthy individual's potential for developing a certain disease | Body mass index | Anthropomorphic | Hypertension, diabetes |
| | Systolic blood pressure | Physiologic | Stroke |
| | Serum CRP | Molecular | Cardiovascular disease |
| | Polymorphism in APOE gene | Genomic | Late-onset Alzheimer's disease |
| | Mutations in BRCA1/2 genes | Genomic | Breast cancer |
| To diagnose a disease/condition or identify individuals with a subtype of the disease | Ejection fraction | Imaging | Cardiomyopathy |
| | Bone mineral density | Imaging | Osteoporosis |
| | Troponin T | Molecular | Myocardial injury |
| | Blood glucose | Molecular | Type 2 diabetes |
| | Urinary hCG | Molecular | Pregnancy |
| | Sweat chloride | Molecular | Cystic fibrosis |
| To determine disease extent and prognosis | Tumour grade score | Histologic | Cancer |
| | SUV measured by $^{18}$F-FDG PET/CT | Imaging | Hodgkin's lymphoma |
| | Serum β2-microglobulin | Molecular | Multiple myeloma |
| | Chromosome 17p deletions | Genomic | Chronic lymphocytic leukaemia |
| To identify the likelihood of a | Mutations in BRCA1/2 genes in | Genomic | Response to PARP inhibitors |

(*Continued*)

**TABLE 1.3 (Continued)**

| Context of use | Biomarker | Characteristic | Target condition or state |
|---|---|---|---|
| favourable or unfavourable effect from exposure to a medicinal product or environmental agent | patients with platinum-sensitive ovarian cancer | | |
| | Serum LDL cholesterol in patients with atherosclerosis | Molecular | Stroke, myocardial infarction, or death |
| To monitor disease status or exposure to xenobiotic substances | Body temperature | Physiologic | Fever |
| | Serum PSA | Molecular | Prostate cancer |
| | Serum CRP | Molecular | Acute pancreatitis |
| | Blood lead | Chemical | Exposure to lead |
| | Urinary tobacco-specific nitrosamines | Chemical | Exposure to tobacco smoke |
| | DNA–protein crosslinks in PBMCs | Genomic | Exposure to carcinogenic agents |
| To show that a biological response has occurred in an individual exposed to a medicinal product or environmental agent | Serum LDL cholesterol | Molecular | Response to lipid-lowering therapy |
| | Circulating B-lymphocytes in patients with lupus erythematosus | Cellular | Response to B-lymphocyte stimulator inhibitors |
| To assess the safety of medical intervention or environmental agent | Hepatic aminotransferases and bilirubin | Molecular | Hepatotoxicity |
| | Serum creatinine | Molecular | Nephrotoxicity |
| | Sperm count | Cellular | Reproductive toxicity |
| | Duration of the QT interval on ECG | Physiologic | Fatal cardiac arrhythmia |

*ECG*, electrocardiography; *FDG*, fluorodeoxyglucose; *hCG*, human chorionic gonadotropin; *LDL*, low-density lipoprotein; *PARP*, poly(ADP-ribose) polymerase; *PBMC*, peripheral blood mononuclear cell; *PET/CT*, positron emission tomography/computed tomography; *PSA*, prostate-specific antigen; *SUV*, standardised uptake value.
Source: Based on the FDA-NIH Biomarker Working Group "BEST (Biomarkers, EndpointS, and other Tools) Resource" (2016).

### 1.3.1 Biomarker discovery

The first recorded use of a biochemical test to diagnose disease can be traced back to the mid-1800s when the presence of the Bence Jones protein in a urine sample was linked to multiple myeloma (Jones, 1848). Throughout the 1900s, significant efforts were expended to identify biomarkers for the diagnosis of common illnesses, such as cancer and cardiovascular disease; however, the progress was hindered by limited technology. The recent advent of high-throughput genomic, transcriptomic, proteomic, and metabolomic tools has made it possible to obtain, in a single experiment, detailed molecular characterisation of biospecimens from individuals with a disease of interest. Biospecimens for biomarker discovery may derive from cell lines, animal models, samples from patients enrolled in ongoing clinical trials, or archived samples from finished prospective studies or biobanks (Quezada et al., 2017).

Genomic techniques, such as real-time polymerase chain reaction, DNA microarrays, and next-generation sequencing, allow researchers to study the genetic make-up of a person to identify individual mutations and/or genetic variants underpinning disease development or susceptibility. Transcriptomic studies use gene expression microarrays or RNA sequencing to quantitatively assess the total complement of RNA transcripts in a cell, tissue, or organism at any given time. For biomarker discovery, transcriptomes derived from 'diseased' and corresponding 'non-diseased' subjects can be compared to highlight the genes that show altered expression in disease. This may be supplemented with proteomic analyses involving immunoassays, microarrays, and mass spectrometry, which enable large-scale screening of patient samples for proteins that could predict clinical diagnosis. Global analyses of metabolites, such as lipids, carbohydrates, and nucleotides, are also possible thanks to advances in mass spectroscopy, nuclear magnetic resonance spectroscopy, and chromatography. Metabolomic studies generate additional insights into biochemical pathways that underlie health and disease to identify early metabolic biomarkers for disease and help understand its progression (Zhang et al., 2015). Technological progress in all those 'omics' fields makes it possible to detect and quantify molecular biomarkers on an increasingly larger and more accurate scale.

In general, new candidate biomarkers can be identified through hypothesis-based approaches or exploratory studies. Hypothesis-based biomarker discovery integrates experimentally obtained data with our existing understanding of molecular mechanisms of disease and drug action, and involves targeted assessment of predefined genetic sequences, molecular interactions, and analytes. For example, the knowledge that diabetes mellitus produces a sustained elevation of blood glucose concentration prompted the identification of glycosylated haaemoglobin as a diagnostic biomarker for the disease. The exploratory approach to biomarker discovery is untargeted and focuses on identifying changes in the presence or relative abundance of molecular species that are statistically correlated with the target disease or condition. A practical illustration of this method is using differential gene expression patterns to classify cancers into molecular subsets to facilitate targeted treatment.

The alternative to de novo biomarker discovery is 'repurposing' of conventional biomarkers and existing tests for new clinical applications.

### 1.3.2 Assay validation

Once a promising new biomarker has been identified, it is necessary to develop an assay that allows it to be accurately and reliably measured in a given sample type. Considerations to be taken into account include safety and ease of deployment in a clinical setting, cost, technical expertise required, stability of the sample and reagents, analytical sensitivity and specificity, and potential sources of variability, such as daily and seasonal variations in analyte concentration, patient characteristics, sample preparation technique, and choice of analytical platform. The analytical procedure should be standardised to maximise repeatability and reproducibility and ensure that results obtained in different centres can be compared in a meaningful way. Common assay parameters that are investigated at this stage of biomarker development are listed in Table 1.4.

### 1.3.3 Evaluation of clinical validity

Biomarker testing relies upon the assumption that biomarker levels are markedly different (usually higher) in individuals with the disease or condition of interest than in the corresponding healthy population. Typically, the outcome variable is dichotomous, for example, 'diseased' or 'non-diseased', but it can also be ordinal, for example, 'definitely normal', 'probably normal', 'uncertain', 'probably abnormal', or 'definitely abnormal'. The diagnostic threshold impacts the ability of the test to discriminate between the different subcategories; its choice is guided by the available data, clinical context, and ethical, financial, and practical considerations. Even if the cutoff value is chosen to maximise the test's diagnostic performance, unavoidably, some individuals will be misclassified, resulting in potential harm. For dichotomous outcomes, this can be depicted graphically as shown in Fig. 1.8.

Results of a diagnostic test with a binary outcome can be presented in a $2 \times 2$ contingency table, which reflects the agreement between the test result and the true disease state (Table 1.5). To determine the true disease state, a 'gold standard' reference test must be available that allows unambiguous classification of test subjects as having or not having the target disease (sensitivity and specificity equal to 100%). However, in many cases, a gold

**TABLE 1.4 Common assay parameters and their definitions.**

| Term | Meaning |
|---|---|
| Precision | The extent to which independent test results obtained under prescribed conditions agree with one another; composed of repeatability, intermediate precision, and reproducibility |
| Repeatability (intra-assay variability) | The degree to which a given assay can obtain consistent results under the same operating conditions and over a short time period |
| Intermediate precision (intralaboratory variability) | The degree to which the same laboratory can obtain consistent results for a given assay when the assay is carried out on different days, by different analysts, or using different equipment |
| Reproducibility (interlaboratory variability) | The degree to which the results of a given assay can be replicated in another laboratory |
| Accuracy | Agreement between the obtained test result and the true or conventionally accepted result |
| Limit of detection (LoD) | The lowest concentration of an analyte that can be reliably distinguished from 'analytical noise', but is not necessarily quantitated as an exact value (Armbruster and Pry, 2008) |
| Limit of quantitation (LoQ) | The lowest concentration of an analyte that can be determined quantitatively with suitable precision and accuracy (Armbruster and Pry, 2008) |
| Analytical sensitivity | The ability of an assay to differentiate between two very close concentrations of an analyte; often wrongly used as a synonym for LoD (Lozano and Cantero, 1997) |
| Analytical specificity | The ability of an assay to distinguish target from non-target analytes in a sample |

standard test is imperfect or not available. An example of such a scenario is the diagnosis of periprosthetic joint infection following hip replacement. The lack of generally accepted criteria by which to diagnose a condition produces bias, which may lead to erroneous conclusions regarding test performance and hinder meaningful comparisons of study outcomes. The methods that can be used to evaluate a medical test in the absence of a gold standard have been reviewed by Umemneku Chikere et al. (2019).

Once the number of true-positive, true-negative, false-positive, and false-negative results is known, the performance of the test can be assessed by calculating its sensitivity and specificity, positive and negative predictive values, and positive and negative likelihood ratios (Table 1.6). While useful

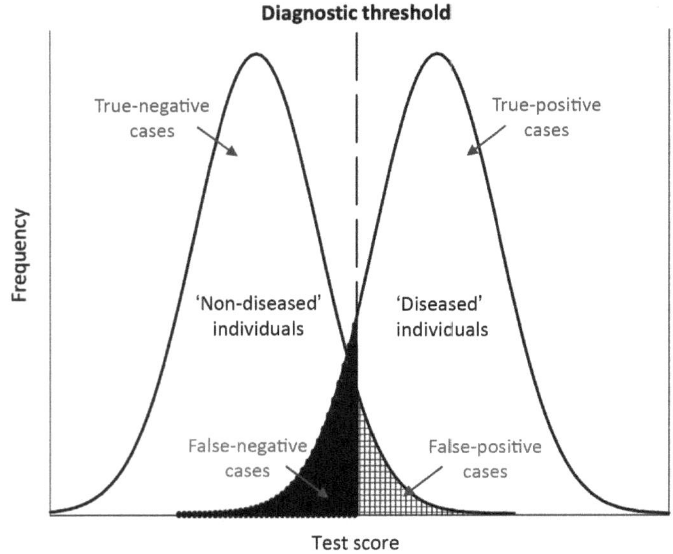


**FIGURE 1.8** Gaussian curves describing the distribution of biomarker levels in individuals with and without the disease/condition of interest.

**TABLE 1.5 The possible outcomes of a dichotomous test.**

| | | Disease status (as determined by reference test) | |
|---|---|---|---|
| | | **Target disease present** | **Target disease absent** |
| Biomarker test | A positive result (suggestive of target disease) | Number of 'diseased' patients with a positive test (true positive) | Number of 'non-diseased' patients with a positive test (false positive) |
| | A negative result (not suggestive of target disease) | Number of 'diseased' patients with a negative test (false negative) | Number of 'non-diseased' patients with a negative test (true negative) |

as basic summaries, none of these parameters validly represent the ability of a biomarker to distinguish patients with and without the target disease. Consequently, paired indicators are not the best way to compare the performance of alternative diagnostic tests, especially if one test does not outperform the other on both accounts. Single-number indicators, such as diagnostic accuracy, area under the receiver operating characteristic curve

**TABLE 1.6 Common descriptors of diagnostic test performance and their interpretation.**

| | Meaning | Test outcome | Result | Rule-of-thumb interpretation |
|---|---|---|---|---|
| | **Paired measures of test performance** | | | |
| Sensitivity (true-positive rate) | The proportion of positive tests in individuals in whom the disease is truly present Defined as TP/(TP + FN) | Dichotomous, polychotomous | 0%–100% | 100%: perfect test $\leq$50%: no discriminatory power |
| Specificity (true-negative rate) | The proportion of negative tests in individuals in whom the disease is truly absent Defined as TN/(TN + FP) | Dichotomous, polychotomous | 0%–100% | 100%: perfect test $\leq$50%: no discriminatory power |
| Positive predictive value (PPV) | The proportion of 'diseased' subjects in those with a positive test result Defined as TP/(TP + FP) | Dichotomous | 0%–100% | 100%: perfect test $\leq$50%: no discriminatory power |
| Negative predictive value (NPV) | The proportion of 'non-diseased' subjects in those with a negative test result Defined as TN/(TN + FN) | Dichotomous | 0%–100% | 100%: perfect test $\leq$50%: no discriminatory power |
| Positive likelihood ratio ($LR_+$) | How much more likely a positive test result is to be a true positive than a false positive Defined as sensitivity/(1−specificity) | Dichotomous, continuous, polychotomous | 1–$\infty$ | Evidence to rule in diagnosis: >10: strong 5–10: moderate 1–4: weak |
| Negative likelihood ratio ($LR_-$) | How much more likely a negative test result is to be a | Dichotomous, continuous, polychotomous | 0–1.0 | Evidence to rule out diagnosis: <0.1: strong |

*(Continued)*

**TABLE 1.6 (Continued)**

| | Meaning | Test outcome | Result | Rule-of-thumb interpretation |
|---|---|---|---|---|
| | false negative than a true negative<br>Defined as (1−sensitivity)/specificity | | | 0.1−0.2: moderate<br>0.3−1.0: weak |
| | **Single measures of test performance** | | | |
| Overall diagnostic accuracy | Proportion of individuals correctly classified as 'diseased' or 'non-diseased'<br>Defined as TP + TN/(TN + TP + FN + FP) | Dichotomous | 0%–100% | > 90%: excellent<br>75%–90%: good<br>50%–75%: poor<br>< 50%: no discriminatory power |
| The area under the ROC curve | The probability that the test correctly discriminates a randomly chosen pair of 'diseased' and 'non-diseased' individuals | Continuous, polychotomous | 0–1.00 | > 0.90: excellent<br>0.75–0.90: good<br>0.50–0.75: poor<br>< 0.50: no discriminatory power |
| Diagnostic odds ratio | The ratio of the odds of positivity in the 'diseased' relative to the odds of positivity in the 'non-diseased'<br>Defined as<br>1) $LR_+/LR_-$<br>2) sensitivity × specificity/(1 − sensitivity) (1 − specificity)<br>3) PPV × NPV/(1 − PPV)(1−NPV) | Dichotomous, continuous, polychotomous | 0–∞ | > 1: useful test<br>1: uninformative test<br>< 1: improper test interpretation (more negative tests among the 'diseased') |

*FN*, false negative; *FP*, false positive; *ROC*, receiver operating characteristic; *TN*, true negative; *TP*, true positive.

(AUC), and diagnostic odds ratio (DOR), are more useful for this purpose (Table 1.6). The following paragraphs discuss the merits and drawbacks of each of the above-mentioned parameters.

### 1.3.3.1 Sensitivity and specificity

Sensitivity and specificity are the most commonly reported indices of biomarker accuracy. Highly sensitive tests ($>90\%$) yield few false negatives, so a negative result constitutes strong evidence to rule out or exclude the target disease. Conversely, highly specific tests ($>90\%$) generate a very low number of false positives, so a positive result strongly implies that the target disease/condition is present. The main advantage of sensitivity and specificity is that they are not influenced by the underlying disease prevalence, that is, percentage of cases in the study population at a single point in time. They are, however, vulnerable to bias: the sensitivity of a test falls with decreasing disease severity while the specificity is diminished by the existence of other possible causes for a positive test result (differential diagnoses) (Fischer et al., 2003). Furthermore, the two metrics are inversely correlated with each other for the diagnostic threshold chosen, such that if the threshold is raised, there will be an increase in specificity at the cost of decreased sensitivity. Combining individual biomarkers into panels that better capture the complexity of disease may result in higher cumulative sensitivity and specificity than with single molecules.

One aspect of sensitivity and specificity to consider is that they provide clinical information in a counterintuitive way: if an individual does or does not have the target disease, what is the probability that the test result is positive (sensitivity) or negative (specificity), respectively? This is confusing because knowing that a particular disease is present would normally render a diagnostic test aimed at the detection of that disease unnecessary (Gallagher, 2003).

### 1.3.3.2 Predictive values

Predictive values are more useful than sensitivity and specificity because they answer the 'right' question: given a positive or negative test result, what is the probability that the patient does or does not have the target disease, respectively? However, unlike sensitivity and specificity, predictive values are intimately linked to the underlying disease prevalence. For any given test, the less common the disease, the weaker the positive predictive value (PPV) (more false positives for every true positive) and the stronger the negative predictive value (NPV) (more true negatives for every false negative). Analogous to this, as the underlying prevalence of the target disease rises, the PPV of the test increases while the NPV of the test drops. In other words, the same diagnostic test can appear to perform better or worse simply because the prevalence of the target disorder is different (Gallagher, 2003).

For this reason, predictive values reported in a study (particularly one that artificially sets the prevalence at 50%, as is the case in 1:1 case–control studies) cannot be readily generalised to other populations and must be interpreted carefully.

#### *1.3.3.3 Likelihood ratios*

The likelihood ratio (LR) is the probability that a specific test result is obtained in patients with the disease divided by the probability of obtaining the same test result in patients without the disease. Similar to predictive values, LRs provide clinical information in an intuitive format by quantifying the change in the certainty of the 'diagnosis' conferred by test results. The odds that the target disease is present can be calculated by multiplying the LR of the test result by the pre-test odds of the diagnosis, which represents the clinician's level of suspicion. LR values exceeding 1 increase disease probability, those ranging from more than 0 to less than 1 diminish it, and an LR of 1 leaves the post-test disease probability unchanged from pre-test estimates. In contrast to predictive values, LRs are reproducible and tend to remain stable between populations unless changes in disease prevalence and severity happen concurrently (Gallagher, 2003). Another advantage of LRs is that they can be calculated for tests with continuous or ordinal outcomes, which helps to retain useful information that would otherwise be lost if the test results were simply dichotomised.

#### *1.3.3.4 Diagnostic accuracy*

Diagnostic accuracy is calculated as the sum of the true positive and true negative results divided by the total number of tests performed. Essentially, it measures how effective a diagnostic test is at discriminating between the 'diseased' and 'non-diseased' subjects, and is proportional to the AUC. Like predictive values, diagnostic accuracy depends on the prevalence of the target disorder in the studied population whenever the sensitivity and specificity are not equal, which means that it will favour classifiers that always predict negative outcomes for rare events (Glas et al., 2003).

#### *1.3.3.5 Receiver operating characteristic curves*

The performance of a test with continuous or ordinal outcomes can be evaluated using ROC curve analysis (Mandrekar, 2010). ROC curves plot the true-positive rate against the false-positive rate to illustrate the trade-off between sensitivity and specificity across all possible threshold values (Fig. 1.9) (Hajian-Tilaki, 2013). The plot can be used to compare two or more candidate biomarkers for the same condition—a test with a curve that lies wholly above the curve of another will be more clinically useful. Once the clinical benefit of a test is established, only a portion of the ROC curve

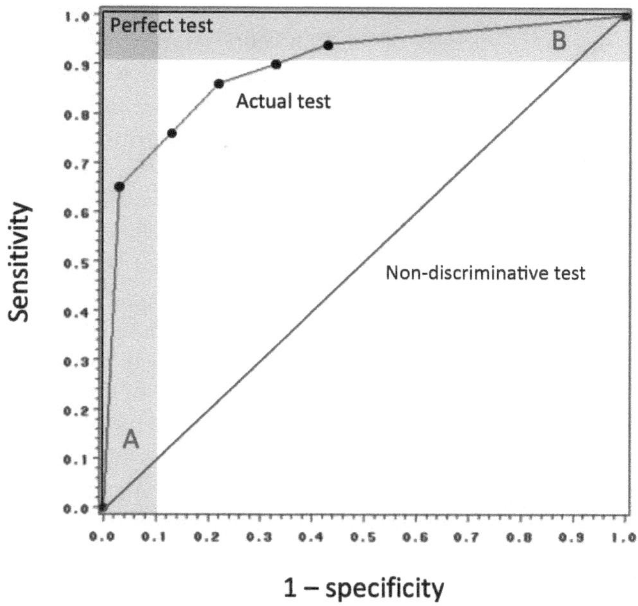


**FIGURE 1.9** Receiver operating characteristic curve showing the relationship between sensitivity and specificity for every possible cut-off of the test under evaluation. The closer the curve approximates the upper left corner of the plot, the better the diagnostic ability. The diagonal line is the line of equivalence, where the sensitivity is equal to the proportion of false positives. Cut-off values of the curve falling into rectangle A (specificity >90%) or B (sensitivity >90%) are useful for ruling in or ruling out disease, respectively.

is usually of interest, such as the region reflecting high specificity or high sensitivity.

The AUC is an effective way to describe the overall test performance. The AUC can be thought of as the average sensitivity over all possible values of specificity (or vice versa), or the probability that a randomly chosen 'diseased' individual has a result indicating greater suspicion than a randomly chosen 'non-diseased' individual (Hanley and McNeil, 1982). The latter interpretation is based on non-parametric Mann–Whitney U statistic that is used in calculating the AUC (Hanley and McNeil, 1982). A test with an AUC of 1.0 (sensitivity and specificity of 100%) correctly classifies the true disease state of all individuals, while one with an AUC of 0.5 is no better than a coin toss. The metric can be used to directly compare the diagnostic ability of alternative biomarkers for the same disease or condition. Correlated or independent AUCs can be compared using the methods proposed by DeLong et al. (1988) and Hanley and McNeil (1982, 1983).

Because the AUC is derived from the sensitivity and specificity, it is dependent on the patients' characteristics and disease spectrum but invariant to the underlying disease prevalence. The latter feature makes it useful for evaluating classifiers predicting rare events. Importantly, since the AUC summarises the entire location of the ROC curve rather than depending on a specific diagnostic threshold, it cannot distinguish tests with high sensitivity and low specificity from tests with low sensitivity and high specificity, and provides limited information on how well a test performs at a specific cut-off value.

Depending on the clinical scenario, different approaches can be used to define the most appropriate diagnostic threshold for a biomarker. If sensitivity and specificity are deemed to be equally important, for example, when false-positive and false-negative diagnoses would be equally costly, it is reasonable to choose a cut-point that achieves an optimum mathematical balance between sensitivity and specificity. A common way to do this is to use the maximum height of the ROC curve above the line of equivalence, which represents the maximum sum of sensitivity and specificity (the Youden index method) (Youden, 1950). In the case of a rare disease with a high associated cost of false-positive diagnosis, the threshold may be raised to maximise specificity. Conversely, if a disease has a high prevalence and the consequences of missed diagnosis outweigh those of overdiagnosis, the threshold is likely to be lowered to boost sensitivity (Hajian-Tilaki, 2013).

#### *1.3.3.6 Diagnostic odds ratios*

An alternative way to compare the diagnostic performance of multiple tests is by means of the DOR. The DOR describes the odds of a positive test in those with the target disease relative to the odds of a positive test in those without the target disease, and higher values indicate better test performance. The measure is relatively independent of changes in disease prevalence and severity, and tends to be reasonably constant irrespective of the diagnostic threshold—these features make it particularly useful in meta-analyses, which combine results from different studies (often using different cut-off values) into more precise summary estimates. A chief limitation of the DOR is that it does not differentiate between false-positive and false-negative results, so it is unable to distinguish tests with high sensitivity and low specificity from tests with low sensitivity and high specificity, and cannot be used in isolation to rule in or rule out a diagnosis (Glas et al., 2003).

### 1.3.4 Characteristics of an ideal biomarker

A new biomarker must fulfil several criteria to be clinically applicable (Selvaskandan et al., 2020). First of all, it must be biologically plausible and provide useful information about the underlying disease mechanism. Second, the assay must be safe, inexpensive, and simple to perform using a rapid and widely available platform, preferably on easily accessible samples, such as urine or blood. Third, it must be validated—both within a single assay system and in different laboratories—and applicable across diverse populations, that is, patients of different ages, sex, ethnicity, and geographical origin. Finally, the test must be highly specific for the target disease/condition and sensitive enough to facilitate its early detection and timely treatment. Typically, greater emphasis is placed on high specificity, that is, low false-positive rate, to avoid subjecting healthy individuals to potentially expensive or invasive further tests. It is important to emphasise that high accuracy does not imply clinical utility. To be clinically useful, a new biomarker must facilitate clinical decision-making and lead to improvements in health outcomes.

### 1.3.5 Biomarkers of hip implant function and toxicity

Hip implants represent an unusual route of toxic exposure whereby the source of the hazardous substance resides within the body. While environmental and occupational exposure sources of heavy metals have been well documented in the literature, relatively little is known about the biological effects of implant-derived wear debris and their relationship to clinically relevant sequalae. Implant degradation can be assessed by focusing on the products of wear and corrosion (wear particles, corrosion products, and metal ions) or on molecular mediators that reflect the biological consequences of wear and corrosion. Elevated circulating levels of cobalt, chromium, and titanium may help identify patients with unexpectedly high-wearing devices and those at risk of ALTR and implant failure, while pro-inflammatory cytokines and indices of bone turnover may inform on the quality of implant fixation to bone and the risk of aseptic loosening. Investigations for implant related-systemic cobalt toxicity (arthroprosthetic cobaltism) consider circulating levels of cobalt as well as biomarkers of end-organ function. Periprosthetic infection, which can severely impact the function of a hip implant, relies on a host of serum, synovial fluid, and periprosthetic tissue biomarkers for diagnosis, including acute-phase reactants, inflammatory mediators, and host antimicrobial proteins.

Chapters 4–7 discuss the existing and emerging biomarkers of implant wear, peri-implant osteolysis, aseptic loosening, periprosthetic infection, and systemic cobalt toxicity, while Chapter 8 summarises how current guidelines from regulatory bodies have incorporated biomarker research into clinical practice.

## References

Ackerman, I., Bohensky, M., Zomer, E., et al., 2019. The projected burden of primary total knee and hip replacement for osteoarthritis in Australia to the year 2030. BMC Musculoskelet. Disord. 20, 90.

Armbruster, D., Pry, T., 2008. Limit of blank, limit of detection and limit of quantitation. Clin. Biochem. Rev. 29, S49–S52.

Ball, S.T., le Duff, M.J., Amstutz, H.C., 2007. Early results of conversion of a failed femoral component in hip resurfacing. J. Bone. Joint. Surg. Am. 89, 735–741.

Banerjee, S., Cherian, J.J., Bono, J.V., Kurtz, S.M., Geesink, R., Meneghini, R.M., et al., 2015. Gross trunnion failure after primary total hip arthroplasty. J. Arthroplasty 30, 641–648.

Bayliss, L., Culliford, D., Monk, A., Glyn-Jones, S., Prieto-Alhambra, D., Judge, A., et al., 2017. The effect of patient age at intervention on risk of implant revision after total replacement of the hip or knee: a population-based cohort study. Lancet 389, 1424–1430.

Bijukumar, D.R., Segu, A., Souza, J.C.M., Li, X.J., Barba, M., Mercuri, L.G., et al., 2018. Systemic and local toxicity of metal debris released from hip prostheses: a review of experimental approaches. Nanomedicine 14, 951–963.

Boutin, P., 1972. Arthroplastie totale de la hanche par prothese en alumina fritte: etude experimentale et premieres applications cliniques. Rev. Chir. Orthop. 58, 229–246.

Califf, R., 2018. Biomarker definitions and their applications. Exp. Biol. Med. 243, 213–221.

Carli, A., Patel, A., Cross, M., Mayman, D., Carroll, K., Pellicci, P., et al., 2020. Long-term performance of oxidized zirconium on conventional and highly cross-linked polyethylene in total hip arthroplasty. SICOT J. 6, 10.

Catelas, I., Wimmer, M.A., 2011. New insights into wear and biological effects of metal-on-metal bearings. J. Bone Joint. Surg. Am. 93, 76–83.

Caton, J., Prudhon, J.L., 2011. Over 25 years survival after Charnley's total hip arthroplasty. Int. Orthop. 35, 185–188.

Charnley, J., 1961. Arthroplasty of the hip: a new operation. Lancet 1, 1129–1132.

Chen, X., Huang, S., Kerr, D., 2011. Biomarkers in clinical medicine. IARC Sci. Publ. 163, 303–322.

Choi, H.-I., Hong, J.A., Kim, M.-S., Lee, S.E., Jung, S.-H., Yoon, P.W., et al., 2018. Severe cardiomyopathy due to arthroprosthetic cobaltism: report of two cases with different outcomes. Cardiovasc. Toxicol. 19, 82–89.

Cobelli, N., Scharf, B., Crisi, G.M., Hardin, J., Santambrogio, L., 2011. Mediators of the inflammatory response to joint replacement devices. Nat. Rev. Rheumatol. 7, 600–608.

Culliford, D., Maskell, J., Judge, A., Cooper, C., Prieto-Alhambra, D., Arden, N., 2015. Future projections of total hip and knee arthroplasty in the UK: results from the UK Clinical Practice Research Datalink. Osteoarthritis. Cartilage 23, 594–600.

Cuthbert, R., Wong, J., Mitchell, P., Kumar, J., 2019. Dual mobility in primary total hip arthroplasty: current concepts. EFORT Open. Rev. 4, 640–646.

D'Antonio, J., Capello, W., Ramakrishnan, R., 2012. Second-generation annealed highly cross-linked polyethylene exhibits low wear. Clin. Orthop. Relat. Res. 470, 1696–1704.

Daniel, J., Pynsent, P.B., McMinn, D.J.W., 2004. Metal-on-metal resurfacing of the hip in patients under the age of 55 years with osteoarthritis. J. Bone. Joint. Surg. Br. 86-B, 177–184.

Dargel, J., Oppermann, J., Brüggemann, G.-P., Eysel, P., 2014. Dislocation following total hip replacement. Dtsch. Arztebl. Int. 111, 884–890.

De Martino, I., Assini, J.B., Elpers, M.E., Wright, T.M., Westrich, G.H., 2015. Corrosion and fretting of a modular hip system: a retrieval analysis of 60 Rejuvenate stems. J. Arthroplasty 30, 1470–1475.

de Villiers, D., Richards, L., Tuke, M., Collins, S., 2020. Ceramic resurfacing: the future and challenges. Ann. Joint. 5, 12.

DeLong, E., DeLong, D., Clarke-Pearson, D., 1988. Comparing the areas under two or more correlated receiver operating characteristic curves: a nonparametric approach. Biometrics 44, 837–845.

Eltit, F., Wang, Q., Wang, R., 2019. Mechanisms of adverse local tissue reactions to hip implants. Front. Bioeng. Biotechnol. 7, 1–17.

Evans, J.T., Evans, J.P., Walker, R.W., Blom, A.W., Whitehouse, M.R., Sayers, A., 2019. How long does a hip replacement last? A systematic review and meta-analysis of case series and national registry reports with more than 15 years of follow-up. Lancet 393, 647–654.

FDA-NIH Biomarker Working Group. 2016. BEST (Biomarkers, EndpointS, and other Tools) Resource. Silver Spring (MD): Food and Drug Administration (US).

Fischer, J., Bachmann, L., Jaeschke, R., 2003. A readers' guide to the interpretation of diagnostic test properties: clinical example of sepsis. Intensive Care Med. 29, 1043–1051.

Fox, A., Bedi, A., Rodeo, S., 2009. The basic science of articular cartilage: structure, composition, and function. Sports Health 1, 461–468.

Gallagher, E., 2003. Evidence-based emergency medicine/editorial. The problem with sensitivity and specificity. Ann. Emerg. Med. 42, 298–303.

Gill, I.P.S., Webb, J., Sloan, K., Beaver, R.J., 2012. Corrosion at the neck-stem junction as a cause of metal ion release and pseudotumour formation. J. Bone Joint. Surg. Br. 94-B, 895–900.

Glas, A., Lijmer, J., Prins, M., Bonsel, G., PM, B., 2003. The diagnostic odds ratio: a single indicator of test performance. J. Clin. Epidemiol. 56, 1129–1135.

Hajian-Tilaki, K., 2013. Receiver operating characteristic (ROC) curve analysis for medical diagnostic test evaluation. Caspian. J. Intern. Med. 4, 627–635.

Hallab, N.J., Jacobs, J.J., 2009. Biologic effects of implant debris. Bull. NYU Hosp. Jt. Dis. 67, 182–188.

Hanley, J., McNeil, B., 1982. The meaning and use of the area under a receiver operating characteristic (ROC) curve. Radiology 143, 29–36.

Hanley, J., McNeil, B., 1983. A method of comparing the areas under receiver operating characteristic curves derived from the same cases. Radiology 148, 839–843.

Hart, A., Ilo, K., Underwood, R., 2011. The relationship between the angle of version and rate of wear of retrieved metal-on-metal resurfacings. J. Bone Joint Surg. Br. 93-B, 315–320.

Hart, A., Shiraz, S., Henckel, J., Lloyd, G., Skinner, J., 2015. Lessons learnt from metal-on-metal hip arthroplasties will lead to safer innovation for all medical devices. Hip Int. 25, 347–354.

Jacobs, J., Campbell, P., Konttinen, Y., 2008. How has the biologic reaction to wear particles changed with newer bearing surfaces? J. Am. Acad. Orthop. Surg. 16, S49–S55.

Jacobs, J.J., Cooper, H.J., Urban, R.M., Wixson, R.L., Della Valle, C.J., 2014. What do we know about taper corrosion in total hip arthroplasty? J. Arthroplasty 29, 668–669.

Jenkins, P., Clement, N., Hamilton, D., Gaston, P., Patton, J., Howie, C., 2013. Predicting the cost-effectiveness of total hip and knee replacement: a health economic analysis. Bone Joint. J. 95-B, 115–121.

Jones, H., 1848. On a new substance occuring in the urine of a patient with mollities ossium. Philos. Trans. Royal Soc. 138, 55–62.

Ko, L.M., Chen, A.F., Deirmengian, G.K., Hozack, W.J., Sharkey, P.F., 2016. Catastrophic femoral head-stem trunnion dissociation secondary to corrosion. J. Bone Joint. Surg. Am. 98, 1400–1404.

Kremers, H., Larson, D., Crowson, C., Kremers, W., Washington, R., Steiner, C., et al., 2015. Prevalence of total hip and knee replacement in the United States. J. Bone Joint Surg. Am. 359, 1386–1397.

Krishnan, H., Krishnan, S.P., Blunn, G., Skinner, J.A., Hart, A.J., 2013. Modular neck femoral stems. Bone Joint. J. 95 B, 1011–1021.

Kurtz, S.M., Lau, E., Ong, K., Zhao, K., Kelly, M., Bozic, K.J., 2009. Future young patient demand for primary and revision joint replacement: national projections from 2010 to 2030. Clin. Orthop. Relat. Res. 467, 2606–2612.

Kurtz, S., Gawel, H., Patel, J., 2011. History and systematic review of wear and osteolysis outcomes for first-generation highly crosslinked polyethylene. Clin. Orthop. Relat. Res. 469, 2262–2277.

Lam, S., Amies, V., 2015. Hip arthritis presenting as knee pain. BMJ Case Rep. 2015, 2014–2016.

Langlois, J., Hamadouche, M., 2020. Recent update on crosslinked polyethylene in total hip arthroplasty. SICOT J. 6, 13.

Lim, S., Yeo, I., Park, C., Lee, K.-J., Min, B.-W., Park, Y.-S., 2019. High survivorship of highly cross-linked polyethylene in revision total hip arthroplasty: a minimum 10-year follow-up study. Arthroplasty 1,, 16.

Lozano, D., Cantero, M., 1997. Difference between analytical sensitivity and detection limit. Am. J. Clin. Pathol. 107, 619–620.

Lübbeke, A., Silman, A., Barea, C., Prieto-Alhambra, D., Carr, A., 2018. Mapping existing hip and knee replacement registries in Europe. Health Policy 122, 548–557.

Mandrekar, J., 2010. Receiver operating characteristic curve in diagnostic test assessment. J. Thorac. Oncol. 5, 1315–1316.

Marchiò, C., Dowsett, M., Reis-Filho, J., 2011. Revisiting the technical validation of tumour biomarker assays: how to open a Pandora's box. BMC Med. 9, 41.

Massin, P., Lopes, R., Masson, B., Mainard, D., 2014. Does biolox delta ceramic reduce the rate of component fractures in total hip replacement? Orthop. Traumatol. Surg. Res. 100, S317–S321.

Matusiewicz, H., 2014. Potential release of in vivo trace metals from metallic medical implants in the human body: from ions to nanoparticles - a systematic analytical review. Acta Biomater. 10, 2379–2403.

McMinn, D.J.W., Treacy, R., Lin, K., Pynsent, P.B., 1996. Metal on metal surface replacement of the hip. experience of the McMinn prothesis. Clin. Orthop. Relat. Res. 329, S89–S98.

Mellon, S., Grammatopoulos, G., Andersen, M., 2015. Optimal acetabular component orientation estimated using edge-loading and impingement risk in patients with metal-on-metal hip resurfacing arthroplasty. J. Biomech. 48, 318–323.

Merle, C., Grammatopoulos, G., Waldstein, W., Pegg, E., Pandit, H., Aldinger, P., et al., 2013. Comparison of native anatomy with recommended safe component orientation in total hip arthroplasty for primary osteoarthritis. J. Bone Joint. Surg. Am. 95, e172.

Morlock, M.M., Dickinson, E.C., Günther, K.-P., Bünte, D., Polster, V., 2018. Head taper corrosion causing head bottoming out and consecutive gross stem taper failure in total hip arthroplasty. J. Arthroplasty 33, 3581–3590.

Nakahara, I., Takao, M., Sakai, T., Nishii, T., Yoshikawa, H., N, S., 2011. Gender differences in 3D morphology and bony impingement of human hips. J. Orthop. Res. 29, 333–339.

National Joint Registry, 2020. 17th Annual Report.

Oral, E., Muratoglu, O., 2011. Vitamin E diffused, highly crosslinked UHMWPE: a review. Int. Orthop. 35, 215–223.

Piconi, C., De Santis, V., Maccauro, G., 2017. Clinical outcomes of ceramicized ball heads in total hip replacement bearings: a literature review. J. Appl. Biomater. Funct. Mater. 15, e1–e9.

Pisanu, F., Doria, C., Andreozzi, M., Bartoli, M., Saderi, L., Sotgiu, G., et al., 2018. Pleomorphic clinical spectrum of metallosis in total hip arthroplasty. Int. Orthop. 43, 85–96.

Pivec, R., Meneghini, R.M., Hozack, W.J., Westrich, G.H., Mont, M.A., 2014. Modular taper junction corrosion and failure: how to approach a recalled total hip arthroplasty implant. J. Arthroplasty 29, 1–6.

Quezada, H., Guzmán-Ortiz, A., Díaz-Sánchez, H., Valle-Rios, R., Aguirre-Hernández, J., 2017. Omics-based biomarkers: current status and potential use in the clinic. Bol. Med. Hosp. Infant. Mex. 74, 219–226.

Sedel, L., Raould, A., 2007. Engineering aspect of alumina on alumina hip prosthesis. Proc. Inst. Mech. Eng. H. 221, 21–27.

Selvaskandan, H., Shi, S., Twaij, S., Cheung, C., Barratt, J., 2020. Monitoring immune responses in IgA nephropathy: biomarkers to guide management. Front. Immunol. 11, 572754.

Shi, H., Magaye, R., Castranova, V., Zhao, J., 2013. Titanium dioxide nanoparticles: a review of current toxicological data. Part Fibre Toxicol. 10, 1–33.

Umemneku Chikere, C., Wilson, K., Graziadio, S., Vale, L., Allen, A., 2019. Diagnostic test evaluation methodology: a systematic review of methods employed to evaluate diagnostic tests in the absence of gold standard - an update. PLoS One 14, e0223832.

World Health Organization & International Programme on Chemical Safety, 1993. Biomarkers and risk assessment: concepts and principles / published under the joint sponsorship of the United Nations environment Programme, the International Labour Organisation, and the World Health Organization.

Widmer, K., Zurfluh, B., 2004. Compliant positioning of total hip components for optimal range of motion. J. Orthop. Res. 22, 815–821.

Xia, Z., Ricciardi, B., Liu, Z., von Ruhland, C., Ward, M., Lord, A., et al., 2017. Nano-analyses of wear particles from metal-on-metal and non-metal-on-metal dual modular neck hip arthroplasty. Nanomedicine 13, 1205–1217.

Youden, W., 1950. Index for rating diagnostic tests. Cancer 3, 32–35.

Zhang, A., Sun, H., Yan, G., Wang, P., Wang, X., 2015. Metabolomics for biomarker discovery: moving to the clinic. Biomed. Res. Int. 2015, 354671.

Chapter 2

# Degradation of metal hip implants

**Andrew R. Beadling, Anne Neville and Michael G. Bryant**
*Institute of Functional Surfaces (iFS), School of Mechanical Engineering, University of Leeds, Leeds, United Kingdom*

## 2.1 Introduction to metallic biomaterials

A biomaterial is 'a nonviable material used in a medical device, intended to interact with biological systems' (Williams, 1987). The scope of biomaterials is wide and varied, ranging from materials used in joint replacements to stents, drug delivery methods, and artificial scaffolds. This section will be limited to materials relevant to total hip replacement (THR) and hip resurfacing (HR) devices.

One of the critical concerns for a joint replacement, and a measure of the success of the device, is the restoration of natural joint function. Metals and their alloys have higher ductility, elastic modulus, and yield strength than other biomaterials, making them a compelling choice for weight-bearing applications (Brunski, 2013).

### 2.1.1 Iron-based alloys

Stainless steels were some of the first metals to be used in total joint replacements, with the original McKee–Farrar and Charnley prosthesis composed of a stainless-steel femoral head and stem (Charnley, 1972; McKee and Watson-Farrar, 1966). The most common types of stainless steel today are 316 and 316 L (the low-carbon variant of 316), with the latter grade defined by ASTM F138 for surgical applications.

The 316 stainless steel is an iron-based alloy with chromium, nickel, and molybdenum constituents. The key function of the chromium constituent is to form a dense, approximately 5-nm thick surface oxide film that enhances the corrosion resistance of the material by acting as a barrier to charge transfer. Stainless steels are susceptible to localised forms of corrosion, such as

**Biomarkers of Hip Implant Function. DOI: https://doi.org/10.1016/B978-0-12-821596-8.00006-9**

pitting and crevice corrosion, where a breakdown in the film can cause high levels of material loss (Fontana and Greene, 1993).

### 2.1.2 Cobalt-based alloys

Cobalt-based alloys have a long history of use in orthopaedics, with the cobalt-chromium-molybdenum alloy (CoCrMo) being the most widely employed. Early trade names for such alloys include Vitallium and Haynes-Stellite 21. CoCrMo alloy composition is defined by ASTM F75 (wrought alloy) and ASTM F1537 (cast alloy).

CoCrMo alloys have a high Young's modulus ($\approx 200$ GPa) and hardness (200 HV10), which enables a high degree of polishing critical for lubrication and allows them to perform well as self-mating sliding materials. Thus, they are typically used in the manufacture of bearing surface components. CoCrMo alloys are not often selected for load-bearing applications, e.g., femoral stems, as the high modulus can lead to complications such as stress-shielding (Sumner and Galante, 1992).

CoCrMo alloy is available in both low-carbon ($<0.15\%$) and high-carbon ($>0.15\%$) variants. High-carbon CoCrMo is harder and, thus, more suitable for sliding bearing surfaces. Like stainless steels, CoCrMo owes its corrosion resistance to a passive oxide layer comprised predominantly of chromium oxides.

### 2.1.3 Titanium-based alloys

Commercially pure titanium and extra-low interstitial Ti-6Al-4V alloy are the two most commonly utilised titanium-based biomaterials. These alloys typically feature in femoral stems, acetabular shells, and modular components and are not used as bearing surfaces due to their poor self-mating tribological properties. Compared to stainless steel and cobalt-based alloys, the theoretical advantages of titanium alloys are their lower modulus of elasticity, reducing the shielding effect, and high corrosion resistance—both owing to a 10-nm thick surface oxide film. Although the mechanical properties of titanium alloys are regarded as superior to those of other biomedical alloys, titanium alloys are often associated with blackening of the adjacent tissue due to wear.

## 2.2 Introduction to tribology

The word 'tribology' is derived from the Greek *tribos*, meaning 'to rub'. It is defined as the science and technology of interacting surfaces in relative motion and of related subjects and practices. The term was first coined in the Jost Report (Jost, 1966), a UK government-commissioned report that examined the cost of friction to the industry. Parameters such as wear, friction, and lubrication of articulating systems, including bearings, all fall within the scope of tribology.

### 2.2.1 Contact of surfaces

When two engineering surfaces are brought into contact and loaded, some deformation will always occur at the contact. This deformation may be purely elastic (Hertz theory), although it is also possible that plastic deformation and, thus, damage to the surface occurs. Fig. 2.1 demonstrates this deformation for a ball-on-flat configuration. The radius (R) of the ball at the contact area deforms under load (W) to produce a circular or ellipsoid contact with a half-width of 'a'. Eq. 2.1 describes the contact pressures associated with such a contact based on the effective modulus and radius of the two surfaces (Hertz, 1881).

$$P_{mean} = \frac{2}{3}P_{max} = \frac{1}{2\pi}\left(\frac{3WE'^2}{R'^2}\right) \tag{2.1}$$

where $P_{mean}$ = Mean Hertzian contact pressure (MPa), $P_{max}$ = Maximum Hertzian contact pressure (MPa), $W$ = Normal load (N), $E'$ = Effective elastic modulus, $R'$ = Effective radius.

Nominally flat surfaces are not perfectly flat and consist of valleys and peaks described by the surface roughness. When two surfaces come into contact and a force is applied, as shown in Fig. 2.2, the peaks will deform such that the load is carried by these asperities. The number of asperity contacts for a given interface is expressed by Eq. 2.2.

$$n = N\int_d^{\infty} \phi(z)dz \tag{2.2}$$

where $n$ = Number of asperity contacts, $N$ = Number of peaks, $d$ = Separation of mean surface levels, $\phi(z)$ = Probability density function, $z$ = Peak heights.

It is possible to treat each asperity contact as a separate Hertzian contact; thus, the true contact area between two surfaces is the sum of the 'n' asperity areas across the interface given by Eq. 2.3. In reality, therefore, an engineering contact consists of interlocking and deformed asperities across the two surfaces. This is critical to understanding how the surfaces behave in contact and under relative motion and how wear of the material occurs.

$$A = \sum A_i = N\pi R_s\int_f^{\infty} \phi(z)(z-d)dz \tag{2.3}$$

where $A$ = Real contact area, $A_i$ = Single asperity contact area, $R_s$ = Characteristic summit radius.

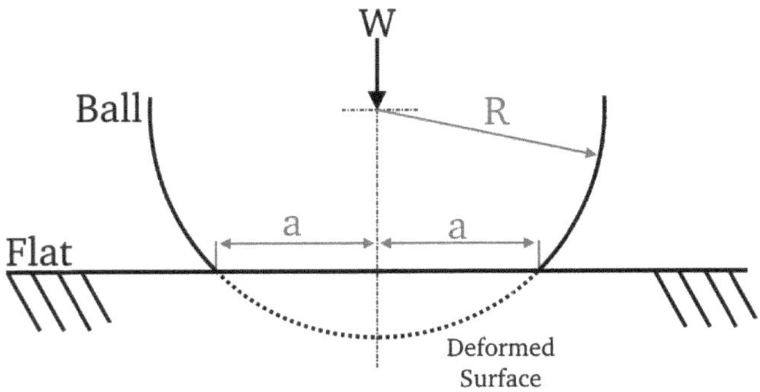


FIGURE 2.1 The contact area between a ball and flat surface under a load.

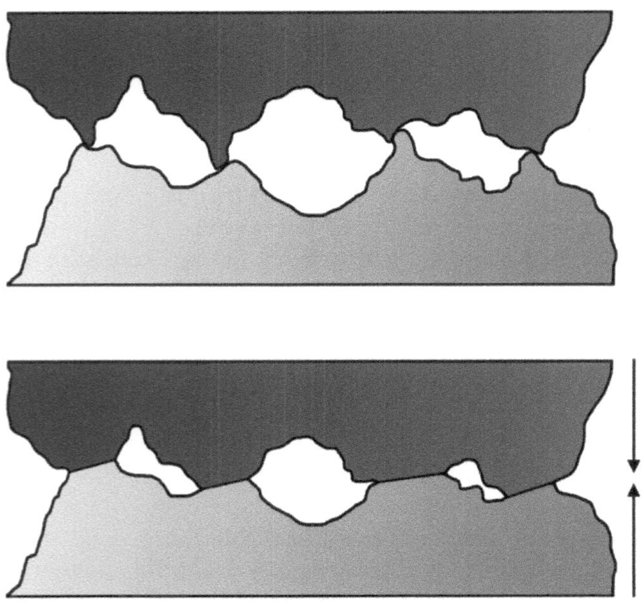

FIGURE 2.2 Asperity deformation in two contacting surfaces.

### 2.2.2 Friction

Friction is the resistance a surface or an object encounters when moving over another. The laws of friction were first stated quantitatively by Guillaume Amontons in 1699 (Williams, 1994). Amontons' First Law states that the resultant frictional force of two surfaces in relative motion is proportional to the normal load (Fig. 2.3).

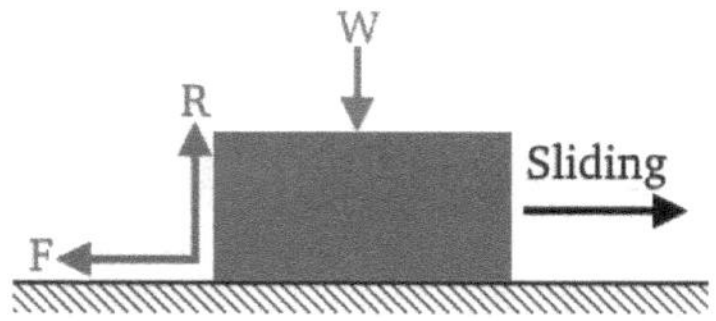


**FIGURE 2.3** Illustration of a body in relative motion over a surface, where W is the normal load (N), R is the reaction force (N), and F is the frictional force (N).

The constant of proportionality was first termed the *coefficient of friction* (μ) by Leonhard Euler, such that:

$$F = \mu W \tag{2.4}$$

where $F$ = Frictional force (N), $\mu$ = Coefficient of friction, $W$ = Normal load (N).

Amontons' Second Law states that the frictional force experienced at the sliding interface of two bodies in contact and the relative motion are independent of the apparent contact area. This is because two surfaces under contact are a series of interlocking and deformed asperities that can be independent of the nominal contact area. The frictional force, in this case, is a function of the true contact area and the specific resistance to shearing of those asperities (Eq. 2.5).

$$F = A \times \tau \tag{2.5}$$

where $F$ = Frictional force (N), $A$ = Real contact area ($m^2$), $\tau$ = Specific resistance to asperity shearing.

### 2.2.3 Wear of materials

Wear is the gradual removal of material from surfaces due to relative motion against each other. Archard proposed that the volumetric wear of a sliding tribosystem in an adhesive wear scenario is proportional to the contact area (Archard, 1953) and is a function of the load, surface hardness, and sliding distance, as shown in Eq. 2.6. This expression can be reduced to normalise wear volume for both the load and sliding distance to produce a wear coefficient (Eq. 2.7), and is often used to represent the wear severity of a system.

$$V = \frac{KWL}{H} \tag{2.6}$$

$$k = \frac{V}{WL} \tag{2.7}$$

where $V$ = Volumetric wear ($m^3$), $K$ = Dimensionless constant, $W$ = Load (N), $L$ = Sliding distance (m), $H$ = Hardness, $k$ = Wear coefficient ($m^3/Nm$).

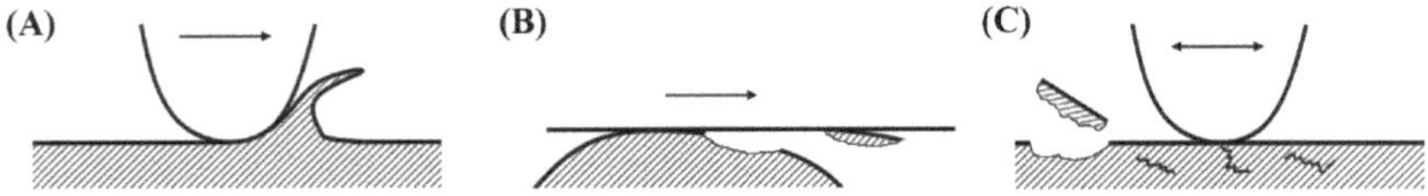


**FIGURE 2.4** The four principal mechanical wear processes: (A) abrasive wear, (B) adhesive wear, (C) fretting/fatigue wear.

There are three principal sliding wear mechanisms typically observed on implanted metallic devices which can act synergistically with corrosive/chemical degradation processes: abrasive wear, adhesive wear, and fretting/fatigue wear. These mechanisms are schematically presented in Fig. 2.4 and discussed below. Other mechanisms of wear exist within the engineering tribological literature albeit they are not linked to biomedical applications. A discussion of these can be found elsewhere (Kato, 2005).

### 2.2.3.1 *Abrasive wear*

Abrasive wear, illustrated in Fig. 2.4A, is a result of asperity contact between a hard and a more ductile surface in relative motion, and occurs particularly when hard third-body particulates are present between the mating surfaces. The most common abrasive wear mechanisms are micro-cutting, ploughing, and grain pull-out.

Attempts to model abrasive wear in the past have seen a good agreement between the theoretical prediction and experimental results (Hokkirigawa and Kato, 1989; Zum Gar, 1987). Eq. 2.8 describes the volumetric wear as a result of single-asperity abrasive wear.

$$V = \alpha\beta\frac{WL}{H} \tag{2.8}$$

where $V$ = Volumetric wear ($m^3$), $\alpha$ = Asperity shape factor, $\beta$ = Degree of wear by abrasive asperity, $W$ = Normal load (N), $L$ = Sliding distance (m), $H$ = Hardness.

### 2.2.3.2 *Adhesive wear*

Adhesive wear, illustrated in Fig. 2.4B, occurs during sliding surface contact where the asperities adhere together due to high local pressure and are 'plucked' from one surface. The debris either remains adhered to the second surface or forms wear particles. Adhesive wear, also termed galling, scoring, or seizing, can increase the surface roughness and, in turn, accelerate the wear of the surface.

Attempts to model adhesive wear have met with difficulty and often do not a produce good agreement with experimental results. However, following on from Archard, adhesive wear volume increases linearly with the load and sliding distance (Kato, 2005).

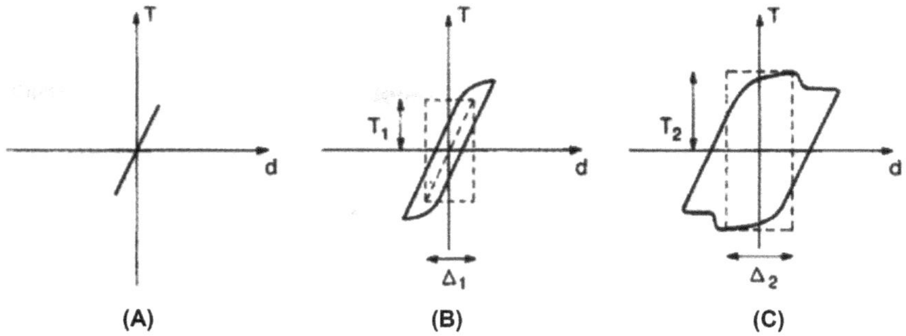


**FIGURE 2.5** Fretting loops under (A) stick, (B) stick-slip, and (C) gross slip (Vingsbo and Soderberg, 1988). *Reproduced with permission from Vingsbo, O., Soderberg, S., 1988. On fretting maps. Wear 126, 131–147, ISSN 0043-1648. https://doi.org/10.1016/0043-1648(88)90134-2.*

### *2.2.3.3 Fretting/fatigue wear*

Fretting, illustrated in Fig. 2.4C, is defined as the reciprocating motion of two contacting surfaces where the displacement amplitude is smaller than the contact area. How a given contact will behave is system and displacement dependent, giving rise to several different fretting 'regimes'. These regimes are described by their fretting loop, calculated by plotting the displacement and tangential force. Examples of the stick, stick-slip, and gross-slip fretting loops can be seen in Fig. 2.5.

In the stick regime (Fig. 2.5A), the conditions and amplitude of displacement result in a relatively low amount of motion. The majority of the displacement is, therefore, elastic deformation of the surfaces, and the contact is static. If lubricated or immersed in an aqueous solution, portions of the contact area will not be exposed to the lubricant.

Under stick-slip (Fig. 2.5B), the loop begins to open as contact areas begin to 'slip'. This will often occur in a ring around the outer edge of the contact, termed *slip annulus*. Then, gross slip occurs with much higher amounts of relative motion and relatively few areas of stick. The area inside the fretting loop (force × displacement) can be used to describe the contact dissipated energy ($E_d$), i.e., the energy put into the surface by fretting.

Fatigue, or 'spalling' (Fig. 2.5C), is a consequence of repeated cyclic sliding. The cyclic stresses induced at the surface result in crack initiation and propagation, which eventually causes material loss. Fatigue wear can lead to high levels of plastic strain at the surface, cause changes in the microstructure of the material, and alter the wear performance of the tribosystem going forward (Williams, 1994). Unlike abrasive or adhesive wear, fatigue wear will have a negligible impact until a critical number of load cycles has been reached.

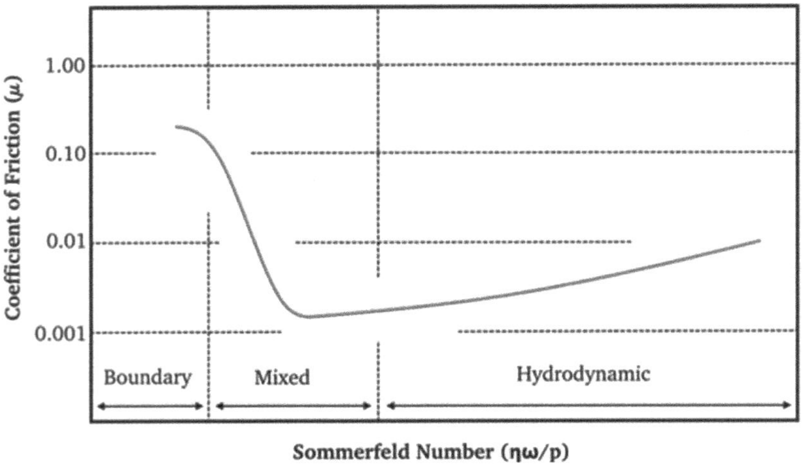


FIGURE 2.6 A typical Stribeck curve.

### 2.2.4 Lubrication

Lubrication of two surfaces in relative motion occurs when a fluid is present within the articulating interface, and can have profound effects on both the friction and wear of the tribological system. Several different lubrication regimes exist that describe how the load is supported at the interface and, ultimately, how the lubrication can affect the wear performance and resulting friction coefficient (Stachowiak and Batchelor, 2005). The transition between regimes can be seen clearly on a Stribeck curve, an example of which is shown in Fig. 2.6.

The Stribeck curve relates the coefficient of friction to the Sommerfeld or Hersey number (Williams, 1994), which is a function of the lubricant viscosity ($\eta$), rotational speed ($\omega$), and load (p). The lubrication regime is linked to the roughness of the sliding surfaces. A modified Stribeck curve that replaces the Sommerfeld number with the lambda ratio ($\lambda$) is often used to describe regimes (Neville and Morina, 2005). The lambda ratio relates the theoretical minimum film thickness generated during sliding to the composite surface roughness (Eq. 2.9).

$$\lambda = \frac{h_{min}}{\sqrt{R_{a1}^2 + R_{a2}^2}} \tag{2.9}$$

where $\lambda$ = Lambda ratio, $h_{min}$ = Theoretical minimum film thickness (m), $R_{a1}$ = Roughness of surface 1, $R_{a2}$ = Roughness of surface 2.

#### *2.2.4.1 Boundary lubrication*

Boundary lubrication is present when the theoretical minimum film thickness is less than the composite surface roughness of the sliding bodies.

The lambda ratio for such a system is less than unity ($0 < \lambda < 1$). Most of the load applied to the interface is supported by asperity contact. Therefore, the frictional force generated during articulation is largely a result of this contact (Stachowiak and Batchelor, 2005). Natural synovial joints tend to operate within the boundary or mixed regime where the wear is dominated by surface interactions (Jin et al., 2000).

#### 2.2.4.2 Fluid-film lubrication

Fluid-film lubrication occurs when the minimum film thickness is much greater than the composite surface roughness of the articulating surfaces, and the associated lambda ratio is much greater than unity ($\lambda > 3$). The exact value of $\lambda$, which indicates full fluid-film lubrication, is system dependent. Under fluid film, the two surfaces are completely separated by the lubricant, which compresses and supports the load applied to the bearing (Neville and Morina, 2005). This type of lubrication is hydrodynamic and the interface frictional force is mostly generated by the viscous properties of the lubricant.

#### 2.2.4.3 Mixed lubrication

Mixed lubrication between the boundary and fluid film exists when the minimum film thickness and composite surface roughness values are comparable. The $\lambda$ ratio is only slightly greater than unity ($1 < \lambda < 3$) in this system (Neville and Morina, 2005). Therefore, the load applied to the bearing is supported by the asperity contact and the lubricant pressure. As the $\lambda$ approaches unity, the regime is termed 'severely' mixed, and a greater proportion of the load relies on asperity interaction. Most hard-on-hard THRs are thought to operate under the mixed regime, with elastohydrodynamic lubrication (EHL) causing elastic deformation of the articulating surfaces due to the high pressures generated in the fluid lubricant (Jin et al., 1997).

#### 2.2.4.4 Lubrication in metal hips

In the case of a metal femoral head articulating with an acetabular cup (Fig. 2.7), the steady-state Hamrock–Dowson equation can be used to calculate the theoretical minimum film thickness for a given bearing couple (Eq. 2.10). This can then be used to calculate the $\lambda$ ratio for that articulating interface and, thus, to determine which lubrication regime the hip is expected to operate in.

$$h_{min} = 2.8R' \left(\frac{\eta u}{E'R'}\right)^{0.65} \left(\frac{W}{E'R'^2}\right)^{-0.21} \tag{2.10}$$

where $h_{min}$ = Theoretical minimum film thickness (m),
$R'$ = Equivalent radius (m) = $\frac{r_1 r_2}{r_2 - r_1}$, *where* $r_1$ = Femoral head radius (m) and $r_2$ = Acetabular cup radius (m),

$\eta$ = Lubricant viscosity (Pa.s),
$u$ = Entraining velocity $\left(\text{m/s}\right) = \frac{\omega r_1}{2}$, *where* $\omega$ = Angular velocity $\left(\text{s}^{-1}\right)$,
$E'$ = Equivalent elastic modulus (Pa) $= 2\left(\frac{1-\nu_1^2}{E_1} + \frac{1-\nu_2^2}{E_2}\right)$,
$W$ = Normal load (N),
$E_1, \nu_1$ = Femoral head elastic modulus and Poisson's ratio,
$E_2, \nu_2$ = Acetabular cup elastic modulus and Poisson's ratio.

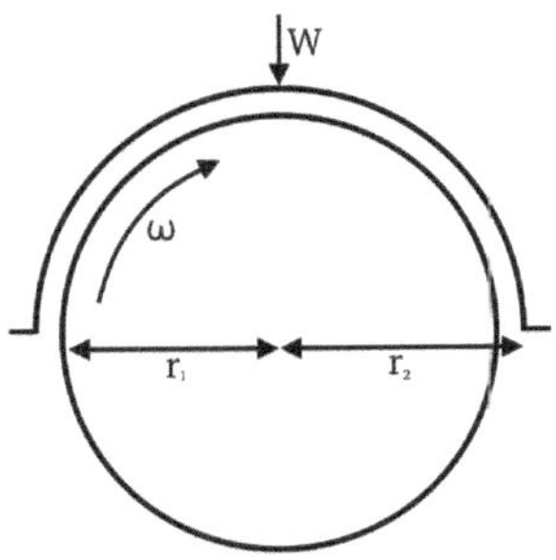


**FIGURE 2.7** Simplified schematic of a femoral head articulating with an acetabular cup. Eq. 2.10 defines variables listed within this figure.

## 2.3 Introduction to corrosion

Corrosion is the destruction or deterioration of a material occurring as a result of a chemical reaction with its environment, and is the sole mechanism of metal ion generation. The environment is typically aqueous, although it can also be gaseous. While corrosion is principally associated with metallic materials, polymers and ceramics are also known to corrode or oxidise. Electrochemistry is defined as the study of chemical reactions that result in the transfer of electrons, typically between an electrical conductor (metal) and an ionic conductor (electrolyte). As the focus of this chapter is metal hip replacements, the principles of corrosion will be discussed in relation to common metals used for biomedical devices in aqueous environments.

### 2.3.1 Thermodynamics and electrochemistry

When in equilibrium, the corrosive process consists of half-cell reactions, namely oxidation and reduction reactions. The dissolution of metal into an aqueous solution (electrolyte), expressed by Eq. 2.11, produces metal ions and free electrons. This loss of electrons is termed oxidation, and the sites at which oxidation reactions occur are known as anodic electrodes (anodes).

$$M \rightarrow M^{z+} + ze^- \tag{2.11}$$

where $M$ = Metal, $M^{z+}$ = Metal ion, $z$ = Valence state, $e^-$ = Electron.

The freed electrons remain within the bulk metal, which creates a potential difference and limits further ion release. A reverse reaction can also occur, i.e., a metal ion rejoining the substrate and forming a dynamic equilibrium. The latter reaction is termed reduction and sites at which it occurs are known as cathodic electrodes (cathodes) (Tait, 1994). Some metals can form a hydroxide layer in an electrolyte which prevents the ions from discharging and returning to the substrate (Talbot and Talbot, 2007).

The liberated electrons can be consumed in other half-cell reduction reactions as oxidising agents come into contact with the metal surface. A common example of a reduction reaction in aerated solutions is the reduction of oxygen, shown in Eq. 2.12. As the electrons are 'mopped up' by oxidising agents, the oxidation reactions continue and corrosive degradation of the metal proceeds at the interface. A layer forms at the metal surface, termed the electrical double layer (EDL), which has both a resistance and capacitance caused by the separation of charge. The EDL is illustrated in Fig. 2.8.

$$\frac{1}{2}O_2 + H_2O + 2e^- \rightarrow 2OH^- \tag{2.12}$$

Oxidising agents, also known as electrochemically active species (EAS), are any species present within the electrolyte that carry a charge. Not all EAS will consume an electron at the metal surface and reduce. Some, such as $Cl^-$ and $SO_4^-$, will form ionic bonds with the released metal ions to produce metal chlorides and sulphates, respectively.

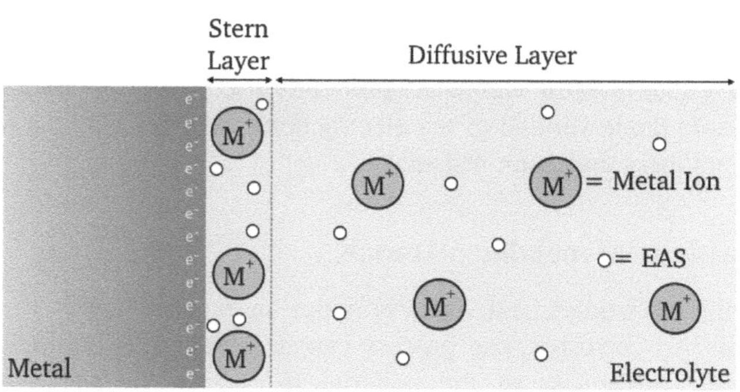


**FIGURE 2.8** Electrical double layer formed by metal ions and electrochemically active species (EAS).

The conditions of equilibrium at a given temperature can be derived from a variant of the Van't Hoff reaction isotherm (Eq. 2.13):

$$\Delta G = \Delta G^0 + RTlnJ \tag{2.13}$$

where $\Delta G =$ Gibbs free energy change, $\Delta G^0 =$ Standard free energy for all reactants, $R =$ Ideal gas constant, $T =$ Temperature, $J =$ Activity quotient

$$J = \frac{[a_{product1}][a_{product2}][etc]}{[a_{reactant1}][a_{reactant2}][etc]}.$$

The Gibbs free energy change ($\Delta$G) is a useful measure of the propensity of a given element to either oxidise or reduce. Elements with $\Delta G > 0$ require energy to oxidise and are, therefore, considered stable. Conversely, oxidation of elements with $\Delta G < 0$ releases energy and may, therefore, happen spontaneously. In the presence of an electrolyte, the ions can diffuse away from the bulk, resulting in material degradation.

At equilibrium, the chemical energies are balanced by the potential acquired at an electrode and can be expressed in the electrical terms shown in Eq. 2.14. This gives rise to the Nernst equation (Eq. 2.15), which mathematically relates the composition of the EDL to the electrical potential (Fontana and Greene, 1993).

$$\Delta G = -zFE \text{ and } \Delta G^0 = -zFE^0 \tag{2.14}$$

$$\Delta E = \Delta E^0 - \frac{RT}{zF} lnJ \tag{2.15}$$

where $F =$ Faraday constant (96,490 C/mol), $E =$ Electrode potential (V), $E^0 =$ Standard electrode potential (V).

The electrochemical series, listed in Table 2.1, ranks common half-cell reactions according to the standard electrode potential. Stable (noble) metals appear towards the top of the table with positive values, whereas metals that are likely to corrode appear towards the bottom with negative values. It is worth noting that metallic elements commonly used within biomedical alloys reside towards the lower end of the electrochemical series, and the high reactivity renders them corrosion resistant.

### 2.3.2 Passivity of metallic materials

Although the electrochemical series provides an excellent guide to a metal's susceptibility to corrosion, the process can be affected by multiple factors, including the propensity of the material to form a surface oxide layer. Chromium, for example, tends to form an oxide ($Cr_2O_3$) spontaneously, both in air and aqueous solutions. This can be seen in Pourbaix diagrams, such as that presented in Fig. 2.9. Given a neutral pH, $Cr^{3+}$ ions will form an oxide across a wide potential range, which is why chromium is often used as an

**TABLE 2.1** The electrochemical series for selected standard electrode potentials at 25°C and pressure of 1 atm.

| Metal | Half-reaction | Standard electrode potential ($E^0$) relative to the standard hydrogen electrode, V |
|---|---|---|
| Gold | $Au^{3+} + 3e^- \rightarrow Au$ | +1.500 |
| Platinum | $Pt^{2+} + 2e^- \rightarrow Pt$ | +1.200 |
| Silver | $Ag^+ + e^- \rightarrow Ag$ | +0.799 |
| Iron | $Fe^{2+} + e^- \rightarrow Fe^{3+}$ | +0.771 |
| Copper | $Cu^+ + e^- \rightarrow Cu$ | +0.520 |
| | $Cu^{2+} + 2e^- \rightarrow Cu$ | +0.337 |
| Hydrogen | $2H^+ + 2e^- \rightarrow H_2$ | 0.000 (by definition) |
| Lead | $Pb^{2+} + 2e^- \rightarrow Pb$ | −0.126 |
| Tin | $Sn^{2+} + 2e^- \rightarrow Sn$ | −0.136 |
| Molybdenum | $MoO_2 + 4H^+ + 4e^- \rightarrow Mo(s) + 2H_2O$ | −0.15 |
| Titanium | $Ti^{2+} + 2e^- \rightarrow Ti$ | −0.163 |
| Nickel | $Ni^{2+} + 2e^- \rightarrow Ni$ | −0.250 |
| Cobalt | $Co^{2+} + 2e^- \rightarrow Co$ | −0.280 |
| Iron | $Fe^{2+} + 2e^- \rightarrow Fe$ | −0.440 |
| Chromium | $Cr^{3+} + 3e^- \rightarrow Cr$ | −0.740 |
| Vanadium | $V^{2+} + 2e^- \rightarrow V$ | −1.13 |
| Aluminium | $Al^{3+} + 3e^- \rightarrow Al$ | −1.660 |
| Magnesium | $Mg^{2+} + 2e^- \rightarrow Mg$ | −2.370 |
| Sodium | $Na^+ + e^- \rightarrow Na$ | −2.714 |

alloying element. Titanium also infers its corrosion resistance from a spontaneously formed oxide, typically $TiO_2$, but the exact chemistry can vary based on the aqueous environment.

The spontaneous formation of a thin oxide layer on a metal's surface, termed 'passivity', can protect the bulk metal from a corrosive environment. The general passivation process can be described by Eq. 2.16.

$$M^{z+} + \frac{z}{2}H_2O \rightarrow \frac{z}{2}MO + zH + ze^- \quad (2.16)$$

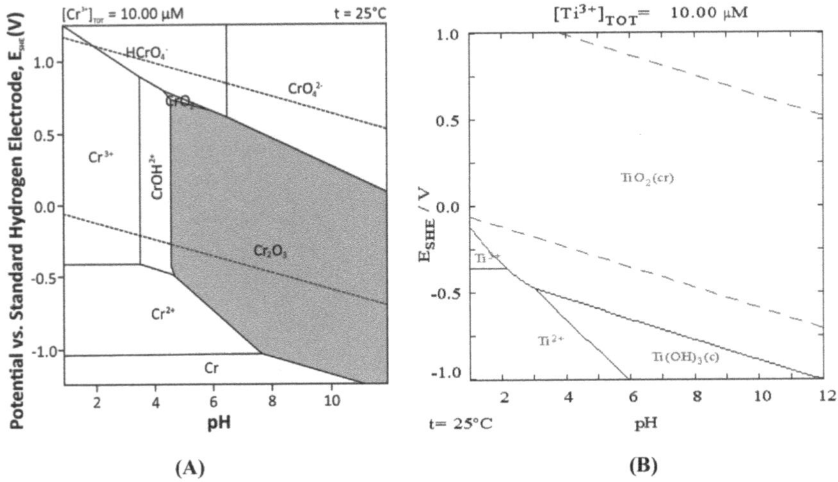


**FIGURE 2.9** Pourbaix diagram for (A) chromium and (B) titanium in pure water (Herdman et al., 2010). *Data taken from Herdman, R.D., Handy, S., Pearson, T., Yamamoto, T., Ishiwata, K., Hara, M., et al., 2010. Chromium alloy coating with enhanced resistance to corrosion in calcium chloride environments. In: US Patent Application, US12/409,629.*

The breakdown of the passive film and corrosion initiation is currently the least understood aspect of the corrosion process. Under static passive conditions, in the absence of any crevices, the breakdown of the passive film is rare and usually only occurs on a tiny, rapid scale. Although the passive layer is usually depicted as an inert barrier covering the underlying metal, in reality, it is much more complicated. In aqueous solutions, the film consists of two layers: an inner oxide layer and an outer hydroxide layer (Borex and Olefjord, 1985). Additionally, the characteristics of the passive film can entirely depend on the composition of the alloy, the environment with which it interacts, and exposure history (Frankel, 1998).

If the oxide layer becomes damaged or detached due to chemical (e.g., reduction in pH or change of driving forces) or mechanical (e.g., wear or stress) forces, corrosion will be accelerated in places where the highly reactive substrate is exposed to the corrosive environment. Although the passivation layer can regenerate itself, not all metal ions liberated from the surface of the metal will contribute to the reformation of the oxide film, and a certain proportion of them will pass into the aqueous solution or environment as free or complex ions. The extent and degree of oxide reformation will depend on several factors that determine re-passivation kinetics: thermodynamic drivers, environment, and, in some cases, mechanical forces.

### 2.3.3 Types of corrosion

A corrosive attack can take on several forms and have different initiation mechanisms. Although Fontana and Greene (1993) described eight main types of corrosion processes, this section will only discuss the most common forms related to biomedical devices.

#### *2.3.3.1 Uniform/general corrosion*

Uniform or general corrosion involves the chemical or electrochemical degradation of a sample that proceeds uniformly over the entire exposed surface. Uniform corrosion is the most commonly associated with metals that do not form a protective oxide layer and is thought to be the most widespread form of corrosion.

#### *2.3.3.2 Galvanic corrosion*

Galvanic corrosion occurs when two dissimilar metals are electrically connected and immersed in, or otherwise exposed to, the same electrolyte. As mentioned in Section 2.3.1, the propensity of a given metal to corrode can be predicted and ranked against that of other metals. In the case of galvanic corrosion, when a metal with a higher tendency to corrode (less noble) is coupled to another metal (more noble), the less noble metal will have a lower potential in a given electrolyte. Therefore, a potential difference is formed between the two metals, and electrons flow from the corroding metal to the other metal. The latter becomes a net cathode and is protected from corrosive attack by the corroding metal, which acts as a net anode. This phenomenon can be exploited to protect metals from corrosive degradation by attaching a sacrificial anode to the system. The rate of galvanic corrosion is affected by several factors, such as the ratio of the anode to cathode surface area and the properties of the electrolyte. The potential of metals for galvanically driven corrosion in a given environment is ranked using the galvanic series, which can be used to predict the preferentially corroding metal for a given couple and guide the selection of metals to be joined (Fontana and Greene, 1993).

#### *2.3.3.3 Crevice corrosion*

Crevice corrosion is a form of localised corrosion that occurs when a volume of electrolyte close to a metal or alloy surface is stagnant. This stagnant volume becomes starved of oxygen and ionic species migrate into the crevice. Crevice corrosion is typically observed in alloys that form a passive surface oxide layer which usually acts as a barrier to charge transfer and protects the bulk alloy from electrochemical degradation.

Once crevice corrosion is initiated, the passive layer is disturbed within the crevice but remains intact outside of it. Several different models for the crevice corrosion initiation have been proposed but once initiated, the reaction can become self-sustaining and rapidly corrode the bulk alloy within the crevice (Fontana and Greene, 1993). Fig. 2.10 illustrates crevice corrosion initiation and propagation processes occurring at the stem–cement interface of a hip implant.

### 2.3.3.4 Pitting corrosion

Pitting corrosion is another form of localised corrosive attack and, like crevice corrosion, is usually associated with alloys that owe their corrosion resistance to a protective oxide film. Interruption of the film creates an anodic site, which is electrically coupled to the remaining passive film outside the pit. The passive film acts as a cathode, creating a local galvanic circuit.

Local passivity can be lost by the actions of aggressive anions, most notably chloride ($Cl^-$). Pits initially go through a metastable phase where

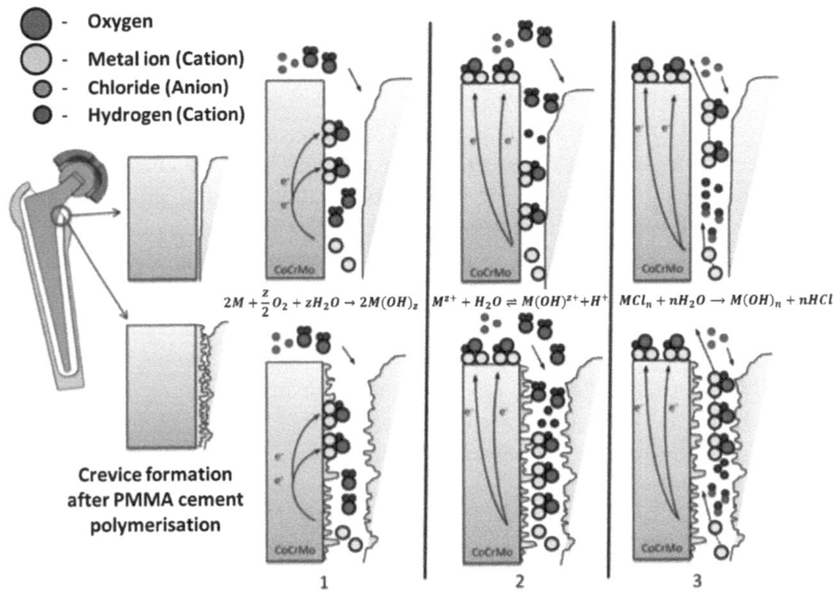


**FIGURE 2.10** Schematic representation of the initiation of crevice corrosion of cemented femoral stems. Stage 1: The cathodic current equals the anodic current within the crevice, reducing oxygen to hydroxyl ions. Stage 2: Metal cations from the alloy pass into the solution and hydrolyse. The metal ion concentration in the crevice increases until the solubility product of one or more of the metal hydroxides is exceeded. Stage 3: Metal hydroxides precipitate from the crevice into the bulk solution due to an electrochemical imbalance. $H^+$ ions that are generated in the process reduce the pH of the crevice solution. To ensure electrochemical equilibrium, $Cl^-$ ions enter the crevice, further reducing the pH through the formation of hydrochloric acid.

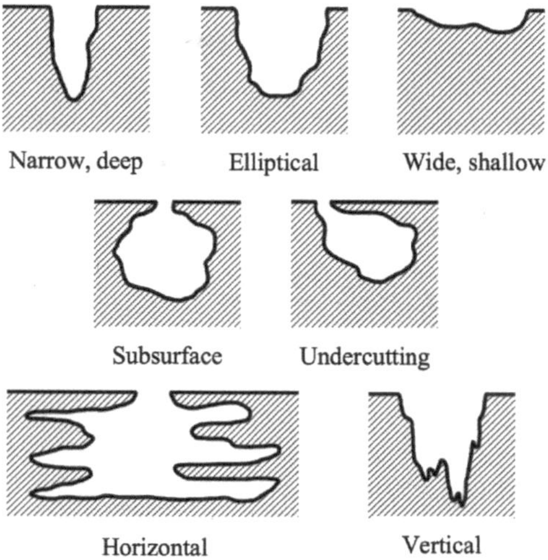


**FIGURE 2.11** The different pit geometries, as defined in ASTM G46 (ASTM International, 2013).

passivity can be temporarily restored before the pit's geometry and chemistry make the pit stable. This only occurs once the local potential passes the pitting potential of the metal and causes rapid material loss within the pit (Pistorius and Burstein, 1992). The film can also be interrupted by physical damage to the surface or, in the case of metal coatings deliberately applied to prevent corrosion, by a defect in the coating layer.

Pits can take on many morphologies, as shown in Fig. 2.11, and one of the most complex forms of corrosion to combat. For a relatively small amount of material loss through corrosive attack, pitting can cause extreme structural weakness within a metal, resulting in the failure of a component.

#### *2.3.3.5 Intergranular corrosion*

Intergranular corrosion is the preferential corrosive attack at, or adjacent to, the metallic grain boundaries of an alloy. In biomedical alloys, chromium is often added to aid in forming a passive oxide film. Chromium carbides ($Cr_{23}C_6$) can precipitate along the grain boundaries of these materials through a process known as sensitisation, which leaves chromium-depleted zones along the grain boundary. The depleted regions have a different potential to that of the bulk alloy and act as anodes, creating a galvanic circuit and leading to electrochemical degradation of the material at the grain boundary.

## 2.4 Tribocorrosion

As mentioned previously, several metals and alloys owe their corrosion resistance to a passive oxide film that forms spontaneously on their surface.

When two surfaces under relative motion (sliding of fretting) come into contact, the passive film can become damaged, exposing the underlying bulk metal to the electrolyte. The dissolution of metal ions can proceed in the wear track, which becomes an anodic site, as depicted in Fig. 2.12. This may form a local galvanic link with the undisturbed film outside the wear scar, which acts as a cathode (Vieira et al., 2012).

Some of the released ions will react with oxygen in the electrolyte to repair the passive film; however, this process is not instantaneous, and the material is free to degrade in the meantime. It must be noted that corrosion processes are the sole source of metal ions, whereas metallic particles arise from the mechanical removal of material.

In some tribological systems, such as the bearing of a metal-on-metal (MoM) hip replacement or modular tapers, mechanical abrasion in the presence of an electrolyte causes the passive film to be continuously disturbed and reformed as a result of mechanical wear and corrosive phenomena. A method for calculating the synergistic effect of these processes, first proposed by Watson et al. (1995), is shown in Eq. 2.17. Watson highlighted that the combined effects of wear and corrosion accelerate the rate of material degradation in such systems. The de-passivation of the oxide film through wear can accelerate corrosion ($C_w$). Conversely, the electrochemical degradation of the surface can increase the surface roughness or weaken grain boundaries and, in turn, facilitate material loss through wear ($W_c$). Depending on the tribological system, these synergistic effects can account for 12%–42% of material loss (Hesketh et al., 2013a; Mathew et al., 2012).

$$T = W_0 + C_0 + S \tag{2.17}$$

where $T$ = Total material loss, $W_0$ = Pure wear (in absence of corrosion), $C_0$ = Pure corrosion (in absence of wear),
$S$ = Synergistic material loss = $W_C + C_W$,
*where* $W_C$ = Wear accelerated by corrosion and $C_W$ = Corrosion accelerated by wear.

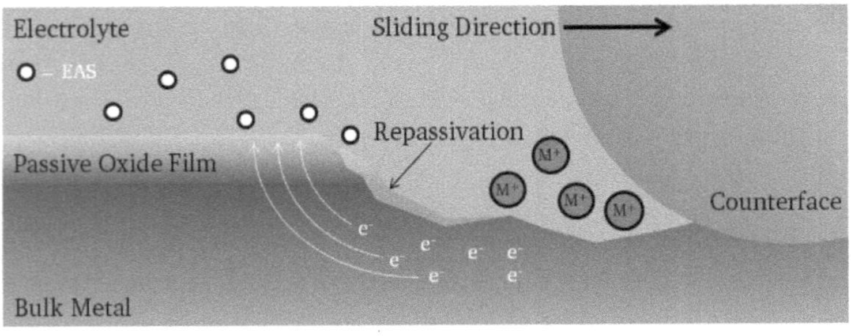


**FIGURE 2.12** De-passivation of a protective oxide film as a result of the sliding action of a counterface. *EAS*, electrochemically active species; $M^+$, metal ion.

## 2.5 Modern hip replacements

Modern MoM hip replacements introduced over the last two decades generally fall into two broad categories: conventional THR, most of which include modular interfaces, and HR implants.

A modular THR, illustrated in Fig. 2.13, consists of a femoral stem, fixed either using cementless osseointegration or conventional bone cement, and a femoral head attached to the stem by a Morse-type taper. The acetabular side of the device is typically comprised of two components, also assembled using an interference-fit fastener. An acetabular shell fixed to the bone holds an acetabular liner that acts as the counterface for the femoral head to articulate against. HR devices forgo a traditional stem and preserve bone stock by using a component designed to simply replace the articulating surface of the femoral head.

This section aims to highlight and discuss the pathways to degradation across the different interfaces for these devices.

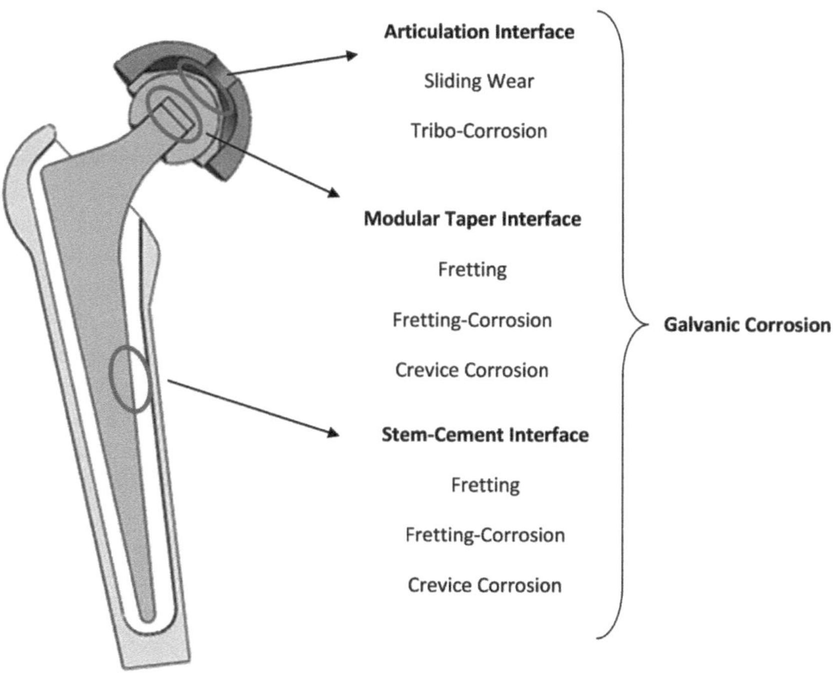


**FIGURE 2.13** Schematic of a typical MoM THR implant with the possible sites of metal debris release and corresponding tribocorrosion mechanisms (Bryant et al., 2014). *Reproduced with permission from Bryant, M., Farrar, R., Freeman, R., Brummitt, K., Nolan, J., Neville, A. 2014. Galvanically enhanced fretting-crevice corrosion of cemented femoral stems. J Mech Behav Biomed Mater. 40, 275-286. https://doi.org/10.1016/j.jmbbm.2014.08.021. Epub 2014 Sep 4. PMID: 25259666.*

### 2.5.1 Sources of degradation

The degradation of orthopaedic biomaterials is a complex phenomenon that depends on multiple factors, such as those related to the surgeon, implant, and patient (Karachalios et al., 2018). The process is undesirable not only because it can compromise the structural integrity of the implant but also because it may elicit an adverse biological reaction in the host through the release of degradation products (Jacobs et al., 1998a,b). Although the degradation of hip implants is rarely the only cause of their failure, the subject has been under much discussion and attracted national press attention. Clinical literature reports that the modular nature of hip implants, fixation, and lubrication regimes underpinning their performance all contribute to mechanical and corrosive material loss.

#### *2.5.1.1 Bearing surfaces*

The introduction of modern MoM bearings addressed the need for long-lasting devices for younger and more active patient subgroups. In vitro hip simulator studies of these implants demonstrated up to a 100-fold reduction in gravimetric/volumetric wear compared to metal-on-polyethylene (MoP) bearings (Firkins et al., 2001). The bearing surface was designed to operate under EHL, with a degree of the bearing's load supported by the fluid pressure in a thin film of entrained lubricant. Evidence of complete bearing surface separation and, thus, no sliding contact was noted in hip simulator studies (Dowson et al., 2000). However, this separation only occurred over portions of the gait cycle and in ideal laboratory hip simulator conditions. Clinically and in vivo, bearings still engage in sliding contact, resulting in the tribocorrosion processes described in Section 2.4. Comparing the wear of MoM bearings with that of MoP bearings neglected the effects of degradation mechanisms specific to sliding metallic interfaces. Mechanical loss of the surface passive oxide film generated high levels of metallic debris and ions in patients with MoM implants (Hart et al., 2010) and contributed to early failures of particular device designs (National Joint Registry, 2015).

Additionally, evidence emerged that MoM bearings in THR and HR implants could be sensitive to the angle of inclination of the acetabular cup (De Haan et al., 2008; Hart et al., 2008; Langton et al., 2008). Higher inclination may lead to 'edge-loading' during articulation, causing higher contact pressures, and compromise the load-supporting ability of the lubricant (Fisher, 2011).

Other studies highlighted a phenomenon termed 'microseparation', whereby the bearing separates during gait (Clarke et al., 2004). The condition could be a result of a number of factors, such as joint laxity or component migration, causing a much more severe contact on heel strike (Liu et al., 2015).

The synovial fluid also plays a role in bearing performance. Its constituents ($\alpha$-globulin, $\beta$-globulin, and hyaluronic acid) can form carbon-rich tribochemical reaction layers that act as solid lubricants, reducing adhesive and

**FIGURE 2.14** The femoral head (left) and acetabular liner (right) from an explanted McKee–Farrar prosthesis displaying visible tribofilm formation on both bearing surfaces (Wimmer et al., 2003). *Reproduced with permission from Wear.*

abrasive wear (Wimmer et al., 2010, 2003) (Fig. 2.14). The layers contain oxygen, calcium, nitrogen, sulphur, and sodium, and their formation is thought to be the result of the denaturation of proteinaceous material in the synovial fluid, possibly due to the high mechanical shear at the interface and interaction with metal ions (Maskiewicz et al., 2010).

### *2.5.1.2 Modular tapers*

Due to the flexibility of modular THR systems, femoral stems with conical Morse-type taper connections for interchangeable heads are often the desired choice of surgeons. Although the change in design philosophy from monobloc systems to modular implants offers intraoperative advantages, it introduces additional interfaces susceptible to wear and corrosion. Thus, these devices have inherent shortcomings, and their performance has been subjected to extensive scrutiny over the past decade. One study revealed that approximately 16%–35% of 138 retrieved THRs showed signs of moderate to severe corrosive attack at the taper connection (Gilbert et al., 1993). This was observed in mixed-metal implants with a titanium-alloy femoral stem and cobalt-alloy femoral head, as well as in devices in which the stem and head were both made of a cobalt alloy. The femoral tapers were subject to crevice, galvanic, and fretting corrosion. Although these processes alone would not have caused the failure of the device, when combined with stress and motion, the results were devastating. Corrosion products generated at the taper connection reportedly migrate into the periprosthetic tissue (Urban et al., 2000) and bearing surfaces, causing further complications (Jacobs et al., 1998a,b).

Fretting-crevice corrosion, also known as mechanically assisted crevice corrosion, is a term coined to describe the complex degradation mechanisms at modular interfaces (Gilbert et al., 1993). The combination of an interfacial crevice and mechanical micromotion creates an aggressive environment that

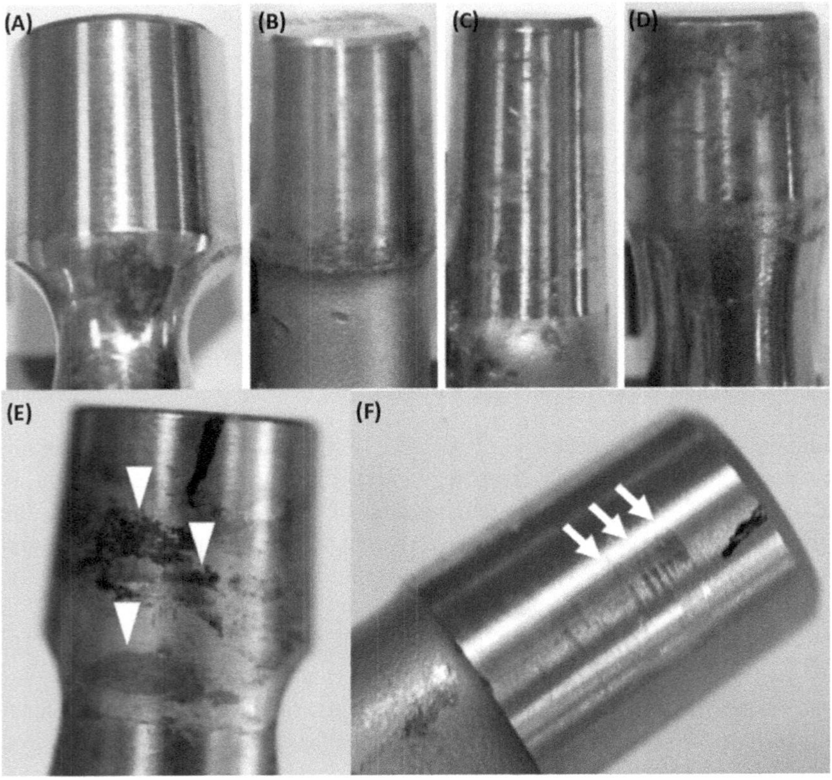


**FIGURE 2.15** Top panels: Representative digital images from grading of fretting and/or corrosion damage of trunnions, including none (grade = 1) (A), mild (grade = 2) (B), moderate (grade = 3) (C), and severe (grade = 4) (D). Bottom panels: Digital images illustrating the differences between fretting and corrosion damage. (E) Severe corrosion in the form of adherent black oxide, as well as focal discolouration without gross evidence of mechanical damage at the base of the taper (arrowheads). (F) Fretting in the form of mild mechanical damage in the middle of the taper (arrows) (Siljander et al., 2018). *Reproduced with permission from the Jorunal of Arthroplasty.*

accelerates the degradation of metallic surfaces in vivo (Fig. 2.15). Similar degradation mechanisms have also been observed at the acetabular shell–liner interfaces and femoral head–stem tapers in bimodular implants. To date, there is little consensus as to which design variables affect fretting-crevice corrosion at modular taper interfaces.

### *2.5.1.3 Stem–cement interface*

Degradation of the cemented portions of stemmed metal devices as a result of fretting-crevice corrosion is commonly observed in retrieval studies.

Although it rarely necessitates intervention, this mechanism was implicated in the unacceptable revision rates of some femoral stem designs.

A study in the 1970s examined a cohort of 28 cemented Müller straight titanium-alloy femoral stems that were mainly revised for pain and discomfort (Willert et al., 1996). The pain, which differed from that in aseptic loosening, occurred approximately 14.5 months after implantation. Revisions were performed at an average of 25.5 months after implantation. Upon retrieval, it was noted that abraded metallic particles were dominant in the joint capsule, whereas corrosion products impregnated the cement–bone interface.

More recently, degradation at the stem–cement interface has been associated with the early failure of DePuy Ultima TPS MoM hip replacements, which featured cemented CoCr-alloy stems. Of the 90 hips examined by Donell et al. (2010), 17 required revision for periprosthetic fracture, with early dislocation in 3 and late dislocation in 16 of the 90 hips. Nine hips were infected. In addition, symptomatic, peri-articular soft-tissue necrosis was found in 44 hips, of which 35 had normal plain radiographs. Dramatic corrosion of generally well-fixed femoral stems was frequently observed on the cemented part of the retrieved femoral component (Fig. 2.16). Necrosis

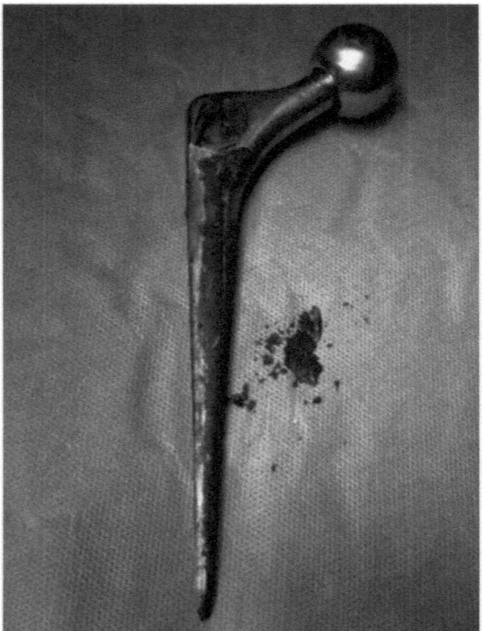

**FIGURE 2.16** Explanted femoral stem with extreme corrosion at the stem–cement interface. *Image provided by John Nolan.*

of implant-adjacent tissue was attributed to the release of metal ions, such as cobalt and chromium ions, from the stem surface. Similar observations have been made by Bolland et al. (2011), who revealed high corrosion of the cemented portions of Zimmer CPT CoCr-alloy femoral stem. Again, elevated metal ion levels and extensive soft-tissue necrosis were observed at revision.

### 2.5.2 Adverse reaction to metal debris

Adverse reaction to metal debris (ARMD) and adverse local tissue reaction (ALTR) are general terms used to describe soft-tissue conditions thought to be related to the release of metal debris and ions from the implant into the joint (Willert et al., 2005). The exact definitions of the different conditions are unclear and are sometimes used interchangeably. They include aseptic lymphocyte-dominated vasculitis-associated lesion (Watters et al., 2010), metallosis (Korovessis et al., 2006), and pseudotumours (Campbell et al., 2010).

ARMD is generally diagnosed by pain reported by the patient and has also been linked to other failure mechanisms, such as osteolysis and aseptic loosening (Park et al., 2005; Willert et al., 2005). Upon revision surgery, aspirated synovial fluid is often cloudy and dark in colour while periprosthetic soft tissue can be stained grey and be necrotic (von Schewelov and Sanzen, 2010). ARMD can also be diagnosed by elevated levels of metal ions in the blood of otherwise asymptomatic patients. The blood/serum metal levels associated with different types of hip implants, and guidelines for managing patients with metal implants, are discussed in Chapter 4 and Chapter 8, respectively.

### 2.5.3 Assessing material loss from metal hip implants

During preclinical testing in hip simulators, implants undergo a gravimetric and volumetric assessment of wear as set out in the ISO 14242 standard (International Organization for Standardization, 2016).

Gravimetric assessment involves simply weighing the bearing components at set intervals before, during, and after simulation. Any change in mass is ascribed to material lost due to the wear of the parts, and the results can be converted to volumetric loss using a simple density conversion. The method comes with several complications that must be mitigated in order to achieve an accurate measurement. First, the simulator must be designed in such a way that the components can be removed and replaced at each weigh-point without affecting the interface or orientation of the component when the test is resumed. Secondly, the balance used to weigh the components must be of sufficient accuracy to register the small mass changes expected (in the order of micrograms). The ISO 14242–2 standard calls for accuracy of at least $\pm$ 100 μg (International Organization for Standardization, 2016), although for metal components, the magnitude of mass change can be so small that a five- or six-point balance may be required. The components must also undergo a set cleaning procedure before each weigh-point to remove the wear debris and deposits from the surface. With the above factors well controlled, gravimetric assessment is often considered as the 'gold standard' for evaluating implant wear.

Under volumetric or 'dimensional change' assessment, the surface of the bearing component is mapped using a coordinate measuring machine (CMM). This can be done using either a standard, contact-based CMM with a ruby stylus or an optical/white light interferometry-based instrument. Like in gravimetric assessment, the components are removed from the simulator at set intervals before being cleaned and re-measured. To control for volume fluctuations, the components must be allowed to acclimatise at the experimental temperature before the test is run. The generated surface maps can be compared to previous measurements to calculate the dimensional change at the surface and, thus, the volume lost. Unlike gravimetric analysis, volumetric analysis can be used to assess the wear of retrieved implants or components with no before-articulation measurements by assuming a perfect sphere. Determining the wear scar can often be difficult, however, and is user subjective. Bergiers et al. (2020) found a good correlation between gravimetric and volumetric assessment when using custom automated software to determine worn data points and calculate volume loss.

Gravimetric and volumetric methods were largely developed around MoP bearings and only determine the total mass or volume loss; thus, they do not provide insight into the wear mechanisms or pathways to degradation at MoM interfaces. Furthermore, they are not useful for assessing the wear of devices in vivo.

It is possible to obtain an in situ corrosion measurement by equipping hip simulators with a three-electrode electrochemical cell that measures the electrochemical reactions taking place at the metal surfaces during sliding (Beadling et al., 2017; Bingley et al., 2018; Hesketh et al., 2013b). Alternatively, volume loss due to corrosion (sliding corrosion at the bearing surface or corrosion of metallic wear debris) can be assessed by quantifying the metal ions released into the test medium using methods such as inductively coupled plasma–mass spectrometry (ICP-MS). Kretzer et al. (2010) found the technique useful during simulation of ultra low-wearing MoM hip replacements where gravimetric assessment lacked the sensitivity to determine mass losses.

ICP-MS can also measure the ions generated by an implant in vivo by assessing their levels in the patient's blood or urine. However, systemic levels of implant-derived metal ions and the degree of patient sensitivity to them are highly variable, meaning that the relationship between the systemic metal load, implant wear, and severity of adverse reactions is not straightforward.

### 2.5.4 Studying metal deposits in tissue

Implant degradation gives rise to metal debris which, depending on its physicochemical properties and the severity of wear, can become incorporated into the periprosthetic tissue or accumulate in distant tissues and organs. Probing the distribution and chemical form of the metal deposits can shed light on the implant degradation mode and provide clues as to how metal debris is metabolised by the body and the likely toxicity mechanisms at play. With the advent of modern and accessible electron and X-ray microscopy, techniques such as transmission electron microscopy, micro-X-ray fluorescence, and micro-X-ray absorption spectroscopy are being increasingly adopted by researchers to better understand the nature and origin of implant-derived debris. Table 2.2 shows examples of methodologies that can be used to study metals in tissue.

#### *2.5.4.1 Periprosthetic tissue*

Periprosthetic tissue samples can be harvested during revision surgery and probed for metal distribution and speciation. Synchrotron analyses revealed that the predominant species in tissues surrounding MoM implants was chromium phosphate ($CrPO_4$), with cobalt present occasionally in areas of high chromium concentration (mainly as metallic cobalt or CoCr particles) (Hart et al., 2010). However, it has also been noted that the chemical speciation of the debris varies depending on the nature of the interfaces present within the THR system.

Di Laura et al. (2017) reported an abundance of $CrPO_4$ and $Cr_2O_3$ within the tissues adjacent to MoM HR devices, where sliding tribocorrosion can be expected. Soft tissues local to bi-modular THRs, where fretting-crevice corrosion mechanisms have been reported, were rich in $Co^{2+}$, metallic cobalt, $CrPO_4$, $Cr_2O_3$, and $TiO_2$ (anatase, rutile, or amorphous forms).

Hart et al. (2012) examined two types of failed MoM devices—Ultima TPS modular implants and HR—and reported that visible corrosion was only observed in the former implant design. Additionally, retrieved tissues adjacent to Ultima hips contained relatively more cobalt than chromium. The authors ascribed these findings to different dominant degradation mechanisms taking place (stem corrosion in Ultima hips vs bearing surface wear in HR implants) and hypothesised that the preferential loss of cobalt over chromium from Ultima stems was caused by the acidic environment created by tissue inflammation. Although $CrPO_4$ and $Cr_2O_3$ have a certain toxic potential, the aforementioned observations concluded that the soft-tissue reactions seen in patients with MoM hips are primarily caused by free cobalt ions ($Co^{2+}$).

More recently, Xia et al. (2017) employed a combination of histological and electron microscopy techniques to investigate the nature of wear

**TABLE 2.2** Examples of techniques used to study metal composition of human tissue samples, and the information they can provide (Swiatkowska et al., 2018).

| Technique | Metal mapping in tissue | Metal mapping in cells | Type of metal | Oxidation state | Surrounding atoms | Sample preparation | Destructive? | Reference |
|---|---|---|---|---|---|---|---|---|
| Light microscopy | Yes | Yes | No | No | No | H&E staining | Yes | Di Laura et al. (2017), Hart et al. (2010), Huber et al. (2009) |
| Laser ablation ICP-MS | Yes | No | Yes | No | No | Dewaxing | Yes | Swiatkowska et al. (2018) |
| TEM | No | Yes | Yes | Yes | Yes | Fixation, dehydration, resin embedding, semithin sectioning, staining | Yes | Case et al. (1994), Xia et al. (2017) |
| SEM/EDXA | No | Yes | Yes | No | No | Fixation, dehydration, drying, metal coating | Yes | Huber et al. (2009), Urban et al. (2004) |
| μ-XRF | Yes | No | Yes | No | No | Dewaxing | No | Abdel-Gadir et al. (2016), Hart et al. (2012, 2010) |
| μ-XAS | No | No | Yes | Yes | Yes | Dewaxing | No | Abdel-Gadir et al. (2016), Wu et al. (2016) |

*H&E*, haematoxylin and eosin; *ICP-MS*, inductively coupled plasma–mass spectrometry; *TEM*, transmission electron microscopy; *SEM/EDXA*, scanning electron microscopy with energy dispersive X-ray analysis; *μ-XRF*, micro-X-ray fluorescence; *μ-XAS*, micro-X-ray absorption spectroscopy.
*Source*: Adapted with permission from the Journal of Trace Metals in Medicine and Biology.

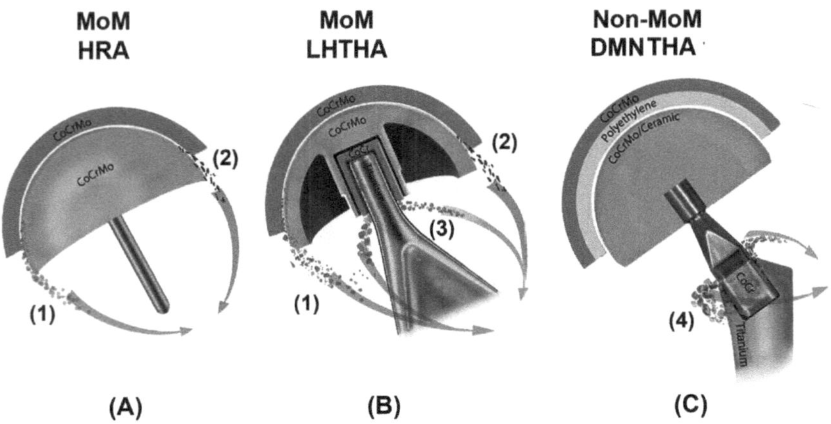


**FIGURE 2.17** Three common classes of hip implants associated with ALTR. Left: Metal-on-metal hip resurfacing arthroplasty (MoM HRA). Metallic particles are produced at the bearing surface by sliding tribocorrosion (1) and edge-loading (2). Middle: Metal-on-metal large-head total hip arthroplasty (MoM LHTHA). Metal particles are produced at the bearing surface by sliding tribocorrosion (1) and edge-loading (2), and at the metallic stem–adapter sleeve interface by fretting and crevice corrosion and, possibly, abrasion (3). Right: Non-MoM bearing surface (metal-on-polyethylene or ceramic-on-polyethylene) THA with a CoCrMo dual-modular femoral neck (Non-MoM DMNTHA). Metal particles are produced by fretting and crevice corrosion at the neck–stem interface (4) (Xia et al., 2017). *Reproduced with permission from Nanomedicine: Nanotechnology, Biology and Medicine.*

particles within tissues surrounding three implant classes, illustrated in Fig. 2.17. Table 2.3 summarises the findings of the study. Debris of varying shape, size, and composition was identified depending upon the source interface. Despite the lowest periprosthetic particulate burden, patients with bi-modular implants had the highest amounts of lymphocytes and tissue destruction and the shortest time to implant failure. Thus, metal debris generated at the femoral neck–stem junction may be more immunogenic and cytotoxic than bearing wear debris, leading to a quicker onset and higher severity of ALTR. These conclusions were later corroborated by Lehtovirta et al. (2018).

### *2.5.4.2 Organ tissue*

Obtaining systemic tissue specimens for analysis is less straightforward because biopsy procedures are invasive, and post-mortem organ samples are not readily available for research purposes. Thus, the current evidence is not always robust.

Post-mortem studies using electron microscopy detected increased metal levels in the lymph nodes, spleen, and liver samples from patients with THA, with the highest metal loads associated with loose or worn prostheses and a

**TABLE 2.3 Summary of size, morphology, and composition of implant-derived particulate debris (Xia et al., 2017).**

| | | MoM resurfacing | Large-diameter MoM THR | MoP/CoP bi-modular THR |
|---|---|---|---|---|
| Size | Nano- | < 10 nm<br>10–800 nm | < 10 nm<br>10–2250 nm | < 10 nm, rare<br>10–400 nm |
| | Micro- | Not present | 1–30 μm | 1–30 μm |
| Shape | | Circular, irregular, small, needle-like | Circular, irregular, small and large, needle-like | Circular nanoparticles and large agglomerations |
| Composition | | Cr >> Co, Mo, P, O | Cr > Co, Mo, Ti, V, P, O<br>Not co-localised | Cr, Co, Mo, Ti, V, P, O<br>Co-localised |

*MoM*, metal-on-metal; *MoP*, metal-on-polyethylene; *CoP*, ceramic-on-polyethylene; *THA*, total hip arthroplasty.
*Source*: Adapted with permission from Nanomedicine: Nanotechnology, Biology and Medicine.

history of multiple revision surgeries (Case et al., 1994; Urban et al., 2004). Case et al. (1994) showed that heavy accumulation of wear debris caused fibrotic and necrotic changes in the lymph nodes but a similar effect was not seen in the spleen or liver, possibly due to the larger size of these organs.

To date, only a handful of groups attempted to determine the exact chemical form of metal debris in the distant organs of patients with hip replacements. Abdel-Gadir and Berber et al. (2016) reported on a 44-year-old individual with catastrophic third-body wear of a CoCr-alloy femoral head and highly elevated blood cobalt and chromium levels. Synchrotron analysis of the patient's liver biopsy revealed an abundance of highly co-localised cobalt and chromium signals in hepatic macrophages. Despite the high particulate load, the results of liver function tests were normal, suggesting that the deposits were well tolerated, at least at the time of assessment.

Swiatkowska et al. (2018) examined necropsy samples of cardiac, hepatic, and splenic tissue from patients with MoP THRs. Cobalt was typically present in the divalent oxidation state, whereas titanium was found exclusively as titanium dioxide, in either anatase or rutile form. High oxidation states of chromium were also identified, uncovering the possibility that these potentially toxic forms may arise in the vital organs of THR recipients. Further work is needed to ascertain whether ex vivo findings in post-mortem tissue can be extrapolated to living patients.

Metal toxicity in the setting of THA is discussed in more detail in Chapters 3 and 7.

## 2.6 Summary and future directions

The degradation of metallic biomaterials is a nuanced, system-dependent, multifactorial process. In vivo, THR interfaces are subjected to harsh chemical environments promoted by device geometry, complex physiological loading, and articulation patterns, and produce wear debris and ions that go on to affect the system. How devices with metal surfaces in sliding interfaces perform tribologically and lose material through those processes dictates the implant's success or failure.

Following the failures of the recent generation of MoM devices, much effort has gone into understanding the clinical indications for revision of an implant. As seen in patient studies, metal hips appear sensitive to high inclination or malposition/component migration. The modular taper connections on some devices have also been highlighted as a compounding factor and may have been the main source of material loss. This has led to certain adaptions in preclinical device testing, such as additions to the ISO 14242 standard. ISO 14242−4:2018 (Part 4) introduces steep cup inclination angles and dynamic separation to hip simulation methods to replicate the adverse conditions found clinically.

Little has been added to the preclinical analysis of MoM bearings that fully elucidates the degradation mechanisms or goes beyond simple wear comparisons to MoP bearings. More is to be done in understanding and representing the adverse daily living activities metal hips are subjected to in vivo. This includes the development of new sensing techniques to better interrogate the degradation mechanisms in situ and in real time. With the recent European Union's Medical Device Regulations stipulating the need for all manufacturers to assess any nanoparticulate debris derived from a medical device, there is a renewed driver to better understand and explore the nature of implant-derived debris. Doing so will be critical going forward, not only to ensure that the degradation of devices with metallic sliding interfaces is fully understood, but also to inform the treatment of patients with such devices already implanted.

## References

Abdel-Gadir, A., Berber, R., Porter, J.B., Quinn, P.D., Suri, D., Kellman, P., et al., 2016. Detection of metallic cobalt and chromium liver deposition following failed hip replacement using T2* and R2 magnetic resonance. J. Cardiovasc. Magn. Reson. 18, 29.

Archard, J.F., 1953. Contact and rubbing of flat surfaces. J. Appl. Phys. 24, 981−988.

ASTM International, 2013. ASTM G46−94—Standard Guage for Examination and Evaluation of Pitting Corrosion.

Beadling, A.R., Bryant, M.G., Dowson, D., Neville, A., 2017. A link between the tribology and corrosive degradation of metal-on-metal THRs. Tribol. Int. 113, 354−361.

Bergiers, S., Hothi, H., Richards, R., Dall'Ava, L., Henckel, J., Hart, A., 2020. Quantifying material loss from the bearing surfaces of retrieved hip replacements: method validation. Tribol. Int. 142, 105975.

Bingley, R., Martin, A., Manfredi, O., Nejadhamzeeigilani, M., Oladokun, A., Beadling, A.R., et al., 2018. Fretting–corrosion at the modular tapers interface: inspection of standard ASTM F1875-98. Proc. Inst. Mech. Eng. H. 232, 492–501.

Bolland, B.J., Culliford, D.J., Langton, D.J., Millington, J.P., Arden, N.K., Latham, J.M., 2011. High failure rates with a large-diameter hybrid metal-on-metal total hip replacement: clinical, radiological and retrieval analysis. J. Bone Joint Surg. Br. 93, 608–615.

Borex, B., Olefjord, I., 1985. Preferential dissolution of iron during the polarization of stainless steels in acids. J. Inst. Met. 84, 134.

Brunski, J.B., 2013. Metals: basic principles. Biomaterials Science: An Introduction to Materials in Medicine. Elsevier, London.

Bryant, M., Farrar, R., Freeman, R., Brummitt, K., Nolan, J., Neville, A., 2014. Galvanically enhanced fretting-crevice corrosion of cemented femoral stems. J. Mech. Behav. Biomed. Mater. 40, 275–286. Available from: https://doi.org/10.1016/j.jmbbm.2014.08.021.

Campbell, P.A., Ebramzadeh, E., Nelson, S., Takamura, K., De Smet, K.A., Amstutz, H.C., 2010. Histological features of pseudotumor-like tissues from metal-on-metal hips. Clin. Orthop. Relat. Res. 468, 2321–2327.

Case, C.P., Langkamer, V.G., James, C., Palmer, M.R., Kemp, A.J., Heap, P.F., et al., 1994. Widespread dissemination of metal debris from implants. J. Bone Joint Surg. 76, 701–712.

Charnley, J., 1972. The long-term results of low-friction arthroplasty of the hip performed as a primary intervention. J. Bone Joint Surg. Br. 54, 61–76.

Clarke, M.T., Lee, P.T.H., Rayment, A., Villar, R.N., Rushton, N., 2004. Bearing microseparation during the gait cycle with metal-on-metal and metal-on-polyethylene bearings. J. Bone Joint Surg. Br. 86-B, 79–80.

De Haan, R., Pattyn, C., Gill, H.S., Murray, D.W., Campbell, P.A., De Smet, K., 2008. Correlation between inclination of the acetabular component and metal ion levels in metal-on-metal hip resurfacing replacement. J. Bone Joint Surg. Br. 90-B, 1291–1297.

Di Laura, A., Quinn, P., Panagiotopoulou, V., Hothi, H., Henckel, J., Powell, J., et al., 2017. The chemical form of metal species released from corroded taper junctions of hip implants: synchrotron analysis of patient tissue. Sci. Rep. 7, 10952.

Donell, S.T., Darrah, C., Nolan, J.F., Wimhurst, J., Toms, A., Barker, T.H., et al., 2010. Early failure of the Ultima metal-on-metal total hip replacement in the presence of normal plain radiographs. J. Bone Joint Surg. Br. 92, 1501–1508.

Dowson, D., McNie, C.M., Goldsmith, A.A.J., 2000. Direct experimental evidence of lubrication in a metal-on-metal total hip replacement tested in a joint simulator. Proc. Inst. Mech. Eng. C. 214, 75–86.

Firkins, P.J., Tipper, J.L., Ingham, E., Stone, M.H., Farrar, R., Fisher, J., 2001. Influence of simulator kinematics on the wear of metal-on-metal hip prostheses. Proc. Inst. Mech. Eng. H. 215, 119–121.

Fisher, J., 2011. Bioengineering reasons for the failure of metal-on-metal hip prostheses. J. Bone Joint Surg. Br. 93-B, 1001–1004.

Fontana, M.G., Greene, N.D., 1993. Corrosion Engineering, second ed. McGraw-Hill, Japan.

Frankel, G.S., 1998. Pitting corrosion of metals: a review of the critical factors. J. Electrochem. Soc. 145, 2186–2198.

Gilbert, J.L., Buckley, C.A., Jacobs, J.J., 1993. In vivo corrosion of modular hip prosthesis components in mixed and similar metal combinations. The effect of crevice, stress, motion, and alloy coupling. J. Biomed. Mater. Res. 27, 1533–1544.

Hart, A.J., Buddhev, P., Winship, P., Faria, N., Powell, J.J., Skinner, J.A., 2008. Cup inclination angle of greater than 50 degrees increases whole blood concentrations of cobalt and chromium ions after metal-on-metal hip resurfacing. Hip Int. 18, 212–219.

Hart, A., Quinn, P., Sampson, B., Sandison, A., Atkinson, K., Skinner, J., et al., 2010. The chemical form of metallic debris in tissues surrounding metal-on-metal hips with unexplained failure. Acta Biomater. 6, 4439–4446.

Hart, A., Quinn, P., Lali, F., Sampson, B., Skinner, J.A., Powell, J.J., et al., 2012. Cobalt from metal-on-metal hip replacements may be the clinically relevant active agent responsible for periprosthetic tissue reactions. Acta Biomater. 8, 3865–3873.

Herdman, R.D., Handy, S., Pearson, T., Yamamoto, T., Ishiwata, K., Hara, M., et al., 2010. Chromium alloy coating with enhanced resistance to corrosion in calcium chloride environments. In: US Patent Application, US12/409,629.

Hertz, H., 1881. On the contact of elastic solids. J. Reine. Angew. Math. 92, 156–171.

Hesketh, J., Beadling, A., Dowson, D., Neville, A., 2013a. The influence of continuous and intermittent sliding on the release of ions from cobalt chromium surfaces, In: 5th World Tribology Congress, WTC 2013, Torino, Italy, September 8–13, 2013.

Hesketh, J., Hu, X., Yan, Y., Dowson, D., Neville, A., 2013b. Biotribocorrosion: some electrochemical observations from an instrumented hip joint simulator. Tribol. Int. 59, 332–338.

Hokkirigawa, K., Kato, K., 1989. Theoretical estimation of abrasive wear resistance based on microscopic wear mechanism. Wear Mater. 1, 1–8.

Huber, M., Reinisch, G., Trettenhahn, G., Zweymüller, K., Lintner, F., 2009. Presence of corrosion products and hypersensitivity-associated reactions in periprosthetic tissue after aseptic loosening of total hip replacements with metal bearing surfaces. Acta Biomater. 5, 172–180.

International Organization for Standardization, 2016. ISO 14242–2:2016, Implants for Surgery—Wear of Total Hip-Joint Prostheses—Part 2: Methods of Measurement.

Jacobs, J.J., Gilbert, J.L., Urban, R.M., 1998a. Current concepts review—corrosion of metal orthopaedic implants. J. Bone Joint Surg. 80, 268–282.

Jacobs, J., Skipor, A., Patterson, L., Hallab, N., Paprosky, W., Black, J., et al., 1998b. Metal release in patients who have had a primary total hip arthroplasty. A prospective, controlled, longitudinal study. J. Bone Joint Surg. 80 (A), 1447–1458.

Jin, Z.M., Dowson, D., Fisher, J., 1997. Analysis of fluid film lubrication in artificial hip joint replacements with surfaces of high elastic modulus. Proc. Inst. Mech. Eng. H. 211, 247–256.

Jin, Z.M., Firkins, P., Farrar, R., Fisher, J., 2000. Analysis and modelling of wear of cobalt-chrome alloys in a pin-on-plate test for a metal-on-metal total hip replacement. Proc. Inst. Mech. Eng. H. 214, 559–568.

Jost, P., 1966. Lubrication (Tribology)—A Report on the Present Position and Industry's Needs. H. M. Stationary Office, London, UK.

Karachalios, T., Komnos, G., Koutalos, A., 2018. Total hip arthroplasty: survival and modes of failure. EFORT Open. Rev. 3, 232–239.

Kato, K., 2005. Classification of wear mechanisms/models. In: Stachowiak, G.W. (Ed.), Wear: Materials, Mechanisms and Practice. John Wiley & Sons, Hoboken, pp. 9–20.

Korovessis, P., Petsinis, G., Repanti, M., Repantis, T., 2006. Metallosis after contemporary metal-on-metal total hip arthroplasty. Five to nine-year follow-up. J. Bone Joint Surg. Am. 88, 1183–1191.

Kretzer, J.P., Krachler, M., Reinders, J., Jakubowitz, E., Thomsen, M., Heisel, C., 2010. Determination of low wear rates in metal-on-metal hip joint replacements based on ultra trace element analysis in simulator studies. Tribol. Lett. 37, 23–29.

Langton, D.J., Jameson, S.S., Joyce, T.J., Webb, J., Nargol, A.V., 2008. The effect of component size and orientation on the concentrations of metal ions after resurfacing arthroplasty of the hip. J. Bone Joint Surg. Br. 90, 1143–1151.

Lehtovirta, L., Reito, A., Parkkinen, J., Pera, S., Eskelinen, A., 2018. Association between periprosthetic tissue metal content, whole blood and synovial fluid metal ion levels and histopathological findings in patients with failed metal-on-metal hip replacement. PLoS One 13, 1–13.

Liu, F., Williams, S., Fisher, J., 2015. Effect of microseparation on contact mechanics in metal-on-metal hip replacements – a finite element analysis. J. Biomed. Mat. Res. B 103, 1312–1319.

Maskiewicz, V.K., Williams, P.A., Prates, S.J., Bowsher, J.G., Clarke, I.C., 2010. Characterization of protein degradation in serum-based lubricants during simulation wear testing of metal-on-metal hip prostheses. J. Biomed. Mat. Res. 94B, 429–440.

Mathew, M.T., Jacobs, J.J., Wimmer, M.A., 2012. Wear-corrosion synergism in a CoCrMo hip bearing alloy is influenced by proteins. Clin. Orthop. Relat. Res. 470, 3109–3117.

McKee, G.K., Watson-Farrar, J., 1966. Replacement of arthritic hips by the McKee-Farrar prosthesis. J. Bone Joint Surg. Br. 48-B, 245–259.

National Joint Registry. 12th Annual Report, 2015.

Neville, A., Morina, A., 2005. Wear and chemistry of lubricants. In: Stachowiak, G.W. (Ed.), Wear: Materials, Mechanisms and Practice. John Wiley & Sons, Hoboken, pp. 71–94.

Park, Y.S., Moon, Y.W., Lim, S.J., Yang, J.M., Ahn, G., Choi, Y.L., 2005. Early osteolysis following second-generation metal-on-metal replacement. J. Bone Joint Surg. Am. 87, 1515–1521.

Pistorius, P.C., Burstein, G.T., 1992. Metastable pitting corrosion of stainless steel and the transition to stability. Phil. Trans. Phys. Sci. Eng. 341, 531–559.

Siljander, M.P., Baker, E., Baker, K.C., Salisbury, M.R., Thor, C.C., Verner, J.J., 2018. Fretting and corrosion damage in retrieved metal-on-polyethylene modular total hip arthroplasty systems: what is the importance of femoral head size? J. Arthroplasty 33, 931–938.

Stachowiak, G.W., Batchelor, A.W., 2005. Engineering Tribology. Elsevier, London.

Sumner, D.R., Galante, J.O., 1992. Determinants of stress shielding: design vs materials vs interface. Clin. Orthop. Relat. Res. 274, 202–212.

Swiatkowska, I., Mosselmans, J.F.W., Geraki, T., Wyles, C.C., Maleszewski, J.J., Henckel, J., et al., 2018. Synchrotron analysis of human organ tissue exposed to implant material. J. Trace Elem. Med. Biol. 46, 128–137.

Tait, W.S., 1994. An Introduction to Electrochemical Corrosion Testing for Practicing Engineers & Scientists. PairODocs Publication, Racine, WI.

Talbot, D.E.J., Talbot, J.D.R., 2007. Corrosion Science and Technology, second ed. CRC Press, Boca Raton, FL.

Urban, R.M., Jacobs, J.J., Tomlinson, M.J., Gavrilovic, J., Black, J., Peoc'h, M., 2000. Dissemination of wear particles to the liver, spleen, and abdominal lymph nodes of patients with hip or knee replacement. J. Bone Joint Surg. Am. 82 (A), 457–476.

Urban, R.M., Tomlinson, M.J., Hall, D.J., Jacobs, J.J., 2004. Accumulation in liver and spleen of metal particles generated at nonbearing surfaces in hip arthroplasty. J. Arthroplasty 19, 94–101.

Vieira, A.C., Rocha, L.A., Papageorgiou, N., Mischler, S., 2012. Mechanical and electrochemical deterioration mechanisms in the tribocorrosion of Al alloys in NaCl and in $NaNO_3$ solutions. Corros. Sci. 54, 26–35.

Vingsbo, O., Soderberg, S., 1988. On fretting maps. Wear 126, 131–147ISSN 0043-1648. Availavle from: https://doi.org/10.1016/0043-1648(88)90134-2.

von Schewelov, T., Sanzen, L., 2010. Catastrophic failure due to aggressive metallosis 4 years after hip resurfacing in a woman in her forties—a case report. Acta Orthop. 81, 402–404.

Watson, S.W., Friedersdorf, F.J., Madsen, B.W., Cramer, S.D., 1995. Methods of measuring wear-corrosion synergism. Wear 181–183, 476–484.

Watters, T.S., Cardona, D.M., Menon, K.S., Vinson, E.N., Bolognesi, M.P., Dodd, L.G., 2010. Aseptic lymphocyte-dominated vasculitis-associated lesion: a clinicopathologic review of an underrecognized cause of prosthetic failure. Am. J. Clin. Pathol. 134, 886–893.

Willert, H.G., Broback, L.G., Buchhorn, G.H., Jensen, P.H., Koster, G., Lang, I., et al., 1996. Crevice corrosion of cemented titanium alloy stems in total hip replacements. Clin. Orthop. Relat. Res. 333, 51–57.

Willert, H.G., Buchhorn, G.H., Fayyazi, A., Flury, R., Windler, M., Köster, G., et al., 2005. Metal-on-metal bearings and hypersensitivity in patients with artificial hip joints. A clinical and histomorphological study. J. Bone Joint Surg. Am. 87, 28–36.

Williams, D.F., 1987. Definitions in Biomaterials. Elsevier, Amsterdam.

Williams, J., 1994. Engineering Tribology. Oxford University Press, Oxford.

Wimmer, M.A., Sprecher, C., Hauert, R., Tager, G., Fischer, A., 2003. Tribochemical reaction on metal-on-metal hip joint bearings. Wear 255, 1007–1014.

Wimmer, M.A., Fischer, A., Buscher, R., Pourzal, R., Sprecher, C., Hauert, R., et al., 2010. Wear mechanisms in metal-on-metal bearings: the importance of tribochemical reaction layers. J. Orthop. Res. 28, 436–443.

Wu, L.E., Levina, A., Harris, H.H., Cai, Z., Lai, B., Vogt, S., et al., 2016. Carcinogenic chromium(VI) compounds formed by intracellular oxidation of chromium(III) dietary supplements by adipocytes. Angew. Chem. Int. Ed. 55, 1742–1745.

Xia, Z., Ricciardi, B., Liu, Z., von Ruhland, C., Ward, M., Lord, A., et al., 2017. Nano-analyses of wear particles from metal-on-metal and non-metal-on-metal dual modular neck hip arthroplasty. Nanomedicine 13, 1205–1217.

Zum Gar, K.H., 1987. Microstructure and Wear of Materials. Elsevier, London.

Chapter 3

# Implant metals and their potential toxicity

Ilona Świątkowska
*Institute of Orthopaedics and Musculoskeletal Science, University College London, Stanmore, United Kingdom*

## 3.1 Hip implant metals and the human health

The human body needs a number of metallic elements to function properly. These can be broadly classified into macrominerals, which serve as constituents of cells and body fluids and must be provided in large quantities in the diet (e.g., potassium, sodium, calcium, and magnesium), and trace minerals, which participate in important cellular and metabolic processes and are only required in minute quantities (e.g., iron, copper, zinc, and molybdenum). Trace elements are usually supplied in adequate amounts from the diet, but deficiency symptoms can develop if their intestinal absorption or transport is impaired, for example, because of genetic defects or chronic illness. Adverse health effects can also occur if trace elements are present in the body in too high amounts. Excessive environmental, occupational, dietary, or medicinal exposure to trace metals may cause them to build up in local tissue (near the exposure site) or elevate their systemic concentrations (in the blood or organs distal from the exposure site), which may lead to clinically relevant toxicity. The adverse effects will be influenced by the magnitude, duration, and route of exposure, the metal's chemical speciation, its tissue distribution and efficiency of elimination, and patient-specific factors, such as genetic make-up and nutritional status. While certain metals are key to optimal health, others have no known biological function and some are highly toxic, even in small amounts.

It is well established that metallic biomaterials shed metal particles and ions into the periprosthetic tissue and bloodstream once implanted. Cobalt, chromium, and titanium have all been detected in the blood of hip implant recipients at levels higher than those that occur in the general population, prompting concerns over the long-term safety of these devices. The

**Biomarkers of Hip Implant Function. DOI: https://doi.org/10.1016/B978-0-12-821596-8.00001-X**

following sections describe the characteristics, biological roles and toxic potential of the common hip implant metals (Table 3.1).

### 3.1.1 Cobalt

Cobalt is a transition metal that occupies a central position in the periodic table of elements. It is a relatively rare constituent of the Earth's crust that does not occur on its own in nature and is generally found in the form of mixed ores that also contain variable amounts of nickel, copper, iron, sulphur, arsenic, and/or silver. The name of the element derives from the word *kobold*, meaning 'goblin'—a devilish mountain-dwelling spirit from German folklore. Medieval miners blamed the creature for the poisonous fumes released from cobalt ore during the silver extraction process (the toxic fumes were later attributed to arsenic and sulphur vapours).

The earliest use of cobalt was as a blue colouring agent in pottery, glass, and jewellery, dating back to ancient Egypt and China. In the eighteenth century, cobalt was successfully isolated and identified as an element, but it was not adopted for industrial applications until the twentieth century. In its pure form, cobalt is a ductile, brittle, grey metal with properties similar to those of iron and nickel. Owing to its durability, high melting point, and resistance to oxidation, it is commonly used to make strong, corrosion- and heat-resistant alloys, such as the cobalt-chromium-molybdenum (CoCrMo) alloy prominent in hip implant manufacture. Other uses of cobalt include permanent magnets, batteries, catalysts, drying agents in paints, fertilisers, and contrast agents in radiology. Industrial activities, engine emissions, volcanic eruptions, and forest fires give rise to a low level of cobalt in urban air ($<2$ ng Co/m$^3$) (Barceloux and Barceloux, 1999a).

Inorganic cobalt is not required by mammals. However, vitamin $B_{12}$ (cobalamin), which features a trivalent cobalt ion at the centre of its structure, is an essential micronutrient with key roles in DNA synthesis and red blood cell production. Like most vitamins, $B_{12}$ cannot be made by the body and must be obtained from the diet. If the recommended daily vitamin $B_{12}$ intake of 2–3 μg (approx. 0.01 μg Co) is not met, vitamin $B_{12}$ deficiency can develop, manifesting as neuropathy and megaloblastic anaemia (Stabler, 2013). The mean daily intake of inorganic cobalt is approximately 12 μg, with most derived from fish, cereals, nuts, and green leafy vegetables (Paustenbach et al., 2013). Although cobalt supplementation is unnecessary, numerous cobalt-containing products are available for purchase in the United States, which are touted to increase the body's antioxidant defence, reduce inflammation, and enhance aerobic performance (Lippi et al., 2006).

At the root of cobalt's perceived performance-enhancing action is its erythropoietic effect, i.e., the ability to stimulate red blood cell production and increase haemoglobin levels. Cobalt salts were historically used to treat

**TABLE 3.1** Metal composition of the three most common orthopaedic alloys.

| Metal alloy | Weight percentage | | | | | | |
|---|---|---|---|---|---|---|---|
| | Co | Cr | Mo | Ni | Ti | Al | V |
| Cobalt-chromium alloy (ASTM F75) | 62–67 | 27–30 | 5–7 | 1 | – | – | – |
| Stainless steel (ASTM F138) | — | 17–20 | 2–4 | 13–15 | – | – | – |
| Titanium alloy (ASTM F136) | – | – | – | – | 89–91 | 5.5–6.5 | 3.5–4.5 |

*Source*: Based on Hallab, N., Merritt, K., Jacobs, J.J., 2001. Metal sensitivity in patients with orthopaedic implants. J. Bone Joint Surg. Am. 83-A, 428–436.

certain types of anaemia (primarily those secondary to chronic renal failure and malignancies), with typical adult doses ranging from 11 to 150 mg of cobalt chloride ($CoCl_2$) per day (Duckham and Lee, 1976). While generally successful, the therapy was associated with adverse effects in adult patients with renal compromise and children and was eventually discontinued.

### 3.1.1.1 Toxicokinetics

In the blood, cobalt occurs primarily in the divalent state ($Co^{2+}$). Following their release from metal implants, $Co^{2+}$ ions bind to proteins in the synovial fluid and surrounding tissue before entering the circulation. In healthy individuals, approximately 88%–95% of blood $Co^{2+}$ is bound to serum albumin and much of the remainder is complexed with smaller biomolecules, such as alpha-lipoic acid and glutathione (Jansen et al., 1996). The chemical similarity of $Co^{2+}$ ions to other divalent cations allows them to 'hijack' various cellular transport mechanisms to gain entry into cells. For example, $Co^{2+}$ can enter red blood cells via the calcium membrane transporter. The process is essentially irreversible as the metal becomes bound to haemoglobin, leading to intracellular cobalt accumulation (Simonsen et al., 2012). Non-protein-bound $Co^{2+}$ ions can cross cell membranes through broad-specificity divalent metal transporters, including the $P2X_7$ ionotropic receptor and divalent metal transporter-1 (DMT-1) (Skorringe et al., 2015). DMT-1 is highly expressed in the neurons and thought to play a key role in the entry of metal ions into the brain (Howitt et al., 2009). The cellular uptake of cobalt may also be mediated by active transport ion pumps, such as the calcium/magnesium pump, and endocytosis (Fig. 3.1).

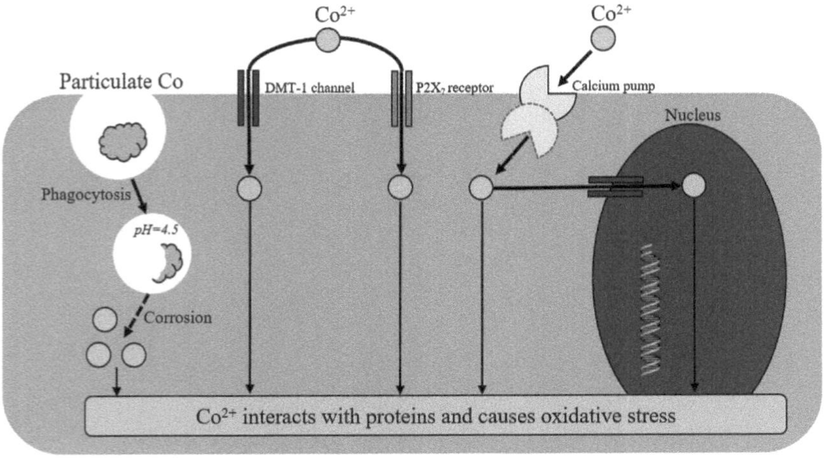


**FIGURE 3.1** Cellular uptake of cobalt. *DMT-1*, divalent metal transporter-1.

During short-term exposure (10–30 h after a single dose), the concentration of cobalt in the serum/plasma is up to two orders of magnitude higher than in the red blood cell fraction (Smith et al., 1972). Chronic exposure leads to slow, continuous uptake of cobalt into erythrocytes, which reflects the time-averaged $Co^{2+}$ exposure during their 120-day lifespan (Walter et al., 2008). The total buffering capacity of human erythrocytes is approximately 1 g of cobalt for a volume of 0.8 L (Simonsen et al., 2011). A greater proportion of cobalt in whole blood than in the plasma suggests that the binding of cobalt to plasma proteins is altered, resulting in a higher ratio of free $Co^{2+}$ to bound $Co^{2+}$ (Catalani et al., 2011). This is of clinical importance because only free $Co^{2+}$ ions can interact with biomolecules and cause toxicity.

Cobalt is highly soluble and readily excreted in the urine; the little that is retained is stored in the liver and kidneys, with smaller amounts in the heart and spleen (Posada et al., 2015). It is estimated that the adult human body contains 1 mg of cobalt, of which 85% is in the form of vitamin $B_{12}$. High acute doses or long-term exposure to cobalt can increase its circulating levels and lead to abnormal deposition of the metal in the vital organs.

### *3.1.1.2 Systemic toxicity*

Cobalt exhibits a range of toxicities that can vary depending on the route of exposure. The most important adverse effects associated with inhalation of cobalt dust are respiratory sequalae, such as wheezing, asthma, and lung fibrosis. Chronic ingestion of cobalt does not seem to trigger pulmonary complications and is instead associated with haematological, thyroid, neurological, and cardiac abnormalities. Mechanisms proposed to underlie these toxic effects include disruption of cellular energy production, generation of reactive oxygen species (ROS), inhibition of thyroidal iodine uptake, displacement of divalent cations from metabolic enzymes, interference with calcium signalling, and induction of a hypoxia-like state in cells.

Reversible hypothyroidism manifesting as goitre, lassitude, lethargy, weakness, poor concentration, and/or diminished reflexes has been observed in individuals receiving $CoCl_2$ treatment, factory workers, and in patients with hip implants made of CoCrMo alloy (Apel et al., 2013; Schirrmacher, 1967; Swennen et al., 2008; Weber et al., 2015).

The neurotoxic effects of cobalt have been demonstrated in numerous animal studies and case reports. Rabbits treated with intravenous cobalt showed histopathological evidence of optic nerve damage and loss of retinal and cochlear cells, the severity of which was related to the dose and time of exposure (Apostoli et al., 2013). Reversible deafness and/or visual deficits have also been noted in recipients of $CoCl_2$ therapy and occupationally exposed persons (Meecham and Humphrey, 1994; Schirrmacher, 1967). In patients with hip implants, cobalt intoxication can manifest as short-term memory loss, inability to focus, cognitive decline, depression, limb paraesthesia, loss of balance, and altered vision and/or hearing (Dahms et al., 2014; Ikeda et al., 2010; Weber et al., 2015).

The link between cobalt and heart damage was first made in the 1950s when $CoCl_2$ tablets were commonly prescribed to treat anaemia. Little and Sunico (1958) and La Grutta et al. (1984) reported on children presenting with rapid-onset cardiomyopathy after treatment with a cobalt-containing supplement. In both cases, the cardiotoxic effects were associated with elevated serum cobalt level (270 μg/L) and resolved upon cessation of cobalt therapy. Cases of ischaemic heart disease and fatal cardiomyopathy were also observed in cobalt-exposed workers (Barborik and Dusek, 1972; Kennedy et al., 1981). In the mid-1960s, a unique cardiomyopathy syndrome was reported in a subset of heavy drinkers of cobalt-laced beer after certain breweries in North America and Europe began using cobalt salts as foam stabiliser. The condition was characterised by rapid-onset heart failure with pericardial effusion, polycythaemia, and thyroid abnormalities, and often had a fast clinical progression to death (Alexander, 1972; Morin et al., 1967). The cobalt doses ingested by the beer drinkers were much lower than those administered to patients with anaemia, so it is likely that the cardiotoxic effects resulted from a combination of cobalt's depressant action on the heart, direct effects of alcohol, and co-existing malnutrition (Alexander, 1969; Rona, 1971). As the clinical and industrial use of cobalt salts waned and measures to reduce occupational exposure in cobalt plants have been implemented, the incidence of subacute cobalt-related cardiomyopathy plummeted. Recent years have seen a resurgence of reports describing the classical features of the condition among recipients of metallic hip replacements. The affected persons typically presented with shortness of breath, palpitations, and exertional chest tightness and had extremely elevated circulating cobalt levels (typically $>100$ μg Co/L) (Packer, 2016).

The symptoms of systemic cobalt toxicity in the setting of hip arthroplasty and the underlying molecular mechanisms are reviewed in detail in Chapter 7.

In vitro studies have shown that cobalt metal and its salts induce DNA strand breaks and crosslinks and inhibit DNA repair, which may play a role in the initiation of cancer. Cobalt-induced tumours have been observed in laboratory animals, albeit at doses markedly higher than those hip implant recipients are usually exposed to (Shabaan et al., 1977). Owing to a lack of convincing in vivo evidence in humans, cobalt is currently classified as 'possibly carcinogenic to humans' (Group 2B) by the International Agency for Research on Cancer (IARC).

### 3.1.2 Chromium

Chromium is a relatively abundant element in the Earth's crust and is present naturally in air, water, soil, and plants. The word 'chromium' originates from the Greek word *khrôma*, meaning 'colour', which reflects the metal's ability to impart intense colouration to minerals, such as orange in potassium (VI) dichromate, red in chromium (VI) oxide, and green in chromium (III) oxide. The colourful nature of chromium compounds made them popular as pigments for paint, ink, textile dyes, and porcelain glaze. In the elemental form, chromium is a hard, brittle, silvery metal with a high melting point (2000°C) and shiny appearance. Owing to its hardness, corrosion resistance and ability to take on a high-shine polish, metallic chromium is used as an alloying agent in cast iron, stainless steel, and CoCrMo alloy, and as a surface coating on other metals. Other uses of chromium compounds include catalysts, fungicides, wood preservatives, and leather tanning agents. In the environment, chromium is predominantly found in the trivalent ($Cr^{3+}$) or hexavalent ($Cr^{6+}$) state, with the latter originating almost exclusively from anthropogenic (man-made) sources.

The general population may be exposed to chromium through ingestion of chromium-containing food, supplements, and drinking water, or inhalation of cigarette smoke. Exposure may also occur through skin contact with certain consumer products, such as detergents or chromium-tanned leather. For three decades, chromium (as $Cr^{3+}$) was accepted as an essential micronutrient required for normal carbohydrate, protein, and fat metabolism. However, recent years have seen this view contested and the European Food Safety Authority no longer considers chromium essential to human health. Current evidence supports a pharmacological rather than nutritional role for chromium (Vincent, 2017). Chromium supplements, including chromium (III) picolinate and chromium (III) nicotinate, demonstrated insulin-mimetic action in rodent models and are marketed for their potential ability to improve glycaemic control and reduce insulin resistance in persons with type 2 diabetes and metabolic syndrome. Their purported beneficial effects on lean body mass, muscle growth, and athletic performance, though not necessarily backed by scientific research, have also made them popular among athletes (Vincent, 2017).

#### 3.1.2.1 Toxicokinetics

Cell membranes are relatively impermeable to free $Cr^{3+}$ ions; thus, synthetic chromium supplements often contain hydrophobic ligands to facilitate the ion's passage into cells. Moreover, $Cr^{3+}$ ions are kinetically inert and slow to form complexes with surrounding ligands (Saha et al., 2011). Trivalent chromium ions released from hip implants tend to combine with ubiquitous phosphate ions and accumulate in extra-synovial tissues as chromium phosphate ($CrPO_4$). Corrosion products such as chromium oxide ($Cr_2O_3$) and chromium hydroxide ($Cr[OH]_3$) become deposited in the synovial environment (Polyzois et al., 2012). Particles of $Cr_2O_3$ or CoCr alloy are taken up by macrophages and gradually dissolved within the acidic environment of lysosomes to release free $Cr^{3+}$ ions. In intestinal epithelial cells, $Cr^{3+}$–transferrin complexes can be taken up by endocytosis (Beyersmann and Hartwig, 2008; Saxena et al., 2018).

In physiological environments, $Cr^{6+}$ predominantly exists as the chromate anion, $[CrO_4]^{2-}$, which readily crosses cellular and nuclear membranes via non-specific anionic channels such as the phosphate and sulphate exchange pathway. Inside the cell, $Cr^{6+}$ undergoes metabolic reduction to $Cr^{3+}$ by intracellular reductants, including glutathione, ascorbate, and cysteine. The final $Cr^{3+}$ species cannot leave the cell and is effectively trapped, usually bound to intracellular proteins or DNA. The reduction process generates short-lived $Cr^{5+}$ and $Cr^{4+}$ intermediates and ROS, which can damage cellular components. If $Cr^{6+}$ reduction takes place inside the nucleus, the reactive intermediates as well as the final $Cr^{3+}$ species may induce genotoxic DNA modifications, including Cr–DNA adducts, strand breaks, and cross-links, which promote genetic mutations (Fig. 3.2). Although cellular environments are generally reducing, chromium (III) complexes can be oxidised in the blood and cells, potentially leading to repeated oxidation–reduction cycles (Levina and Lay, 2008). In fact, evidence is mounting that the insulin-enhancing activity of chromium (III) supplements is mediated by in vivo formation of reactive $Cr^{5+}$ and $Cr^{6+}$ species (Mulyani et al., 2004; Wu et al., 2016).

In the blood, $Cr^{3+}$ is transported by serum transferrin while $Cr^{6+}$ accumulates in red blood cells. Cellular uptake of chromium is often taken as an

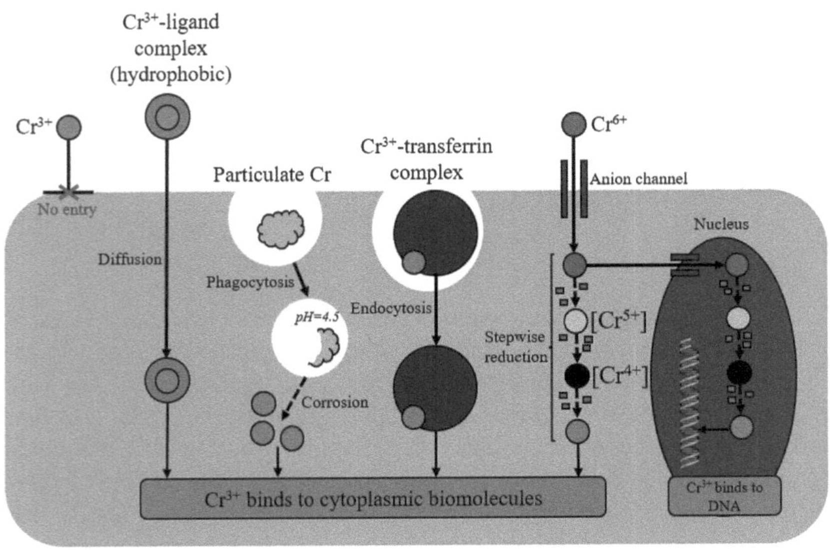


**FIGURE 3.2** Cellular uptake and intracellular fate of chromium ions. Pink rectangles represent toxic by-products of $Cr^{6+}$ reduction.

indication of its valence, and increased chromium concentration in red blood cells serves as indirect evidence of $Cr^{6+}$ exposure (Walter et al., 2008). When workers of a dichromate producing factory were divided into those mainly exposed to $Cr^{3+}$ and those mainly exposed to $Cr^{6+}$, chromium erythrocyte content was seven-fold higher in the latter group (Minoia and Cavalleri, 1988). Red blood cell Cr/serum Cr ratio was also measured in THA recipients to infer whether orthopaedic implants could be a source of hexavalent chromium. Finley et al. (2017) showed that chromium released from MoM hip implants preferentially distributed into the serum, with no associated increase in red blood cell chromium concentration, indicating that the metal was in the relatively non-toxic trivalent state.

Trivalent chromium is excreted in the urine, with approximately 10% lost in the bile and smaller amounts eliminated in the hair, nails, breast milk, and perspiration (Posada et al., 2015). Although trivalent chromium is rapidly cleared from the plasma, its elimination from tissues occurs at a much slower rate. Following revision of a high-wearing MoM implant, blood chromium levels tend to decrease more slowly than cobalt levels, which could be partly explained by the poor water solubility of $Cr^{3+}$ and its tendency to accumulate within the periprosthetic tissue.

#### *3.1.2.2 Systemic toxicity*

The oxidation state of chromium is intimately related to its biochemical activity in humans. Hexavalent compounds, such as chromium trioxide and potassium dichromate, are powerful oxidising agents and known carcinogens, while trivalent chromium is considered fairly innocuous at normal exposure concentrations (Welling et al., 2015). Inhaling hexavalent chromium can cause perforation of the nasal septum, laryngitis, asthma, bronchitis, pneumonitis, and lung tissue damage. Allergic contact dermatitis and ulceration are common with chronic dermal exposure to hexavalent chromium salts, such as that experienced by chrome platers, cement workers, and painters (Shelnutt et al., 2007). Moreover, both trivalent and hexavalent chromium compounds are selectively accumulated in the renal cortex where, in large doses, they can cause kidney damage (Wedeen and Qian, 1991). Adverse renal effects have been reported in humans after inhalation, ingestion, and dermal exposure to chromium. Another potential consequence of $Cr^{6+}$ intoxication is hepatocellular necrosis (Kurosaki et al., 1995).

The main toxicological hazard related to chromium is the carcinogenicity of its hexavalent form. Intracellular reduction of $Cr^{6+}$ to $Cr^{3+}$ generates reactive intermediates and ROS, which can all induce genotoxic DNA modifications. Aside from directly causing DNA damage, ROS liberated during $Cr^{6+}$ reduction activate various mitogen-activated protein kinases (MAPKs) and mitogenic transcription factors, with an overall pro-inflammatory and oncogenic effect (Beyersmann and Hartwig, 2008). Chronic inhalation of hexavalent chromium translates into an elevated risk of lung cancer in factory workers (Nickens et al., 2010), while oral exposure is associated with an increased incidence of stomach, lung, and liver cancer (Beaumont et al., 2008; Linos et al., 2011; Welling et al., 2015). Hexavalent chromium is classified as a human carcinogen (Group 1) by the IARC; however, there is insufficient evidence in humans for the carcinogenicity of chromium metal or trivalent chromium.

It is yet unclear whether chromium released from hip implants could produce clinically relevant systemic toxicity. Reassuringly, current evidence suggests that hip implants are not a source of hexavalent chromium (Hart et al., 2010; Jacobs et al., 1995a,b; Urban and Jacobs, 1994).

### 3.1.3 Molybdenum

Molybdenum is a naturally occurring element that is widely distributed throughout the soil and common foods, such as legumes, nuts, and whole grains. In its elemental form, molybdenum is a hard, ductile, silvery-white metal with one of the highest melting points of all pure elements (2623°C). Its primary application is as an alloying agent in steel and cast iron, where it improves the strength and durability of the material and reduces its corrosion potential. The ability to withstand extreme heat and pressure makes molybdenum particularly well suited for use in furnaces, nuclear reactors, jet engines, and engine lubricants. The metal is also commonly added to the CoCr alloy and stainless steel during manufacture of hip implant components (Barceloux and Barceloux, 1999b).

In the human body, molybdenum serves as a co-factor for at least four key metabolic enzymes, earning it the status of an essential micronutrient. The requirement for molybdenum (34 μg Mo/day) is easily met through diet alone, with an average dietary intake of 100–500 μg molybdenum per day (Barceloux and Barceloux, 1999b). Although molybdenum can be toxic to animals, data on its adverse effects in humans are sparse. In two isolated reports, exposure to high amounts of molybdenum (10–15 mg Mo/day) through food consumption or inhalation of molybdenum dust was associated with hyperuricaemia, aching joints, and gout-like symptoms, without evidence of permanent damage (Kovalskii et al., 1961; Walravens et al., 1979).

#### *3.1.3.1 Toxicokinetics*

Molybdenum exists in the body in either the quadrivalent or hexavalent form. In the blood, it is transported as the molybdate ion, $[MoO_4]^{2-}$, bound to albumin and α2-macroglobulin or loosely associated with red blood cells. The human body contains approximately 70 μg of molybdenum per kilogram of body weight, with the highest concentrations in the liver and kidneys.

Excessive amounts of molybdenum are rapidly eliminated in the urine, and there is little evidence of its retention in tissue (Blanco and Blanco, 2017).

#### *3.1.3.2 Systemic toxicity*

Molybdenum has low perceived toxicity and the potential relationship between molybdenum ions released from hip implants and any local or systemic sequelae remains relatively unexplored. Since molybdenum is only a minor constituent of orthopaedic alloys, molybdenum levels are rarely measured in patients with hip implants.

### 3.1.4 Nickel

Nickel is a hard, ductile, silvery-white metal whose main uses are in the production of alloys for industrial, medical, and domestic applications, electroplating, and as a chemical catalyst. The addition of nickel to stainless steel produces a material with superior strength, ductility, toughness, and resistance to extreme temperatures and corrosive environments. Hip implant components made of stainless steel contain up to 15% of nickel. Other applications of nickel metal include jewellery, coins, rechargeable batteries, electronic equipment, and kitchenware.

Nickel is widely distributed in the air, water, and soil, and the predominant route of exposure is oral ingestion. Foods that contribute the most nickel to the diet are dark chocolate, soybeans, nuts, and oatmeal; stainless steel cookware and water taps may leach additional nickel into food and drinking water. The average dietary intake in the United States ranges from 70 to 160 μg nickel per day (Bennett, 1984). Although nickel is required by plants and many animal species, its nutritional value and essentiality in humans is debated, and the biological consequences of nickel deficiency are not clearly defined.

#### *3.1.4.1 Toxicokinetics*

Although nickel may exist in several oxidation states, the divalent form is the most widespread in biological systems. Like $Co^{2+}$, $Ni^{2+}$ ions are highly soluble and rapidly excreted in the urine, with little accumulation in tissue (Merritt et al., 1992). The total body burden of nickel in a healthy adult is approximately 500 μg (7.3 μg Ni/kg body weight).

#### *3.1.4.2 Systemic toxicity*

Nickel and nickel compounds are well-known occupational carcinogens; however, only the inhalation route is associated with cancer, with the tumours local to the respiratory tract. Consequently, nickel metal and its alloys are classified as 'possibly carcinogenic to humans' (Group 2B) by the IARC.

The main adverse effect associated with nickel is its sensitising quality—approximately 10%–20% of the general population suffer from nickel allergy. The condition disproportionately affects women and is related to prolonged contact with nickel-containing jewellery, piercings, and clothing fasteners. It is believed that corrosion of orthopaedic prostheses can release sufficient amounts of nickel to cause local or systemic allergic reactions, but not enough to provoke systemic toxicity (Barceloux and Barceloux, 1999c). As a result, routine measurement of circulating nickel levels is not considered clinically useful as an indicator of hip implant wear or risk of adverse reactions (Newton et al., 2012).

### 3.1.5 Titanium

Titanium is a lustrous, ductile, silvery-white metal that exhibits high strength and chemical stability. When added to iron, aluminium, vanadium, molybdenum, or tantalum, it forms lightweight, yet durable, alloys commonly used in aerospace, automotive, metallurgic, and biomedical industries. Its excellent mechanical properties, combined with high biocompatibility and ability to integrate with the surrounding bone, have made titanium ideal for use in joint replacement implants, spinal instrumentation, and dental inserts. The most stable oxidation state of titanium is quadrivalent ($Ti^{4+}$), but titanium (III) compounds are also prevalent in the environment.

Human exposure to titanium is largely via titanium (IV) dioxide ($TiO_2$)—a naturally occurring mineral that exists in three main crystallographic structures: rutile, anatase, and brookite. Owing to its light-reflecting abilities, small amounts of $TiO_2$ are added to certain personal care products, pharmaceutical preparations, and processed foods to enhance their white colour and opacity, and help block ultraviolet rays. Examples of such products include toothpaste, sunscreen, deodorant, chewing gum, and candy (Weir et al., 2012). When used as a pigment, $TiO_2$ may be referred to as Titanium White, Pigment White 6, CI 77891, or additive E171. Ingestion of pigment-grade $TiO_2$ is particularly commonplace in westernised populations: in the United Kingdom, the median adult intake is approximately 2.5 mg per day (Lomer et al., 2004). In addition to its brightening and opacifying properties, $TiO_2$ acts as a chemical catalyst when exposed to ultraviolet light, lending to its use in self-cleaning glass and ethylene-removing food packaging (Awalgaonkar et al., 2020; Parkin and Palgrave, 2005). Although titanium is relatively abundant in the human body (10–20 mg total body burden), its biological function, if any, is unclear as no titanium-requiring biomolecules have ever been identified (Buettner and Valentine, 2012).

#### *3.1.5.1 Toxicokinetics*

Despite the widespread use of $TiO_2$ in processed foods, skincare products, and pharmaceuticals, information on the absorption, distribution, and toxicity of titanium in humans is limited. Implant-derived soluble titanium ions

($Ti^{4+}$) bind to serum transferrin, which allows them to be transported to different parts of the body. The transferrin–titanium complex can enter cells via endocytosis, though the intracellular activities of $Ti^{4+}$ and how it exerts its effects remain largely unknown (Saxena et al., 2018; Tinoco et al., 2008). The non-protein-bound fraction of implant-derived $Ti^{4+}$ is likely complexed with small molecular anions such as citrate (Silwood and Grootveld, 2005). At the pH of blood (7.4), many Ti (IV) compounds dissociate and transform into $TiO_2$ (Saxena et al., 2018). Although $TiO_2$ is 'generally recognised as safe' by the US Food and Drug Administration, there is some concern regarding the potential consequences of dermal and intestinal absorption of $TiO_2$ nanoparticles (1–100 nm in diameter); it is unclear if the same risks apply to endogenous exposure. Complicating the safety evaluation of hip implants is the fact that titanium debris can exist in several forms, crystal structures, and particle sizes, which may differ in biological activity and toxicity. In vitro studies suggest that $TiO_2$ nanoparticles are more cytotoxic and more readily taken up by cells than their coarser counterparts, and that anatase is more toxic than rutile (He et al., 2015; Sayes et al., 2006).

Titanium is not readily excreted in the urine and tends to accumulate in tissue. Periprosthetic and systemic titanium deposition, e.g., in the bone marrow, spleen, lungs, skin, nails, and various mucosal surfaces, has been described in animals and humans (Urban et al., 2000; Wu et al., 2009). It is likely that the retained particles act as a sustained source of titanium ions as they undergo gradual dissolution (Urban et al., 2000). The clinical implications of chronic, low-level exposure to titanium ions warrant further study.

### 3.1.5.2 Systemic toxicity

Although cases of titanium-related organ injury have been described (Chen et al., 2015; Liu et al., 2009; Moran et al., 1991; Sheng et al., 2013; Tarpada et al., 2020; Urban et al., 2000), implanted titanium poses relatively low health risk and adverse systemic effects associated with titanium release from implants have been relatively mild and infrequently reported. There is a likely connection between titanium and the 'yellow nail syndrome' (YNS)—a rare medical condition characterised by nail changes (thickening of the nail plate, yellow discolouration, and slowed growth), lymphedema, and/or respiratory manifestations such as chronic sinusitis, cough, bronchiectasis, and pleural effusions (Samman and White, 1964). Berglund and Carlmark (2011), who first pointed to the involvement of titanium in YNS, found high levels of titanium in nail clippings from individuals with one or more symptoms of the condition (median, 5 μg/g; range, 1.1–170 μg/g; normal value, 0 μg/g). The two affected groups were persons exposed to $TiO_2$ from medications or confectionery and patients with titanium implants and concurrent gold or amalgam dental work. The authors hypothesised that the dominant cause of YNS symptoms in the latter group was galvanic interaction between gold/amalgam and titanium, accelerating the release of titanium ions.

More recently, Tarpada et al. (2020) described a case of systemic titanium toxicity triggered by polyethylene liner fracture and consequent abrasive wear of a titanium acetabular cup in a CoP THR. The patient reported headaches, weakness, fatigue, and blurred vision, without any of the classic features of YNS. The symptoms were associated with a massive periprosthetic metallosis, pseudotumour formation, and extremely elevated circulating titanium levels (460 μg/L).

Titanium intoxication secondary to THA failure may also cause renal insufficiency (Fabi et al., 2011). While the case report evidence for titanium-related nephrotoxicity is weak, a recent study demonstrated that administration of $TiO_2$ nanoparticles to rats caused marked histological alterations in the renal tissues and affected kidney function (Al-Doaiss et al., 2019).

Titanium is another redox-active metal that can enhance ROS production and induce oxidative stress. Studies in mice have shown that $TiO_2$ nanoparticles have the potential to convert benign tumour cells to malignant ones through the generation of ROS (Onuma et al., 2009). Based on experimental evidence from animal inhalation studies, $TiO_2$ nanoparticles are classified as 'possibly carcinogenic to humans' (Group 2B) by the IARC and as an 'occupational carcinogen' by the National Institute for Occupational Safety and Health. It is unknown if these results can be extrapolated to human exposure and implanted titanium is currently non-classifiable as to its carcinogenicity to humans.

## 3.1.6 Vanadium

Vanadium is a naturally occurring element that exists in six different oxidations states, of which trivalent ($V^{3+}$), quadrivalent ($V^{4+}$), and pentavalent ($V^{5+}$) are the most environmentally relevant. Pure vanadium is a soft, ductile, silver-grey metal that has been widely adopted in the fabrication of steel and non-ferrous alloys for jet engines, automobile parts, and construction tools. The addition of a small amount of vanadium to steel refines the grain structure, increasing the strength, toughness, and wear resistance of the material. Vanadium is also an important beta-phase stabiliser for orthopaedic titanium alloys; the Ti-6Al-4V alloy contains 4% vanadium. Other uses of vanadium include superconductive magnets, photographic developers, pigments for the ceramic and textile industries, and chemical catalysts.

Even though most foods are low in vanadium, diet is the chief route of vanadium exposure for the general population, with the estimated daily intake ranging from 10 to 60 μg in the United States. The highest levels of vanadium are found in black pepper, parsley, mushrooms, and shellfish; additionally, smokers may inhale up to 0.4 μg vanadium per cigarette (Barceloux and Barceloux, 1999d). Although its biological role in humans is unclear, vanadium is regarded as an essential trace element. Vanadium salts such as vanadyl (IV) sulphate mimic the action of insulin and decrease blood glucose levels when administered to rats or patients with diabetes (Domingo and Gómez, 2016). In addition to its antidiabetic effects, vanadium shows promise in the treatment of parasitic diseases, infections, and malignant tumours (Rehder, 2015). Despite the large number of potential therapeutic uses, no vanadium compound has yet proven to be efficacious and safe enough for long-term use in humans.

### 3.1.6.1 Toxicokinetics

Once in the bloodstream, vanadium species bind to plasma proteins, particularly transferrin and albumin, with a smaller fraction complexed with negatively charged small molecules such as citrate (Treviño et al., 2019). Vanadium is rapidly distributed around the body and stored in tissue. The total body burden of vanadium is estimated at 100 μg, with the highest amount in the kidneys, liver, and bone. Although vanadium can exist in a number of valence states, the pentavalent form predominates in extracellular body fluids, while the quadrivalent form is common inside cells (Barceloux and Barceloux, 1999d). Depending on factors such as local pH, oxygen concentration, and availability of cellular reductants, vanadium may undergo reduction, oxidation, or redox cycling, which may influence its bioavailability and biological effects (ATSDR, 2012). The principal route of elimination is through the kidneys, with only a minor amount excreted in the faeces.

### 3.1.6.2 Systemic toxicity

In experimental animals, vanadium toxicity can manifest as haematological and metabolic alterations, gastrointestinal distress, kidney damage, and reproductive, developmental, and neurobehavioural abnormalities (Domingo and Gómez, 2016). In general, the toxicity of vanadium compounds increases with their valence and solubility, such that pentavalent compounds are the most hazardous. The only clearly documented toxic effects of vanadium in humans are eye irritation and respiratory distress (rhinitis, wheezing, cough, sore throat, and chest pain) following occupational exposure (ATSDR, 2012). Inhalation of vanadium dust is also suspected of causing neurobehavioural impairments affecting the attention and visuospatial abilities (Barth et al., 2002). Acute oral exposure may produce gastrointestinal distress, including abdominal cramping and diarrhoea.

Based on the evidence of lung cancer in vanadium-exposed mice, the metal is thought to be a possible human carcinogen, but only in the form of vanadium pentoxide.

The small amount of vanadium released from well-functioning hip implants is unlikely to pose a toxicological risk for humans (Catalani et al., 2013); however the long-term consequences of chronic 'internal' vanadium generation and its potential accumulation in tissue have not been studied. There has been at least one report of implant-related vanadium neurotoxicity (sensory-motor axonal neuropathy and bilateral sensorineural hearing loss) in a patient with a Ti-6Al-4V alloy femoral stem (Moretti et al., 2012).

### 3.1.7 Aluminium

Aluminium is a lightweight, malleable, corrosion-resistant metal, and the most abundant metallic element in the Earth's crust. The combination of low density, easy workability, high resistance to corrosion, and excellent conductivity makes aluminium suitable for a wide variety of applications, including construction, manufacture of car and aeroplane parts, electronics, packaging, and cookware. Aluminium in the form of aluminium oxide (alumina) is the cornerstone of modern ceramic hip arthroplasty. The metal is also added to commercially pure titanium to enhance its mechanical properties as an implant biomaterial.

Because of its ubiquity in the environment and consumer products, the general population is constantly exposed to aluminium in everyday life. The main routes of exposure are ingestion, inhalation, and dermal contact. Aluminium or its salts occur naturally in some foods and cigarette smoke, as well as being present in many food additives, pesticides, detergents, paints, cosmetics, antiperspirants, and pharmaceuticals. Some of the biggest sources of aluminium are antacids and antidiarrhoeal and anti-ulcerative medications. Uptake of vaccines containing aluminium adjuvants results in acute exposure to the metal. Although most adults ingest as much as 10 mg of aluminium per day, the metal is not required for human metabolism and has no place in nutrition.

#### *3.1.7.1 Toxicokinetics*

While aluminium can exist in several different forms in vivo, the predominant and most toxicologically relevant species is $Al^{3+}$. In the blood, $Al^{3+}$ is distributed equally between the plasma and red blood cells, with 90% of the plasmatic fraction bound to transferrin and the remainder complexed with citrate, phosphate, and hydroxide ions. After absorption, aluminium is distributed unequally to all tissues; the estimated total body burden of aluminium is 30–50 μg, of which 50% is stored in bone and 1% in the brain (Yokel and McNamara, 2001). Approximately 95% of absorbed aluminium is removed in the urine, with the remainder excreted in the faeces, bile, breast milk, sweat, hair, and nails. The heavy reliance on urinary excretion means that inadequate renal function will increase the risk of aluminium accumulation and toxicity.

#### *3.1.7.2 Systemic toxicity*

Continuous exposure to aluminium causes it to gradually build up in the body. Accumulation of aluminium in mammalian tissues is highly disruptive and associated with widespread pathological effects, including haematological, cardiovascular, gastrointestinal, endocrine, hepato-renal, pancreatic, musculoskeletal, reproductive, and developmental abnormalities (Igbokwe et al., 2019). The toxic effects of aluminium emanate from the pro-oxidant and pro-inflammatory action of $Al^{3+}$, which leads to oxidative damage to cellular proteins, lipids, and DNA, inhibition of enzymatic activity and energy production, and, ultimately, cell death (Exley, 2016). Aluminium is a known human neurotoxin, with an established causal role in dialysis encephalopathy and likely involvement in other neurodegenerative disorders, such as Alzheimer's disease, Parkinson's disease, and multiple sclerosis. Since accumulation of aluminium in the body is slow and its effects wide ranging and non-specific, aluminium intoxication may take several years to become apparent and the symptoms may not be easily attributable to their root cause.

It is unclear whether patients with hip implants are at a higher risk of systemic aluminium toxicity than the general population. While CoC bearings are not associated with appreciable ion generation (Grübl et al., 2006), implant components made of Ti-6Al-4V alloy release aluminium ions into the local tissue and bloodstream as they corrode. The extent of this is likely low if the prosthesis is functioning optimally and the resulting contribution to the total aluminium body load is probably insignificant. However, local adverse effects and hypersensitivity reactions are still possible. The toxicological relevance of chronic, low-level exposure to 'endogenous' aluminium certainly deserves consideration, particularly in younger patients, who are expected to have their implants for several decades.

### 3.1.8 Reproductive toxicity

The cytotoxic and mutagenic effects of metal ions have sparked concerns over potential reproductive and developmental toxicity in patients with metal implants or in their offspring. While varying degrees of reproductive toxicity have been observed in male mice dosed with $Cr^{6+}$, $Co^{2+}$, $Ni^{2+}$, $V^{5+}$, and $Al^{3+}$ ions, there is insufficient evidence to link metal release from hip arthroplasties to such effects. Nikolaou et al. (2013) showed that males with blood cobalt concentrations typical of well-functioning MoM implants (2–11 μg/L) exhibited no sperm abnormalities. Further, cobalt and chromium ions are known to cross the placenta in patients with MoM hips (Oppermann et al., 2015; Ziaee et al., 2007), but there has not been a report of birth defects or physiological aberrations in the offspring as a result of maternal MoM arthroplasty. Oppermann et al. (2015) described a patient with a unilateral MoM implant and high blood cobalt and chromium levels (103 and 47 μg/L, respectively), who gave birth to a male infant with minor hypospadias. The boy's blood cobalt and chromium measurements at 9 weeks of age were 10 and 6.7 μg/L, respectively, but he did not exhibit any adverse effects and the hypospadias could not be conclusively linked to raised metal ions. In Fritzsche et al. (2012), a patient with bilateral MoM hips and highly elevated blood cobalt levels throughout pregnancy (138–143 μg/L) gave birth to a healthy male infant at 38 weeks of gestation. The authors noted that the child's development after 14 weeks was 'uneventful' despite abnormal circulating cobalt concentration (13 μg/L) at 8 weeks of age. Similarly, Holly (1955) reported no developmental abnormalities in human foetuses born to mothers receiving 75–100 mg $CoCl_2$ per day for up to 6 months during pregnancy. A recent systematic review of relevant human and animal studies concluded that implanted devices containing cobalt are not a reproductive hazard (Monnot et al., 2021).

Despite the lack of conclusive evidence for a teratogenic risk of high cobalt or chromium levels, the possibility of implant-related pregnancy complications cannot be dismissed. In the United Kingdom, women with MoM replacements are advised to postpone pregnancy planning for at least 2 years after implantation (MHRA, 2010).

### 3.1.9 Genotoxicity and carcinogenicity

Ionic cobalt, chromium, titanium, nickel, vanadium, and aluminium are all mutagenic in cultured cells. The genotoxic effects are thought to be mediated via free radical-induced DNA damage and/or interference with DNA repair mechanisms (Daley et al., 2004).

Significant increases in DNA aberrations (aneuploidy and chromosomal translocations) have been observed in peripheral lymphocytes of patients with MoM prostheses and those undergoing revision of CoCr hip implants (Doherty et al., 2001; Dunstan et al., 2008); however, the clinical relevance of these findings has not been demonstrated. Interestingly, similar effects were noted in patients with CoC implants, which do not appreciably elevate systemic metal levels. It is possible that the reported DNA alterations were caused by a non-specific inflammatory response secondary to the presence of an implant rather than by specific actions of cobalt or chromium (Christian et al., 2014).

Cases of malignant tumours at sites of metal orthopaedic implants have been reported in humans and animals (Sunderman, 1989), and some analyses suggest that patients with MoM devices have a higher risk of soft-tissue sarcoma than those with non-MoM implants and the general population (Mäkelä et al., 2012). Large-scale epidemiological analyses have not shown a causal link between hip replacement and increased overall cancer risk (Mäkelä et al., 2012; Mathiesen et al., 1995; Onega et al., 2006; Visuri et al., 2010). In fact, joint implant recipients display lower mortality from all causes and lower cancer incidence than the general population (Smith et al., 2012). While certain studies point to higher incidence of prostate cancer, bladder cancer, endometrial cancer, melanoma, and hematopoietic cancers in individuals with hip implants than in the general population (Brewster et al., 2013; Levašič et al., 2018; Tharani et al., 2001), these excess risks are likely a result of various biases and lack of information on confounding comorbidities in the studied cohorts. Notably, particle-mediated cancers can have long latency times, with some solid malignant tumours requiring between 20 and 40 years to develop. It is possible that the relatively short follow-up time in the previous studies (up to 25 years) was not long enough to detect any malignant changes. As a result, it may not be possible to extrapolate these findings to younger patients, who are expected to use their hip prostheses for several decades.

## 3.2 Metal hypersensitivity

Metal ions released into systemic circulation bind to serum proteins to form haptens or hapten-like complexes, which can act as antigens for circulating T lymphocytes and elicit hypersensitivity reactions. The resulting immune response has the characteristics of a delayed-type hypersensitivity reaction (type IV), where the antigenic complexes are first processed and presented to $CD4^+$ Th1 lymphocytes by antigen presenting cells (APCs) within the synovial tissue (Akil et al., 2018). Activated T cells generate cytokines that attract macrophages and other inflammatory cells to the implant site. The subsequent release of pro-inflammatory cytokines, including interleukin (IL)-1, IL-2, IL-6, tumour necrosis factor (TNF)-$\alpha$, and interferon (IFN)-$\gamma$, stimulates an adaptive immune response that may damage local tissues and result in symptoms such as periprosthetic joint pain, swelling, effusions, and, less frequently, cutaneous reactions such as eczematous dermatitis, urticaria, and/or vasculitis. Eruptions typically occur on the skin overlaying the implant site, but generalised skin rashes have also been described (Engelhart and Segal, 2017; Hosoki et al., 2016; Steens et al., 2006; Tower, 2010). While the role of metal hypersensitivity in prosthetic loosening is poorly characterised, many investigators believe that it may contribute to implant failure (Hallab et al., 2001). The histological features of aseptic lymphocytic vasculitis-associated lesion (ALVAL) may represent a systemic T lymphocyte-mediated delayed-type hypersensitivity reaction. It has been suggested that in patients with high-wearing implants, the metal debris attracts macrophages and leads to an expanding necrotic zone, while patients with hypersensitivity will react to even low amounts of particles, leading to T-lymphocyte infiltration and ALVAL (Paukkeri et al., 2016).

Nickel is the most potent and common metal sensitiser, with an estimated 8%–25% of the general population affected by nickel sensitivity (Hallab and Jacobs, 2009). In addition, approximately 2.4%–3% of the general population is allergic to cobalt and 1% to chromium (Schäfer et al., 2001). Dermal sensitivity to titanium, vanadium, and aluminium is less frequently reported (Engelhart and Segal, 2017; Granchi et al., 2005; Lalor et al., 1991; Thomas et al., 2006). In line with these findings, there are generally more case reports of hypersensitivity reactions to stainless steel and CoCr-alloy components than to titanium-based implants.

While there is no standard way to diagnose metal hypersensitivity, dermal patch testing, lymphocyte transformation testing, modified lymphocyte stimulation testing, and leukocyte migration inhibitory testing are all useful. Patch testing is the 'gold standard' for in vivo evaluation for delayed-type hypersensitivity reactions (Schalock et al., 2012). However, its applicability to the study of immune responses in deep tissue is highly questioned, because the skin's APCs (the Langerhans cells) are not present within the periprosthetic environment (Wawrzynski et al., 2017). Previous studies have shown that the prevalence of cutaneous metal sensitivity is higher in patients with well-functioning metallic implants (25%) than in the general population (10%–15%), and may be as high as 60% in those with malfunctioning or failed prostheses (Hallab et al., 2001). Unfortunately, the relationship between contact dermatitis and deep tissue allergy is not straightforward, and it is unclear whether metal hypersensitivity is a cause or consequence of implant failure. Patients testing positive for metal sensitivity preoperatively often do not react to their metal prostheses and vice versa, which suggests that alternate mechanisms, such as genetic autoimmunity, may be implicated. It also remains a subject of debate whether pre-existing metal hypersensitivity should impact treatment decisions and implant choice in prospective THA recipients. There is no evidence to suggest that individuals with a positive skin patch test result but no history of reaction to metallic materials are more likely to develop a reaction to an implanted device (Hallab et al., 2001). The current consensus is that 'hypoallergenic' materials (ceramic or ceramic-coated implants) should only be considered in persons with a documented allergy to a metal in the standard implant, e.g., from dermal contact with a metallic watch strap or costume jewellery (Wawrzynski et al., 2017).

## 3.3 Local effects of metal debris

In addition to potential systemic effects and end-organ toxicity, soluble and particulate metallic debris can provoke adverse biological reactions in the vicinity of the implant. The clinical spectrum of these reactions ranges from small, asymptomatic soft-tissue lesions to extensive peri-implant osteolysis, necrosis, effusion, and growing cystic or solid masses that can cause secondary pathological effects by pushing or displacing surrounding tissues (Pisanu et al., 2019). The terms used to describe the local responses to metallic debris include metallosis (macroscopic black or grey staining of periprosthetic tissues), pseudotumours (benign soft-tissue masses), adverse reaction to metal debris, adverse local tissue reaction (ALTR), and ALVAL (Langton et al., 2010; Pandit et al., 2008; Willert et al., 2005). The forms of implant debris most commonly investigated in association with local toxicity are cobalt and titanium ions, particles of CoCr alloy, and $TiO_2$ particles.

Following their generation, most CoCr wear particles are phagocytosed by tissue-resident macrophages and dissolved in the acidic environment of their lysosomes to liberate $Co^{2+}$ and $Cr^{3+}$ ions. Free cobalt ions are cytotoxic to macrophages in vitro, with lower doses causing apoptosis (programmed cell death) and higher doses leading to necrosis (uncontrolled cell death due to cell injury) (Catelas et al., 2001; Huk et al., 2004; Kwon et al., 2009). Cobalt ions can also activate Toll-like receptor 4 on macrophages and monocytes, which stimulates the inflammatory cascade by triggering the expression of a wide array of cytokines (TNF-$\alpha$, IL-1, IL-6, and IL-8), growth factors (macrophage colony stimulating factor-1), and chemokines (MCP-1, MIP-1a), as well as activating a myriad of downstream signalling pathways, including nuclear factor-$\kappa$B, protein kinase B (AKT/PKB), and MAPK pathways (Posada et al., 2015). Although the interplay between these inflammatory mediators and pathways has not been fully elucidated, it is known that TNF-$\alpha$, IL-1, and IL-6 induce the differentiation and maturation of osteoclasts (bone-resorbing cells) while inhibiting the proliferation of osteoblasts (bone-forming cells) and stimulating the release of matrix-resorbing enzymes. The net result is periprosthetic osteolysis, compromised implant fixation, and prosthetic loosening. Soluble and particulate titanium can affect bone homeostasis in a similar way, enhancing bone resorption around the implant (Cadosch et al., 2009; Choi et al., 2005; Yao et al., 2017).

Although ALTRs are most frequently reported in association with CoCr alloy, titanium debris has also been linked to inflammation (Jacobs et al., 1995a,b), pain (Hallam et al., 2004), and pseudotumour formation (McPherson et al., 2014; Sakamoto et al., 2016; Tarpada et al., 2020) in patients with hip implants. In Kwon et al. (2010), the incidence of pseudotumours in patients with MoM hip resurfacing devices was associated with elevated serum metal levels and excessive wear of the femoral component, suggesting that pseudotumours represent a local biological reaction to high localised concentration of metal debris. Confusingly, pseudotumours can also develop in the vicinity of MoP and CoP bearings, which release less metal ions than MoM couples (Clyburn, 2013; Hsu et al., 2012; Mao et al., 2012). Finally, pseudotumours are not exclusively associated with CoCr alloy and were also seen to form adjacent to titanium-based prostheses (Hsu et al., 2012; McPherson et al., 2014; Sakamoto et al., 2016; Tarpada et al., 2020). While the exact pathogenesis of pseudotumour development is unknown, it is thought to involve both chronic inflammation and a delayed-type hypersensitivity response (Sakamoto et al., 2016).

Because of the destructive nature of ALTRs, accurate diagnosis and timely revision of a problematic implant are the key to improving patient outcomes. Chapter 4 describes how measurement of circulating metal levels can be used to assess implant degradation and potential for local adverse reactions in hip implant recipients.

## References

Akil, S., Newman, J.M., Shah, N.V., Ahmed, N., Deshmukh, A.J., Maheshwari, A.V., 2018. Metal hypersensitivity in total hip and knee arthroplasty: current concepts. J. Clin. Orthop. Trauma. 9, 3–6.

Al-Doaiss, A.A., Ali, D., Ali, B.A., Jarrar, B.M., 2019. Renal histological alterations induced by acute exposure of titanium dioxide nanoparticles. Int. J. Morphol. 37, 1049–1057.

Alexander, C.S., 1969. Cobalt and the heart. Ann. Intern. Med. 70, 411–314.

Alexander, C.S., 1972. Cobalt-beer cardiomyopathy. A clinical and pathologic study of twenty-eight cases. Am. J. Med. 53, 395–417.

Apel, W., Stark, D., Stark, A., O'Hagan, S., Ling, J., 2013. Cobalt-chromium toxic retinopathy case study. Doc. Ophthalmol. 126, 69–78.

Apostoli, P., Catalani, S., Zaghini, A., Mariotti, A., Poliani, P.L., Vielmi, V., et al., 2013. High doses of cobalt induce optic and auditory neuropathy. Exp. Toxicol. Pathol. 65, 719–727.

ATSDR, 2012. Toxicological Profile for Vanadium. Agency for Toxic Substances and Disease Registry, Atlanta, GA.

Awalgaonkar, G., Beaudry, R., Almenar, E., 2020. Ethylene-removing packaging: basis for development and latest advances. Compr. Rev. Food Sci. Food Saf. 19, 3980–4007.

Barborik, M., Dusek, J., 1972. Cardiomyopathy accompaning industrial cobalt exposure. Br. Heart J. 34, 113–116.

Barceloux, D.G., Barceloux, D., 1999a. Cobalt. J. Toxicol. Clin. Toxicol. 37, 201–216.

Barceloux, D.G., Barceloux, D., 1999b. Molybdenum. J. Toxicol. Clin. Toxicol. 37, 231–237.

Barceloux, D.G., Barceloux, D., 1999c. Nickel. J. Toxicol. Clin. Toxicol. 37, 239–258.

Barceloux, D.G., Barceloux, D., 1999d. Vanadium. J. Toxicol. Clin. Toxicol. 37, 265–278.

Barth, A., Schaffer, A., Konnaris, C., Blauensteiner, R., Winker, R., Osterode, W., et al., 2002. Neurobehavioral effects of vanadium. J. Toxicol. Environ. Health 65, 677–683.

Beaumont, J.J., Sedman, R.M., Reynolds, S.D., Sherman, C.D., Li, L.H., Howd, R.A., et al., 2008. Cancer mortality in a Chinese population exposed to hexavalent chromium in drinking water. Epidemiology 19, 12–23.

Bennett, B., 1984. Environmental nickel pathways to man. IARC Sci. Publ. 53, 487–495.

Berglund, F., Carlmark, B., 2011. Titanium, sinusitis, and the yellow nail syndrome. Biol. Trace Elem. Res. 143, 1–7.

Blanco, A., Blanco, G., 2017. Essential minerals. In: Medical Biochemistry. Elsevier, Saint Louis, MO, pp. 715–743.

Beyersmann, D., Hartwig, A., 2008. Carcinogenic metal compounds: recent insight into molecular and cellular mechanisms and cellular mechanisms. Arch. Toxicol. 82, 493–512.

Brewster, D.H., Stockton, D.L., Reekie, A., Ashcroft, G.P., Howie, C.R., Porter, D.E., et al., 2013. Risk of cancer following primary total hip replacement or primary resurfacing arthroplasty of the hip: a retrospective cohort study in Scotland. Br. J. Cancer 108, 1883–1890.

Buettner, K.M., Valentine, A.M., 2012. Bioinorganic chemistry of titanium. Chem. Rev. 112, 1863–1881.

Cadosch, D., Chan, E., Gautschi, O.P., Filgueira, L., 2009. Metal is not inert: role of metal ions released by biocorrosion in aseptic loosening – current concepts. J. Biomed. Mater. Res. A 91, 1252–1262.

Catalani, S., Leone, R., Rizzetti, M.C., Padovani, A., Apostoli, P., 2011. The role of albumin in human toxicology of cobalt: contribution from a clinical case. ISRN Hematol. 2011, 1–6.

Catalani, S., Stea, S., Beraudi, A., Gilberti, M., Bordini, B., Toni, A., et al., 2013. Vanadium release in whole blood, serum and urine of patients implanted with a titanium alloy hip prosthesis. Clin. Toxicol. 51, 550–556.

Catelas, I., Petit, A., Zukor, D.J., Huk, O.L., 2001. Cytotoxic and apoptotic effects of cobalt and chromium ions on J774 macrophages- implication of caspase-3 in the apoptotic pathway. J. Mater. Sci. Mater. Med. 12, 949–953.

Chen, Z., Wang, Y., Zhuo, L., Chen, S., Zhao, L., Luan, X., et al., 2015. Effect of titanium dioxide nanoparticles on the cardiovascular system after oral administration. Toxicol. Lett. 239, 123–130.

Choi, M.G., Koh, H.S., Kluess, D., O'Connor, D., Mathur, A., Truskey, G.A., et al., 2005. Effects of titanium particle size on osteoblast functions in vitro and in vivo. Proc. Natl. Acad. Sci. U.S.A. 102, 4578–4583.

Christian, W.V., Oliver, L.D., Paustenbach, D.J., Kreider, M.L., Finley, B.L., 2014. Toxicology-based cancer causation analysis of CoCr-containing hip implants: a quantitative assessment of genotoxicity and tumorigenicity studies. J. Appl. Toxicol. 34, 939–967.

Clyburn, T., 2013. Pseudotumor in metal-on-polyethylene total hip arthroplasty. Reconstr. Rev. 3, 18–21.

Dahms, K., Sharkova, Y., Heitland, P., Pankuweit, S., Schaefer, J.R., 2014. Cobalt intoxication diagnosed with the help of Dr House. Lancet. 383, 574.

Daley, B., Doherty, A., Fairman, B., Case, C., 2004. Wear debris from hip or knee replacements causes chromosomal damage in human cells in tissue culture. J. Bone Joint Surg. Br. 86, 598–606.

Doherty, A., Howell, R., Ellis, L., Bisbinas, I., Learmonth, I., Newson, R., et al., 2001. Increased chromosome translocations and aneuploidy in peripheral blood lymphocytes of patients having revision arthroplasty of the hip. J. Bone Joint Surg. Ser. B. 83, 1075–1081.

Domingo, J., Gómez, M., 2016. Vanadium compounds for the treatment of human diabetes mellitus: a scientific curiosity? A review of thirty years of research. Food Chem. Toxicol. 95, 137–141.

Duckham, J., Lee, H., 1976. The treatment of refractory anaemia of chronic renal failure with cobalt chloride. Q. J. Med. 45, 277–294.

Dunstan, E., Ladon, D., Whittingham-Jones, P., Carrington, R., Briggs, T.W.R., 2008. Chromosomal aberrations in the peripheral blood of patients with metal-on-metal hip bearings. J. Bone Joint Surg. Am. 90, 517–522.

Engelhart, S., Segal, R., 2017. Allergic reaction to vanadium causes a diffuse eczematous eruption and titanium alloy orthopedic implant failure. Cutis 99, 245–249.

Exley, C., 2016. The toxicity of aluminium in humans. Morphologie 100, 51–55.

Fabi, D., Levine, B., Majoras, N., 2011. Chronic renal failure and catastrophic failure of an acetabular component: a case report. J. Bone Joint Surg. Am. 93, 1–4.

Finley, B., Scott, P.K., Glynn, M.E., Paustenbach, D., Donovan, E., Thuett, K.A., et al., 2017. Chromium speciation in the blood of metal-on- metal hip implant patients. Toxicol. Environ. Chem. 99, 48–64.

Fritzsche, J., Borisch, C., Schaefer, C., 2012. Case report: high chromium and cobalt levels in a pregnant patient with bilateral metal-on-metal hip arthroplasties. Clin. Orthop. Relat. Res. 470, 2325–2331.

Granchi, D., Cenni, E., Trisolino, G., Giunti, A., Baldini, N., 2005. Sensitivity to implant materials in patients undergoing total hip replacement. J. Biomed. Mater. Res. B Appl. Biomater. 77, 257–264.

Grübl, A., Weissinger, M., Brodner, W., Gleiss, A., Giurea, A., Gruber, M., et al., 2006. Serum aluminium and cobalt levels after ceramic-on-ceramic and metal-on-metal total hip replacement. J. Bone Joint Surg. Br. 88, 1003–1005.

Hallab, N.J., Jacobs, J.J., 2009. Biologic effects of implant debris. Bull. NYU Hosp. Jt. Dis. 67, 182–188.

Hallab, N., Merritt, K., Jacobs, J.J., 2001. Metal sensitivity in patients with orthopaedic implants. J. Bone Joint Surg. Am. 83-A, 428–436.

Hallam, P., Haddad, F., Cobb, J., 2004. Pain in the well-fixed, aseptic titanium hip replacement. J. Bone Joint Surg. 86-B, 27–30.

Hart, A., Quinn, P., Sampson, B., Sandison, A., Atkinson, K., Skinner, J., et al., 2010. The chemical form of metallic debris in tissues surrounding metal-on-metal hips with unexplained failure. Acta Biomater. 6, 4439–4446.

He, X., Hartlieb, E., Rothmund, L., Waschke, J., Wu, X., Van Landuyt, K., et al., 2015. Intracellular uptake and toxicity of three different titanium particles. Dent. Mater. 31, 734–744.

Holly, R., 1955. Anemia in pregnancy. Obstet. Gynecol. 5, 562–568.

Hosoki, M., Nishigawa, K., Miyamoto, Y., Ohe, G., Matsuka, Y., 2016. Allergic contact dermatitis caused by titanium screws and dental implants. J. Prosthodont. Res. 60, 213–219.

Howitt, J., Putz, U., Lackovic, J., Doan, A., Dorstyn, L., Cheng, H., et al., 2009. Divalent metal transporter 1 (DMT1) regulation by Ndfip1 prevents metal toxicity in human neurons. Proc. Natl. Acad. Sci. U.S.A. 106, 15489–15494.

Hsu, A., Gross, C., Levine, B., 2012. Pseudotumor from modular neck corrosion after ceramic-on-polyethylene total hip arthroplasty. Am. J. Orthop. 41, 422–426.

Huk, O.L., Catelas, I., Mwale, F., Antoniou, J., Zukor, D.J., Petit, A., 2004. Induction of apoptosis and necrosis by metal ions in vitro. J. Arthroplasty 19, 84–87.

Igbokwe, I., Igwenagu, E., Igbokwe, N., 2019. Aluminium toxicosis: a review of toxic actions and effects. Interdiscip. Toxicol. 12, 45–70.

Jacobs, J., Urban, R., Wall, J., Black, J., Reid, J., Veneman, L., 1995a. Unusual foreign-body reaction to a failed total knee replacement: simulation of a sarcoma clinically and a sarcoid histologically. J. Bone Joint Surg. 77-A, 444–451.

Jacobs, J.J., Urban, R.M., Gilbert, J.L., Skipor, A.K., Black, J., Jasty, M., et al., 1995b. Local and distant products from modularity. Clin. Orthop. Relat. Res. 319, 94–105.

Ikeda, T., Takahashi, K., Kabata, T., Sakagoshi, D., Tomita, K., Yamada, M., 2010. Polyneuropathy caused by cobalt-chromium metallosis after total hip replacement. Muscle Nerve 42, 140–143.

Jansen, H.M.L., Knollema, S., Van Der Duin, L.V., Willemsen, A.T.M., Wiersma, A., Korfand, J., et al., 1996. Pharmacokinetics and dosimetry of cobalt-55 and cobalt-57. J. Nucl. Med. 37, 2082–2086.

Kennedy, A., Dornan, J., King, R., 1981. Fatal myocardial disease associated with industrial exposure to cobalt. Lancet 317, 412–414.

Kovalskii, V., Jarovaja, G., Shmavonyan, D., 1961. Changes in purine metabolism in man and animals in various molybdenum-rich biogeochemical provinces. Zh Obshch. Biol. 22, 179–191.

Kurosaki, K., Nakamura, T., Mukai, T., Endo, T., 1995. Unusual findings in a fatal case of poisoning with chromate compounds. Forensic Sci. Int. 1, 57–65.

Kwon, Y.-M., Glyn-Jones, S., Simpson, D.J., Kamali, A., McLardy-Smith, P., Gill, H.S., et al., 2010. Analysis of wear of retrieved metal-on-metal hip resurfacing implants revised due to pseudotumours. J. Bone Joint Surg. Br. 92, 356–361.

Kwon, Y.-M., Xia, Z., Glyn-Jones, S., Beard, D., Gill, H.S., Murray, D.W., 2009. Dose-dependent cytotoxicity of clinically relevant cobalt nanoparticles and ions on macrophages in vitro. Biomed. Mater. 4, 1–8.

La Grutta, A., Amato, G., Fedele, S., Vitaliti, S., 1984. Lntossicazione cronica da cobalto. Presentazione di un caso clinico di cardiomiopatia. Minerva. Pediatr. 36, 691–697.

Lalor, P.A., Revell, P.A., Gray, A.B., Wright, S., Railton, G.T., Freeman, M.A., 1991. Sensitivity to titanium. A cause of implant failure? J. Bone Joint Surg. Br. 73, 25–28.

Langton, D.J., Jameson, S.S., Joyce, T.J., Hallab, N.J., Natu, S., Nargol, A.V.F., 2010. Early failure of metal-on-metal bearings in hip resurfacing and large-diameter total hip replacement. J. Bone Joint Surg. Br. 92-B, 38–46.

Levašič, V., Milošev, I., Zadnik, V., 2018. Risk of cancer after primary total hip replacement: the influence of bearings, cementation and the material of the stem. Acta Orthop. 89, 234–239.

Levina, A., Lay, P.P.A., 2008. Chemical properties and toxicity of chromium (III) nutritional supplements. Chem. Res. Toxicol. 21, 563–571.

Linos, A., Petralias, A., Christophi, C.A., Christoforidou, E., Kouroutou, P., Stoltidis, M., et al., 2011. Oral ingestion of hexavalent chromium through drinking water and cancer mortality in an industrial area of Greece – an ecological study. Environ. Health 10, 50.

Lippi, G., Franchini, M., Guidi, G.C., 2006. Blood doping by cobalt. Should we measure cobalt in athletes? J. Occup. Med. Toxicol. 1, 18.

Little, J., Sunico, R., 1958. Cobalt-induced goiter with cardiomegaly and congestive failure. J. Pediatr. 52, 284–288.

Liu, H., Ma, L., Zhao, J., Liu, J., Yan, J., Ruan, J., et al., 2009. Biochemical toxicity of nano-anatase TiO2 particles in mice. Biol. Trace Elem. Res. 129, 170–180.

Lomer, M.C.E., Hutchinson, C., Volkert, S., Greenfield, S.M., Catterall, A., Thompson, R.P.H., et al., 2004. Dietary sources of inorganic microparticles and their intake in healthy subjects and patients with Crohn's disease. Br. J. Nutr. 92, 947–955.

Mäkelä, K.T., Visuri, T., Pulkkinen, P., Eskelinen, A., Remes, V., Virolainen, P., et al., 2012. Risk of cancer with metal-on-metal hip replacements: population based study. BMJ Support Palliat. Care 345, e4646.

Mao, X., Tay, G.H., Godbolt, D.B., Crawford, R.W., 2012. Pseudotumor in a well-fixed metal-on-polyethylene uncemented hip arthroplasty. J. Arthroplasty 493, e13–e17.

Mathiesen, E., Ahlbom, A., Bermann, G., Lindgren, J., 1995. Total hip replacement and cancer. A cohort study. J. Bone Joint Surg. Br. 77-B, 345–350.

McPherson, E.J., Dipane, M.V., Sherif, S.M., 2014. Massive pseudotumor in a 28mm ceramic-polyethylene revision THA: a case report. Reconstr. Rev. 4, 11–17.

Meecham, H., Humphrey, P., 1994. Industrial exposure to cobalt causing optic atrophy and nerve deafness: a case report. J. Neurol. Neurosurg. Psychiatry 54, 374–375.

Merritt, K., Margevicius, R.W., Brown, S.A., 1992. Storage and elimination of titanium, aluminum, and vanadium salts in vivo. J. Biomed. Mater. Res. 26, 1503–1515.

MHRA, 2010. Advice From the CSD Expert Advisory Group on the Biological Effects of Metal Wear Debris Generated From Hip Implants. Available from: http://data.parliament.uk/DepositedPapers/Files/DEP2013-0653/149850a.pdf.

Minoia, C., Cavalleri, A., 1988. Chromium in urine, serum and red blood cells in the biological monitoring of workers exposed to different chromium valency states. Sci. Total. Environ. 71, 323–327.

Monnot, A.D., Kovochich, M., Bandara, S.B., Wilsey, J.T., Christian, W.V., Eichenbaum, G., et al., 2021. A hazard evaluation of the reproductive/developmental toxicity of cobalt in medical devices. Regul. Toxicol. Pharmacol. 123, 104932.

Moran, C.A., Mullick, F.G., Ishak, K.G., Johnson, F.B., Hummer, W.B., 1991. Identification of titanium in human tissues: probable role in pathologic processes. Hum. Pathol. 22, 450–454.

Moretti, B., Pesce, V., Maccagnano, G., Vicenti, G., Lovreglio, P., Soleo, L., et al., 2012. Peripheral neuropathy after hip replacement failure: is vanadium the culprit? Lancet 379, 1676.

Morin, Y., Foley, A., Martineau, G., Roussel, J., 1967. Quebec beer-drinkers' cardiomyopathy: forty-eight cases. Can. Med. Assoc. J. 97, 881–883.

Mulyani, I., Levina, A., Lay, P.A., 2004. Biomimetic oxidation of chromium(III): does the anti-diabetic activity of chromium(III) involve carcinogenic chromium(VI)? Angew. Chem. Int. Ed. Engl 43, 4504–4507.

Newton, A.W., Ranganath, L., Armstrong, C., Peter, V., Roberts, N.B., 2012. Differential distribution of cobalt, chromium, and nickel between whole blood, plasma and urine in patients after metal-on-metal (MoM) hip arthroplasty. J. Orthop. Res. 30, 1640–1646.

Nickens, K.P., Patierno, S.R., Ceryak, S., 2010. Chromium genotoxicity: a double-edged sword. Chem. Biol. Interact. 188, 276–288.

Nikolaou, V., Petit, A., Zukor, D., Papanastasiou, C., Huk, O., Antoniou, J., 2013. Presence of cobalt and chromium ions in the seminal fluid of young patients with metal-on-metal total hip arthroplasty. J. Arthroplasty 28, 161–167.

Onega, T., Baron, J., MacKenzie, T., 2006. Cancer after total joint arthroplasty: a meta-analysis. Cancer Epidemiol. Biomarkers Prev. 15, 1532–1537.

Onuma, K., Sato, Y., Ogawara, S., Shirasawa, N., Kobayashi, M., Yoshitake, J., et al., 2009. Nano-scaled particles of titanium dioxide convert benign mouse fibrosarcoma cells into aggressive tumor cells. Am. J. Pathol. 175, 2171–2183.

Oppermann, M., Borisch, C., Schaefer, C., 2015. Hip arthroplasty with high chromium and cobalt blood levels – case report of a patient followed during pregnancy and lactation period. Reprod. Toxicol. 53, 51–53.

Packer, M., 2016. Cobalt cardiomyopathy: a critical reappraisal in light of a recent resurgence. Circ. Heart. Fail. 9, e003604.

Pandit, H., Glyn-Jones, S., McLardy-Smith, P., Gundle, R., Whitwell, D., Gibbons, L.C.M., et al., 2008. Pseudotumours associated with metal-on- metal hip resurfacings. J. Bone Joint Surg. Br. 90, 920–923.

Parkin, I.P., Palgrave, R.G., 2005. Self-cleaning coatings. J. Mater. Chem. 15, 1689–1695.

Paukkeri, E., Korhonen, R., Hämäläinen, M., Pesu, M., Eskelinen, A., 2016. The inflammatory phenotype in failed metal-on-metal hip arthroplasty correlates with blood metal concentrations. PLoS One 11, 1–13.

Paustenbach, D.J., Tvermoes, B.E., Unice, K.M., Finley, B.L., Kerger, B.D., 2013. A review of the health hazards posed by cobalt. Crit. Rev. Toxicol. 43, 316–362.

Pisanu, F., Doria, C., Andreozzi, M., Bartoli, M., Saderi, L., Sotgiu, G., et al., 2019. Pleomorphic clinical spectrum of metallosis in total hip arthroplasty. Int. Orthop. 43, 85–96.

Polyzois, I., Nikolopoulos, D., Michos, I., Patsouris, E., Theocharis, S., 2012. Local and systemic toxicity of nanoscale debris particles in total hip arthroplasty. J. Appl. Toxicol. 32, 255–269.

Posada, O., Tate, R., Meek, R.M., Grant, M., 2015. In vitro analyses of the toxicity, immunological, and gene expression effects of cobalt-chromium alloy wear debris and Co ions derived from metal-on-metal hip implants. Lubricants 3, 539–568.

Rehder, D., 2015. The role of vanadium in biology. Metallomics 7, 730–742.

Rona, G., 1971. Experimental aspects of cobalt cardiomyopathy. Br. Heart J. 33, 171–174.

Saha, R., Nandi, R., Saha, B., 2011. Sources and toxicity of hexavalent chromium. J. Coord. Chem. 64, 1782–1806.

Sakamoto, M., Watanabe, H., Higashi, H., Kubosawa, H., 2016. Pseudotumor caused by titanium particles from a total hip prosthesis. Orthopedics 39, e162–e165.

Samman, P., White, W., 1964. The "yellow nail" syndrome. Br. J. Dermatol. 76, 153–157.

Saxena, M., Loza-Rosas, S.A., Gaur, K., Sharma, S., Pérez Otero, S.C., Tinoco, A.D., 2018. Exploring titanium(IV) chemical proximity to iron(III) to elucidate a function for Ti(IV) in the human body. Coord. Chem. Rev. 363, 109–125.

Sayes, C.M., Wahi, R., Kurian, P.A., Liu, Y., West, J.L., Ausman, K.D., et al., 2006. Correlating nanoscale titania structure with toxicity: a cytotoxicity and inflammatory response study with human dermal fibroblasts and human lung epithelial cells. Toxicol. Sci. 92, 174–185.

Schäfer, T., Böhler, E., Ruhdorfer, S., Weigl, L., Wessner, D., Filipiak, B., et al., 2001. Epidemiology of contact allergy in adults. Allergy 56, 1192–1196. Erratum in: Allergy 2002 57:178.

Schalock, P., Menné, T., Johansen, J., 2012. Hypersensitivity reactions to metallic implants—diagnostic algorithm and suggested patch test series for clinical use. Contact Dermatitis. 66, 4–19.

Schirrmacher, U.O.E., 1967. Case of cobalt poisoning. Analgesia 1, 544–545.

Shabaan, A.A., Marks, V., Lancaster, M.C., Dufeu, G.N., 1977. Fibrosarcomas induced by cobalt chloride (CoCl2) in rats. Lab. Anim. 11, 43–46.

Shelnutt, S., Goad, P., Belsito, D., 2007. Dermatological toxicity of hexavalent chromium. Crit. Rev. Toxicol. 37, 375–387.

Sheng, L., Wang, X., Sang, X., Ze, Y., Zhao, X., Liu, D., et al., 2013. Cardiac oxidative damage in mice following exposure to nanoparticulate titanium dioxide. J. Biomed. Mater. Res. A 101, 3238–3246.

Silwood, C.J.L., Grootveld, M., 2005. Chemical nature of implant-derived titanium(IV) ions in synovial fluid. Biochem. Biophys. Res. Commun. 330, 784–790.

Simonsen, L.O., Brown, A.M., Harbak, H., Kristensen, B.I., Bennekou, P., 2011. Cobalt uptake and binding in human red blood cells. Blood Cells Mol. Dis. 46, 266–276.

Simonsen, L.O., Harbak, H., Bennekou, P., 2012. Cobalt metabolism and toxicology- a brief update. Sci. Total. Environ. 432, 210–215.

Skorringe, T., Burkhart, A., Johnsen, K.B., Moos, T., 2015. Divalent metal transporter 1 (DMT1) in the brain: implications for a role in iron transport at the blood-brain barrier, and neuronal and glial pathology. Front. Mol. Neurosci. 8, 19.

Smith, A.J., Dieppe, P., Porter, M., Blom, A.W., 2012. Risk of cancer in first seven years after metal-on-metal hip replacement compared with other bearings and general population:

linkage study between the National Joint Registry of England and Wales and hospital episode statistics. BMJ 344, e2383.

Smith, T., Edmonds, C., Barnaby, C., 1972. Absorption and retention of cobalt in man by whole-body counting. Health Phys. 22, 359–367.

Stabler, S.P., 2013. Vitamin B12 deficiency. N. Engl. J. Med. 368, 149–160.

Steens, W., von Foerster, G., Katzer, A., 2006. Severe cobalt poisoning with loss of sight after ceramic-metal pairing in a hip- a case report. Acta Orthop. 77, 830–832.

Sunderman, F.W., 1989. Carcinogenicity of metal alloys in orthopedic prostheses: clinical and experimental studies. Fundam. Appl. Toxicol. 13, 205–216.

Swennen, B., Buchet, J.P., Stanescu, D., Lison, D., Lauwerys, R., 2008. Epidemiological survey of workers exposed to cobalt oxides, cobalt salts, and cobalt metal. Occup. Environ. Med. 50, 835–842.

Tarpada, S.P., Loloi, J., Schwechter, E.M., 2020. A case of titanium pseudotumor and systemic toxicity after total hip arthroplasty polyethylene failure. Arthroplast. Today 6, 710–715.

Tharani, R., Dorey, F.J., Schmalzried, T.P., 2001. The risk of cancer following total hip or knee arthroplasty. J. Bone Joint Surg. Am. 83-A, 774–780.

Thomas, P., Bandl, W., Maier, S., Summer, B., Przybilla, B., 2006. Hypersensitivity to titanium osteosynthesis with impaired fracture healing, eczema, and T-cell hyperresponsiveness in vitro: case report and review of the literature. Contact Dermatitis. 55, 199–202.

Tinoco, A.D., Eames, E.V., Valentine, A.M., 2008. Reconsideration of serum Ti(IV) transport: albumin and transferrin trafficking of Ti(IV) and its complexes. J Am Chem Soc. 130, 2262–2270.

Tower, S., 2010. Cobalt toxicity in two hip replacement patients. State Alsk. Epidemiol. Bull. 14, 1.

Treviño, S., Díaz, A., Sánchez-Lara, E., Sanchez-Gaytan, B., Perez-Aguilar, J., González-Vergara, E., 2019. Vanadium in biological action: chemical, pharmacological aspects, and metabolic implications in diabetes mellitus. Biol. Trace Elem. Res. 188, 68–98.

Urban, R.M., Jacobs, J.J., 1994. Migration of corrosion products from modular hip prostheses. Particle microanalysis and histopathological findings. J. Bone Joint Surg. Am. 76, 1345.

Urban, R.M., Jacobs, J.J., Tomlinson, M.J., Gavrilovic, J., Black, J., Peoc'h, M., 2000. Dissemination of wear particles to the liver, spleen, and abdominal lymph nodes of patients with hip or knee replacement. J. Bone Joint Surg. Am. 82-A, 457–476.

Vincent, J., 2017. New evidence against chromium as an essential trace element. J. Nutr. 147, 2212–2219.

Visuri, T., Pulkkinen, P., Paavolainen, P., Pukkala, E., 2010. Cancer risk is not increased after conventional hip arthroplasty: a nationwide study from the Finnish arthroplasty register with follow-up of 24,636 patients for a mean of 13 years. Acta Orthop. 81, 77–81.

Walravens, P., Moure-Eraso, R., Solomons, C., Chappell, W., Bentley, G., 1979. Biochemical abnormalities in workers exposed to molybdenum dust. Arch. Environ. Health 34, 302–308.

Walter, L.R., Marel, E., Harbury, R., Wearne, J., 2008. Distribution of chromium and cobalt ions in various blood fractions after resurfacing hip arthroplasty. J. Arthroplasty 23, 814–821.

Wawrzynski, J., Gil, J.A., Goodman, A.D., Waryasz, G.R., 2017. Hypersensitivity to orthopedic implants: a review of the literature. Rheumatol. Ther. 4, 45–56.

Weber, K.P., Schweier, C., Kana, V., Guggi, T., Byber, K., Landau, K., 2015. Wear and tear vision. J. Neuroopthalmol. 35, 82–85.

Wedeen, R.P., Qian, L., 1991. Chromium-induced kidney disease. Environ. Health Perspect. 92, 71–74.

Weir, A., Westerhoff, P., Fabricius, L., Hristovski, K., Von Goetz, N., 2012. Titanium dioxide nanoparticles in food and personal care products. Environ. Sci. Technol. 46, 2242–2250.

Welling, R., Beaumont, J.J., Petersen, S.J., Alexeeff, G.V., Steinmaus, C., 2015. Chromium(VI) and stomach cancer: a meta-analysis of the current epidemiological evidence. Occup. Environ. Med. 72, 151–159.

Willert, H., Buchhorn, G., Fayyazi, A., Flury, R., Windler, M., Koster, G., et al., 2005. Metal-on-metal bearings and hypersensitivity in patients with artificial hip joints: a clinical and histomorphological study. J. Bone Joint Surg. Am. 87, 28–36.

Wu, J., Liu, W., Xue, C., Zhou, S., Lan, F., Bi, L., et al., 2009. Toxicity and penetration of TiO2 nanoparticles in hairless mice and porcine skin after subchronic dermal exposure. Toxicol. Lett. 191, 1–8.

Wu, L.E., Levina, A., Harris, H.H., Cai, Z., Lai, B., Vogt, S., et al., 2016. Carcinogenic chromium(VI) compounds formed by intracellular oxidation of chromium(III) dietary supplements by adipocytes. Angew. Chem. Int. Ed. Engl 55, 1742–1745.

Yao, J.J., Lewallen, E.A., Trousdale, W.H., Xu, W., Thaler, R., Salib, C.G., et al., 2017. Local cellular responses to titanium dioxide from orthopedic implants. BioResearch 6, 94–103.

Yokel, R., McNamara, P., 2001. Aluminium toxicokinetics: an updated minireview. Pharmacol. Toxicol. 88, 159–167.

Ziaee, H., Daniel, J., Datta, A., Blunt, S., McMinn, D., 2007. Transplacental transfer of cobalt and chromium in patients with metal-on-metal hip arthroplasty: a controlled study. J. Bone Joint Surg. 89, 301–305.

Part II

# Assessing hip implant function and risk of local and systemic adverse reactions

Chapter 4

# Markers of hip implant degradation: analytical considerations and clinical interpretation

**Pascal-André Vendittoli[1], Angela Styhler[1] and Ilona Świątkowska[2]**
*[1]Surgery Department, Hôpital Maisonneuve-Rosemont, Montreal University, Montréal, QC, Canada, [2]Institute of Orthopaedics and Musculoskeletal Science, University College London, Stanmore, United Kingdom*

## 4.1 Introduction

Wear and corrosion of metallic biomaterials liberate metal particles and ions into the surrounding tissue and bloodstream. Despite the widely perceived biocompatibility of bulk medical alloys, the chronic release of metal debris into the periprosthetic tissue can produce adverse local tissue reactions (ALTR), such as inflammation, tissue necrosis, and osteolysis, which may cause the implant to fail prematurely (Campbell and Takamura, 2020; Delaunay et al., 2010). Rarely, patients with high-wearing prostheses may also experience health problems affecting the entire body instead of only the hip joint (discussed in Chapters 3 and 7). The diagnostic markers of implant degradation can be classified into two categories: the products of wear and corrosion (e.g., metal particles and ions) and molecular mediators that reflect the biological consequences of wear and corrosion, such as pro-inflammatory cytokines and indices of bone turnover. This chapter focuses on the former category while the latter is covered in Chapter 5.

The principal metals used in the manufacture of modern hip implants are cobalt, chromium, and titanium. Inorganic cobalt is not required by the human body but vitamin B12 (cobalamin), which features a single cobalt atom at the centre of its structure, is important for DNA synthesis, the regulation of nervous system function, and the formation of red blood cells. Chromium is implicated in insulin, sugar, and lipid metabolism; however, its status as an essential micronutrient has been the subject of growing

**Biomarkers of Hip Implant Function. DOI: https://doi.org/10.1016/B978-0-12-821596-8.00005-7**

controversy (Levina et al., 2016). Titanium has no confirmed biological role and no titanium-requiring biomolecules have been identified. Nevertheless, the body content of titanium is presumed to be relatively high on account of its ubiquity in the environment, industrial applications, and common consumer products (Skocaj et al., 2011; Weir et al., 2012). Regulatory bodies such as the Medicines and Healthcare products Regulatory Agency (MHRA) and US Food and Drug Administration advocate the use of blood cobalt and chromium ion measurements as a screening tool in patients with metal-on-metal (MoM) hip replacement devices (Van Der Straeten et al., 2013a). Despite the widely held view that titanium-based implants are corrosion resistant and not associated with clinically significant toxicity, recent studies suggest that titanium is not exempt from degradation in physiological environments, and cases of ALTR to titanium debris have been reported. Nonetheless, the metal has so far received relatively little attention and actionable thresholds are less clearly defined than for cobalt and chromium.

This chapter addresses the most important analytical considerations for the determination of cobalt, chromium, and titanium levels in biological fluids, and discusses how systemic metal levels can be used to inform clinical decision-making in patients with hip implants.

## 4.2 Mechanisms of implant degradation

### 4.2.1 Passive corrosion

Passive corrosion is a thermodynamic process driven by a material's 'desire' to go toward a lower chemical energy state. The process is exothermic, with minimal activation energy, which enables it to occur spontaneously at a fast or slow rate, depending on the type of metal. First, the surface layer of the metal dissolves in the aqueous environment of the periprosthetic area. In this step, also referred to as oxidation, cations are removed. The remaining electrons are attracted to a differential charge at another point on the surface, thereby generating a current and driving the reduction reaction. Metal oxides or insoluble metal hydroxides (rust) form as by-products of the reaction (Urish et al., 2019).

Corrosion of biomaterials is limited by passivity (Dowson and Neville, 2014), i.e., the ability of a metal to form a thin oxide layer on its surface that acts as a barrier to charge transfer. Breakdown or disruption of the surface oxide film may result in areas with a very high rate of corrosion while the general corrosion rate on the surface remains extremely low, if not negligible. Despite formidable oxide films that form spontaneously on the surface of CoCr and Ti alloys, passive corrosion of exposed metallic implant surfaces has been observed and shown to be a contributing factor to the increased systemic metal load after orthopaedic procedures (Jacobs et al., 1998; Luetzner et al., 2007; Yan et al., 2006b).

Titanium levels in particular reflect passive corrosion of hip implants, as titanium components are used solely for non-articulating surfaces. Titanium release should be directly proportional to the metal area exposed to bodily fluids and tissues; however, the actual surface exposed to corrosion is difficult to estimate in grit-blasted or plasma-sprayed components due to their porous nature (Fig. 4.1). A prospective, randomised study that compared the clinical outcomes of patients with hip resurfacing (HR) and those with MoM total hip replacement (THR) revealed that blood titanium levels were elevated in both cohorts, and that plasma-sprayed titanium coatings on HR implants were associated with higher titanium release than grit-blasted surfaces of the femoral and acetabular THR implants at 2 years postoperatively (1.87 and 1.30 μg/L, respectively) (Vendittoli et al., 2010).

### 4.2.2 Galvanic corrosion

Galvanic corrosion is an electrochemical reaction that occurs when two dissimilar metals or alloys are in contact with one another in the presence of an electrolyte, such as synovial fluid (Osman et al., 2016). The metal with the higher electrode potential acts as an anode, releasing ions into the electrolyte solvent, while the other metal acts as a cathode. The electron current flowing from the anode to the cathode corrodes the anode metal.

Modular hip prostheses involve at least one junction between the components, and the choice of metals used in these couples influences their resistance to electrochemical attack. For example, a CoCr/CoCr couple is less susceptible to corrosion than CoCr combined with titanium or stainless steel (Gilbert et al., 2009; Goldberg et al., 2002). Similarly, Ti/Ti tapers are

**FIGURE 4.1** An acetabular component made of titanium. The porous outer surface encourages osseointegration, but also increases the surface area exposed to corrosion, promoting titanium ion release.

reportedly more corrosion resistant than Ti/CoCr tapers (Hutt et al., 2016). In a retrieval analysis of 231 modular implants, femoral neck and head corrosion scores were markedly higher in mixed-alloy than similar-alloy scenarios, with moderate-to-severe corrosion observed in 42% of mixed-alloy implants and 28% of similar-alloy implants (Goldberg et al. 2002).

The presence of metal impurities in the prosthesis may be sufficient to trigger galvanic corrosion in vivo. A study using scanning electron microscopy and X-ray microanalysis reported that holes found on the surface of retrieved CoCr components had a deviant composition of cobalt, chromium, and molybdenum compared with that of smooth prosthetic surface (decreased Co and enhanced Cr and Mo intensities), and that they contained titanium and aluminium particles (Koerten et al. 2001). The authors proposed that the contaminating metals created a galvanic element with the CoCr alloy, resulting in metal ion release and consequent weakening of the articular surface. These findings highlight the importance of using pure compounds in the manufacture of prosthetic components.

### 4.2.3 Mechanically assisted corrosion

Mechanically assisted corrosion (MAC) is induced when the protective oxide film on the surface of metal components is disrupted by mechanical wear. High cyclic loads and slight vibratory motions (micromotion) promote repetitive depassivation-repassivation cycles that deplete the local oxygen concentration and retard the formation of a new oxide layer (Eltit et al., 2019; Venditti et al., 2019). The excess of positively charged metal ions is balanced by the migration of chloride ions, resulting in the production of hydrochloric acid that perpetuates corrosion within the crevice between the two implant components (Collier et al., 1992). MAC occurs predominantly at mixed-alloy modular junctions and MoM articulating surfaces, where it causes severe mechanical damage and high localised release of metal ions (McTighe et al., 2015; Venditti et al., 2011). Retrieval analyses revealed that corroded CoCr head tapers have a typical macroscopic appearance in which the topography of the male taper is imprinted onto the female taper (Fig. 4.2) (Bishop et al., 2013).

There are many factors that can contribute to the wear and corrosion of modular junctions, such as the type of materials used, femoral neck length, intraoperative assembly technique, presence of a sleeve between the femoral head and stem trunnion, surface topography of the implant, femoral head-neck offset, taper geometry, bearing diameter, junction cleanliness, impaction force at surgery, hip load determined by lateral offset, and weight and activity level of the patient (Rieker and Wahl, 2020). A study comparing wear rates between a perfectly matched interface and one with a large taper mismatch (9.12′) showed that the latter produced a high wear rate (2.960 $mm^3$ per million load cycles) that could have a major negative effect

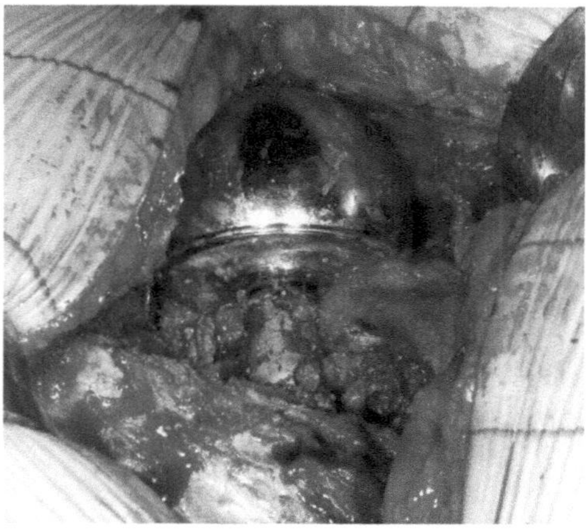

**FIGURE 4.2** Intraoperative image of a revision surgery showing a large-diameter femoral head disimpacted from the femoral stem, with mechanically assisted crevice corrosion evident as black wear debris around the trunnion.

on the clinical outcome of the implant (Ashkanfar et al., 2017). Reducing the taper mismatch to 6′ markedly decreased the magnitude of wear (0.069 $mm^3$ per million load cycles) and achieved better fixation by increasing the contact area between the components.

### 4.2.4 Bearing wear

Bearing wear occurs when the articulating surfaces of an implant come into contact with one another, leading to progressive removal of material and generation of particulate debris. Although MoM bearings have lower wear rates than conventional metal-on-polyethylene (MoP) bearings, the wear particles they generate are much smaller and more numerous, resulting in higher reactivity and potential for local toxicity. Dissolution of these particles in the corrosive environment of the synovium gives rise to measurable increases in systemic cobalt and chromium levels.

The wear of MoM bearings happens in two distinct stages: the bedding-in stage (also referred to as the running-in stage) and the steady-state stage. The bedding-in period, which typically occurs over the first 2 years after implantation (equivalent to the first $1-2 \times 10^6$ cycles in a hip joint simulator) is when the most wear particles are released (Vendittoli et al., 2010; Yan et al., 2006a). Following the bedding-in stage, well-functioning implants experience a slow but steady wear that continues over the subsequent years. In patients with problematic implants, wear rates remain elevated even after the bedding-in phase, leading to further increases in systemic metal levels that can inform the risk of ALTR (Van Der Straeten et al., 2013b).

While bearing wear is mainly adhesive and abrasive, it can also result from material fatigue or a third-body wear process, e.g., when a loose metallic bead gets trapped between two articulating surfaces. One extreme example of third-body wear occurs when a fractured ceramic component is replaced with an MoP bearing (Rambani et al., 2017)—even if a meticulous synovectomy and thorough debridement are performed during revision surgery, residual ceramic fragments can become embedded in the polyethylene liner and abrade the CoCr femoral head.

The amount of metal debris released from MoM bearings is influenced by factors such as implant design, bearing material, diameter, surface finish, and acetabular cup inclination angle (Desy et al., 2011; Hutt et al., 2016; Vendittoli et al., 2010). Clinical investigations, tribological studies, and mathematical models suggest that components with a large bearing size and those with a small clearance limit the amount of wear (Affatato et al., 2006; Dowson et al., 2004; Rieker et al., 2005; Vendittoli et al., 2013). The precision of the surgical technique may also influence the wear rate of MoM bearings. Hart et al. (2008) showed that cup inclination angles between 30° and 50° (the 'safe zone') were associated with mean blood cobalt and chromium levels of 1.6 and 1.9 μg/L, respectively, and that widening the angle beyond 50° increased the cobalt and chromium values to 4.5 and 4.3 μg/L, respectively.

### 4.2.5 Abnormal component contact

Abnormal contact between different components of a hip implant can accelerate their degradation. For example, full penetration of a CoCr femoral head through the polyethylene liner will place it in contact with the metal acetabular shell, causing abrasive wear of the softer material (Fig. 4.3) (Quitmann et al., 2006). Similarly, impingement between the femoral neck and the edge of a ceramic acetabular liner will abrade the former component, resulting in pronounced metal release (Fig. 4.4).

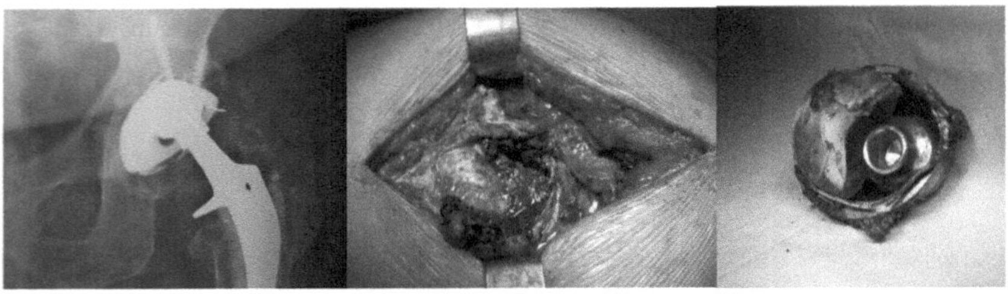

**FIGURE 4.3** Left image: Anteroposterior radiograph of a left hip in an elderly patient with complete polyethylene wear-through of an MoP THR. Middle image: Tissue blackening due to oxidised titanium particles was observed at surgical exposure. Right image: The implant was severely damaged, with a pattern suggestive of direct contact between the CoCr femoral head and titanium acetabular component. Since titanium is softer than CoCr alloy, the acetabular component was more severely damaged.

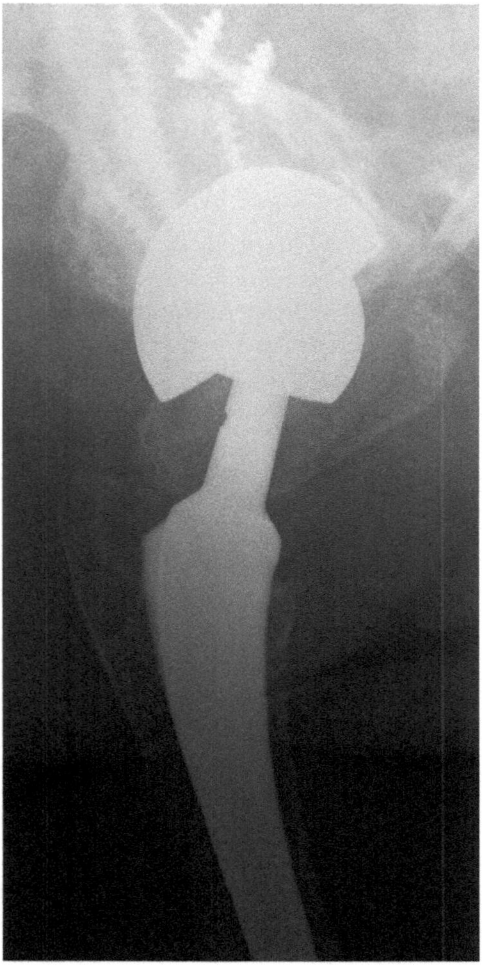

**FIGURE 4.4** A 40-year-old female developed groin pain 4 years after her fourth revision using a ceramic-on-ceramic implant with a titanium stem. The patient's blood titanium levels were elevated at 29 μg/L, while cobalt and chromium were both normal at 0.25 and 0.27 μg/L, respectively. Impingement between the titanium femoral neck and acetabular cup accelerated the wear of the former component (seen as an indentation in the femoral neck), causing a local adverse reaction to metal debris that necessitated a further revision procedure.

## 4.3 Measuring systemic levels of cobalt, chromium, and titanium

The amount of metallic debris released from an implant can inform on implant performance or the need for a revision surgery. High metal content in the synovial fluid is an obvious indicator of a high-wearing device; however, synovial sampling is too invasive to be used in routine patient monitoring, so systemic metal levels are typically measured in the blood or urine. The following sections offer guidance for choosing the appropriate analytical approach.

### 4.3.1 Choice of sample type

#### *4.3.1.1 Urine*

The kidneys are integral to the filtration of harmful substances from the body and measuring cobalt and chromium levels in the urine is the preferred method for assessing occupational exposure to these metals (Delaunay et al., 2010). The results are often expressed in or μg/L, μg/g creatinine, or μg/mmol creatinine. Adjustment to creatinine corrects for variation in sample volume due to urinary water content, which can make the analyte more or less concentrated (Barr et al., 2005). In the general population, normal urinary values of cobalt and chromium are $<2$ μg/g creatinine and $<0.5$ μg/g creatinine, respectively. Since urine metal concentrations are highly dependent on the renal excretion rate, the timing of specimen collection is important. To assess metal ion release from an orthopaedic implant, it is advised to measure metal concentrations in the urine collected over a 12- or 24-hour period. When carried out by the patient, such protocols have a high probability of sample contamination and incomplete collection. Single-timepoint urine sampling is more practical, but likely less reliable because of the variation in specimen volume and concentration of endogenous and exogenous chemicals from void to void. For these reasons, whole blood and serum are the preferred matrices to assess exposure to cobalt and chromium in patients with metal hip implants.

Titanium is less soluble and less readily excreted than cobalt or chromium, and its levels in the urine are often undetectable. Thus, unless titanium release from an implant is extremely high, urine monitoring of titanium levels in orthopaedic patients holds little clinical relevance.

#### *4.3.1.2 Blood*

There is no international consensus on the optimal blood fraction to be used for cobalt and chromium monitoring in patients with hip implants. As a result, some investigators assay whole blood, while others quote serum, plasma, or erythrocyte values. Despite early evidence to the contrary, cobalt and chromium concentrations in the different blood fractions cannot be used interchangeably (Malek et al., 2015). Comparing concurrent whole blood, serum, and erythrocyte samples collected from patients with MoM devices yielded markedly different mean values for both metals (Table 4.1) (Vendittoli et al., 2010). Furthermore, the serum/whole blood ratio for cobalt appears to be concentration dependent, making it impossible to derive a constant conversion factor. This is further confounded by inaccuracies in the

**TABLE 4.1** Mean cobalt, chromium, and titanium concentration in different blood fractions between 1 and 2 years following metal-on-metal hip replacement (Vendittoli et al., 2010).

| | Serum (μg/L) | Whole blood (μg/L) | Erythrocytes (μg/L) | Serum/whole blood ratio | Serum/ erythrocyte ratio |
|---|---|---|---|---|---|
| Co | 0.92 | 0.70 | 0.56 | 1.31 | 1.59 |
| Cr | 1.56 | 1.25 | 0.91 | 1.25 | 1.60 |
| Ti | 4.21 | 1.98 | 0.80 | 1.59 | 6.0 |

reporting of the blood fractions used for analysis and heterogenous methodologies employed by different testing laboratories.

The binding affinity of metal ions for proteins means that metal ions are usually more abundant in the serum than in erythrocytes (Walter et al., 2008). Since whole blood is a combination of the two fractions, its metal content tends to lie in between that of the serum and red blood cells. During short-term exposure, most of the cobalt distributes into the serum/plasma while its red blood cell content remains two orders of magnitude lower. Long-term exposure is associated with a slow, continuous uptake of cobalt into red blood cells, which reflects the time-averaged cobalt exposure during their 120-day lifespan (Delaunay et al., 2010; Ebreo et al., 2011). Studies in volunteers ingesting a cobalt supplement and in patients with MoM implants concluded that whole blood measurements are more stable, potentially offering a better indication of long-term average cobalt exposure (Finley et al., 2013). It appears that the serum cobalt concentration is more representative of recent cobalt exposure, which may vary depending on the patient's physical activity level and bearing surface wear (Simonsen et al., 2012).

An important point to consider is that metal ions with different valence states may distribute into different blood fractions. In contrast to trivalent chromium, hexavalent chromium accumulates in red blood cells, with a much smaller amount found in the serum. Measuring chromium concentration in the serum only would therefore be an incomplete assessment.

Another advantage of whole blood samples is that they require less manipulation following collection, reducing the risk of contamination associated with clotting additives, additional hand-offs between laboratory staff, and transfers between containers (Yao et al., 2020). The complexity of blood matrix components renders whole blood analysis more difficult, however, and the large quantity of iron in red blood cells may interfere with the results.

In the United Kingdom, the MHRA calls for the measurement of cobalt and chromium in whole blood, although the analytical protocol has not been standardised. There are no guidelines concerning the optimal sample type for the determination of titanium levels, but serum/plasma analysis is generally preferred to whole blood analysis because of the simpler sample matrix and higher titanium content, which makes the analyte easier to detect. Future methodological standardisation and alignment with current cobalt and chromium practices may result in a shift to whole blood as the recommended sample type for titanium determination in patients with hip arthroplasty (Yao et al., 2020).

## 4.3.2 Specimen collection and storage

### 4.3.2.1 Urine

For urine testing, a simple midstream sample collected in a urine collection cup, transported at 4°C, and preserved for 1–2 weeks before analysis is appropriate (Delaunay et al., 2010).

### 4.3.2.2 Blood

Whole blood is typically collected at the antecubital fossa. The vein is cannulated with a 22-gauge stainless steel needle and the outer plastic cannula is left in place while the needle is removed. Historically, it was recommended that the first 5 mL of blood withdrawn be discarded to avoid contamination from the metal needle. The current consensus is that this is not necessary to obtain repeatable results (Barry et al., 2013).

Regardless of the analytical method used to quantify cobalt, chromium, and titanium in biological matrices, sample purity must be carefully controlled given the minute levels of ions being measured (Campbell and Estey, 2013; Delaunay et al., 2010). Environmental contamination can be particularly problematic for titanium, which is present in many consumer products. Sampling sites must be thoroughly cleaned, and blood specimens collected into tubes that are certified for trace metal testing. Most manufacturers of blood collection supplies offer specialised trace element tubes with caps that do not release metals and anticoagulant coatings that are also metal free. It is worth noting that, currently, there are no vendor-certified titanium-free collection systems. Thus, the suitability of needles and tubes utilised during the sample collection process should be verified with the reporting laboratory (Yao et al., 2020).

Although it is controversial whether systemic metal ion levels and implant use/activity level are correlated, we suggest that patients avoid modifying their exercise routine or engaging in new or strenuous activities in the week before blood metal level testing to avoid biasing the results (Heisel et al., 2005; Jelsma et al., 2020; Khan et al., 2006).

Following sampling, blood collection tubes should be labelled with the patient's name/identification number and collection date and transported to testing laboratories at 4°C to maximise sample stability. If required, whole blood may be centrifuged at 3000 rpm for 15 minutes to separate the serum and red blood cell fractions, which should then be transferred aseptically to individual polypropylene tubes. Whole blood and serum samples may be stored at 4°C for up to 28 days, or at -20°C indefinitely.

## 4.3.3 Quantification of metal levels

### *4.3.3.1 Sample preparation*

A typical sample preparation procedure for trace metal analysis involves simple dilution, with or without a prior digestion step. Common diluents include nitric acid, hydrochloric acid, and ammonium hydroxide, and a dilution factor between 10 and 50 is usually adequate. Sample digestion can be carried out using strong acids or alkali, either at room temperature or heated in a water bath or high-pressure microwave (Wilschefski and Baxter, 2019). All laboratory glassware and accessories used in the sample preparation process should be presoaked in 2% nitric acid in twice-distilled and deionised water, rinsed in twice-distilled and deionised water, then checked using a nitric acid leaching test to ensure that they do not contain detectable amounts of the relevant trace elements (Savarino et al., 2014). Considering that airborne dust may contain chromium at concentrations 10,000-fold higher than those in the blood of unexposed individuals, sample preparation should be performed in a dedicated room with efficient fume extraction and temperature monitoring; analysis of samples in a class-100 environment is preferred ($\leq$100 particles of size 0.5 μm or larger per cubic foot of air permitted) (Yao et al., 2020).

### *4.3.3.2 Analytical approach*

Although a number of analytical techniques can be used to quantify cobalt, chromium, and titanium levels in biological samples, graphite-furnace atomic absorption spectroscopy (AAS) and inductively coupled plasma–mass spectrometry (ICP-MS) have been the most commonly employed.

Graphite-furnace AAS uses a high-temperature (3000°C) graphite tube to turn liquid samples into free atoms before exposing them to light whose wavelength is specific to the element being analysed, e.g., 241 nm for cobalt or 283 nm for chromium. Upon absorbing the light, the atoms become ionised. Within certain limits, the amount of light energy absorbed can be linearly correlated to analyte concentration in the sample, allowing it to be quantified.

In ICP-MS, the sample is placed in a nebulisation chamber containing an inert gas, such as helium or argon, where it is transformed into an aerosol of very fine droplets. The gas carries the aerosol into the core of inductively coupled argon plasma that generates temperatures capable of dissociating, atomising, and ionising most elements (8000–10,000°C). A differential vacuum system accelerates the plasma ions toward a collection of electrostatic discs that extract the positively charged ions and transport them toward a quadrupole mass filter. Depending on the frequency applied to the quadrupole, only ions with certain mass-to-charge ratio (m/z) are transmitted and detected. Inductively coupled plasma–optical emission spectroscopy (ICP-OES) is a related technique that quantifies analyte levels by measuring the intensity of light emitted by the elements in a sample at appropriate analytical lines.

Owing to its ability to analyse multiple elements/isotopes at the same time, improved sensitivity and specificity, shorter run times, and higher sample throughput, ICP-MS has replaced graphite-furnace AAS in many analytical laboratories and is the preferred means of quantifying systemic metal levels.

#### 4.3.3.3 *Minimising spectral interferences*

ICP-MS instruments fitted with a quadrupole mass analyser have a relatively low resolving power ($R = 300$), and their measurements are affected by a range of polyatomic and isobaric interferences (species with the same nominal mass-to-charge ratio as the measured isotope) that can cause the true analyte concentration to be overestimated. The interferences originate from the sample matrix, plasma gases, reagents used for sample preparation, or entrained atmospheric gasses, and are very common in biological samples, which naturally contain high levels of inorganic and organic compounds (Table 4.2).

Several methods to minimise spectroscopic interferences have been proposed, with cell-based technology being the most common approach.

A dynamic reaction cell (DRC) uses a reactive gas, such as ammonia, oxygen, or hydrogen, to change the analyte or the interfering species to a new form with a different mass-to-charge ratio, or neutralise the interfering species via a charge exchange reaction (Wilschefski and Baxter, 2019).

In a collision reaction cell, inert gas such as helium is used at varying flow rates to lower the kinetic energy of the polyatomic ions and prevent them from reaching the detector (Sarmiento-González et al., 2005). Universal cell technologies that include both collision and DRCs are also available, allowing the analyst to select the best interference control strategy for their samples. The triple-quadrupole ICP-MS platform (ICP-QQQ-MS) contains an octupole-based collision/reaction cell sandwiched between two quadrupole mass analysers. The tandem MS configuration permits greater control of reaction cell chemistry, enabling enhanced removal of polyatomic and isobaric interferences and higher-sensitivity detection of ultra-trace elements.

**TABLE 4.2 Examples of spectroscopic interferences affecting the measurement of cobalt, chromium, and titanium levels in biological samples using ICP-MS (May and Wiedmeyer, 1998).**

| Analyte | Abundance (%) | Interfering species |
|---|---|---|
| $^{59}Co^+$ | 100 | Polyatomic: $^{43}Ca^{16}O^+$, $^{40}Ar^{19}F^+$, $^{24}Mg^{35}Cl^+$, $^{36}Ar^{23}Na^+$ |
| $^{52}Cr^+$ | 83.76 | Polyatomic: $^{40}Ar^{12}C^+$, $^{35}Cl^{16}O^1H^+$, $^{34}S^{18}O^+$ |
| $^{53}Cr^+$ | 9.51 | Polyatomic: $^{37}Cl^{16}O^+$, $^{36}S^{17}O^+$, $^{38}Ar^{15}N^+$ |
| $^{46}Ti^+$ | 7.99 | Polyatomic: $^{32}S^{14}N^+$, $^{14}N^{16}O_2^+$, $^{15}N_2{}^{16}O^+$<br>Isobaric: $^{46}Ca$ |
| $^{47}Ti^+$ | 7.32 | Polyatomic: $^{32}S^{14}N^1H^+$, $^{30}Si^{16}O^1H^+$, $^{12}C^{35}Cl^+$, $^{31}P^{16}O^+$ |
| $^{48}Ti^+$ | 73.98 | Polyatomic: $^{32}S^{16}O^+$, $^{34}S^{14}N^+$, $^{12}C_4^+$, $^{36}Ar^{12}C^+$<br>Isobaric: $^{48}Ca$ |
| $^{49}Ti^+$ | 5.46 | Polyatomic: $^{32}S^{17}O^+$, $^{35}Cl^{14}N^+$, $^{34}S^{15}N^+$, $^{36}Ar^{12}C^1H^+$, $^{31}P^{18}O^+$ |
| $^{50}Ti^+$ | 5.25 | Polyatomic: $^{32}S^{18}O^+$, $^{36}Ar^{14}N^+$, $^{35}Cl^{15}N^+$, $^{36}S^{14}N^+$<br>Isobaric: $^{50}Cr$, $^{50}V$ |

*ICP-MS*, inductively coupled plasma–mass spectrometry.
*Source*: Based on May, T.W., Wiedmeyer, R.H., 1998. A table of polyatomic interferences in ICP-MS. At. Spectrosc. 19, 150–155.

An alternative way to deal with spectroscopic interferences is to use double-focusing sector-field ICP-MS (SF-ICP-MS). The instrument employs electric and/or magnetic fields instead of quadrupoles, which increases its resolving power ($300 \leq R \leq 10{,}000$) and allows for the elimination of nearly all atomic and molecular interferences affecting trace metal analysis. The instrument's limit of detection (LoD) is typically 0.02 μg/L for cobalt, 0.1 μg/L for chromium, and 0.2 μg/L for titanium in whole blood, and its limit of quantification (LoQ)—a more precise estimation of the sensitivity and accuracy of the device—is approximately 3.33 times the LoD (Barry et al., 2013). When operated in a low-resolution mode ($R = 300$), the LoD of SF-ICP-MS is one order of magnitude lower than that offered by single-quadrupole ICP-MS.

Titanium concentration in biological fluids cannot be reliably determined using single-quadrupole ICP-MS because all isotopes of titanium are affected by spectral interferences (Table 4.2). The isobaric overlap between $^{48}Ti$ and $^{48}Ca$ is particularly problematic in view of the high circulating calcium levels. Although SF-ICP-MS can limit the contribution from polyatomic species, the isobaric interferences from Ca, Cr, and V cannot be resolved, making the high-abundance isotopes ($^{46}Ti$, $^{48}Ti$, and $^{50}Ti$) unavailable for analysis and decreasing the analytical sensitivity of the instrument (Koller et al., 2018). Nevertheless, the utility of SF-ICP-MS to quantify trace amounts of titanium in the blood, plasma, and serum samples has been demonstrated (Sarmiento-González et al., 2008; Swiatkowska et al., 2019a). Balcaen et al. (2014) showed that titanium levels can also be reliably measured using ICP-QQQ-MS in MS/MS mode. The method developed by the authors used a mixture of ammonia and helium gas to convert $^{48}Ti^{+}$ to $[Ti(NH_3)_6]^{+}$ cluster ions, providing interference-free conditions and a low LoD (3 ng/L) for the quantification of serum titanium in patients with titanium-based hip implants.

The excellent analytical performance of SF-ICP-MS and ICP-QQQ-MS comes at a price, as both machines incur high running costs and require highly trained operators. As a result, they are not widely available in routine clinical laboratories. The use of ICP-OES may be a more practical and cost-effective solution. In Harrington et al. (2016), ICP-OES displayed a low LoD for serum titanium (0.6 μg/L), but it was not sensitive enough for whole blood specimens, which usually contain lower levels of titanium than the serum.

### *4.3.3.4 Sources of intra- and inter-laboratory variability*

Barry et al. (2013) reported on the reproducibility of cobalt, chromium, and titanium measurement in 78 pairs of whole blood samples collected from patients with MoM THR. The absolute difference between the two sets of samples was greater than the LoQ of the high-resolution ICP-MS instrument for all three metals (0.74 vs 0.07 μg/L for cobalt, 0.84 vs 0.35 μg/L for chromium, and 0.88 vs 0.70 μg/L for titanium), and although the variations were typically very low, in 19%–31% of the cases they were significant enough to risk misinterpretation when assuming that whole blood cobalt levels ≥ 1 μg/L may be used to make decisions on MoM implant performance and/or need for revision surgery.

Particular caution should be exercised when comparing the results of trace metal analyses performed in different laboratories. In Rahme et al. (2014), paired whole blood samples from 46 patients with MoM implants were collected into lavender-top (7.2 mg $K_2$ EDTA) and royal blue-top (10.8 mg $K_2$ EDTA) blood collection tubes before each set of samples was sent to a separate laboratory for analysis. The lower anticoagulant content of lavender-top tubes was associated with significantly higher mean blood cobalt (4.51 μg/L for Lab 1 vs 3.98 μg/L for Lab 2) and chromium (3.04 μg/L for Lab 1 vs 2.27 μg/L for Lab 2). Importantly, the two laboratories also used different ICP-MS instrumentation (SF-ICP-MS in Lab 1 vs ICP-DRC-MS in Lab 2) and employed a different sample preparation protocol (simple dilution in Lab 1 vs acid digestion in Lab 2). In Kerger et al. (2015), detergent dilution of serum samples was associated with numerically greater cobalt recovery, lower variability, and better sensitivity than acid digestion methods.

Inter-laboratory variability also affects titanium measurement (Barry et al. 2020). A recent study by Koller et al. (2018) found that when the same blood samples were analysed by seven different laboratories, the titanium levels ranged from below the instrument's LoD to 25 μg/L depending on the analytical approach used.

In light of these results, it is recommended that sequential samples from the same patient are analysed in the same laboratory to limit the effects of chance, inadvertent errors, and inter-laboratory methodological differences, which can bias the results (Kerger et al., 2015; Saini et al., 2019).

#### *4.3.3.5 The units*

An additional source of confusion when interpreting the results of trace metal analyses is the variety of units used to report the values. Blood/serum metal concentrations are typically expressed in micrograms per litre (μg/L), parts per billion (ppb), or nanograms per millilitre (ng/mL), all of which are equivalent. Some authors choose to use nanomoles per litre (nmol/L); however, this is not interchangeable with the other units. Eq. (4.1) can be used to convert between the different units:

$$x = \frac{Ay}{1000} \tag{4.1}$$

where $x$ = Metal concentration in μg/L, ppb, or ng/mL; $y$ = Measured metal concentration in nmol/L; $A$ = Atomic weight (58.993 for Co, 51.996 for Cr, 47.867 for Ti).

## 4.4 Using systemic metal levels to assess implant degradation and risk of local adverse reactions

Accumulation of metal wear debris in the periprosthetic tissue can trigger local inflammatory reactions, leading to the formation of cystic or solid soft-tissue masses (pseudotumours), metallosis, tissue necrosis, osteolysis, and, ultimately, early implant failure (Mabilleau et al., 2008). The harmful local biological responses to metal particles, corrosion products, and ions are encompassed by the terms 'ALTR' and 'ARMD'. Aseptic lymphocyte-dominated vasculitis-associated lesion (ALVAL) is a histological diagnosis representing a lymphocyte-mediated hypersensitivity reaction, whereby tissue damage is caused by cytotoxic T cells. ALVAL manifests sonographically as synovitis, bursal fluid collection, or osteolysis, and may develop in the presence of a well-performing implant and low metal ion release (Griffiths et al., 1987; Willert et al., 2005).

ARMD lesions are benign, non-infective soft-tissue neoplasms that arise in the vicinity of MoM implants and, occasionally, adjacent to MoP devices. They occur more frequently in females than males and have been reported in up to 69% of patients with MoM THR (Sutphen et al., 2016). Although many are asymptomatic, their presence can lead to a palpable mass or fluid collection near the hip, persistent pain (ranging from simple discomfort to debilitating pain that impacts mobility), local erythema, stiffness, limping, reduced range of motion, and joint instability with or without spontaneous dislocation (Butler and Barrack, 2004; Peacock et al., 2008; Williams et al., 2011). Moreover, ARMD lesions may cause venous and nervous compression, resulting in lower limb oedema and paralysis of the femoral nerve/lateral cutaneous nerve, respectively (Clayton et al., 2008).

The pathogenesis of ARMD implicates macrophage-mediated tissue damage secondary to the increased production of wear debris via component impingement, galvanic corrosion, and/or tribocorrosion (Delaunay et al., 2010). In a small proportion of patients with ARMD, lymphocyte-mediated hypersensitivity to a 'normal' amount of metal debris is observed (Pandit et al., 2008). The factors linked to greater implant wear and/or increased risk of ARMD include the use of large-diameter femoral heads (≥36 mm), recalled implant types, presence of taper junctions, acetabular component malposition, high activity level of the patient, and female sex (Matharu et al., 2015a). The latter finding is ascribed to the higher prevalence of hip dysplasia in women, which may make it more difficult to position the acetabular component accurately, resulting in higher rates of edge-loading and component impingement (Nakano et al., 2017).

Depending on the severity of implant malfunction/malposition, ARMD can manifest within several months to years after the surgery. For example, in patients with mild problems, e.g., slowly progressing trunnionosis, the symptoms can take multiple years to surface. The presence of severe ARMD-related bone and muscle damage (e.g., extensive synovitis, expanded bursae with large volumes of discoloured fluid, femoral neck narrowing, and osteolysis) necessitates a more thorough debridement during revision surgery, which can increase the risk of postsurgical complications and unsatisfactory outcomes (De Smet et al., 2008; Grammatopoulos et al., 2009; Munro et al., 2014). Because of the destructive nature of ARMD, accurate diagnosis and timely revision of a problematic prosthesis are of utmost importance.

Rarely, extremely high blood cobalt levels resulting from accelerated MoM-bearing wear or third-body wear of a CoCr component can bring on systemic toxicity symptoms ranging from mild neurobehavioral abnormalities, polycythaemia, and reversible thyroid dysfunction to peripheral neuropathy, deafness and/or blindness, and lethal cardiomyopathy (Cheung et al., 2016), either with or without concurrent ALTR. Here, we discuss the utility of blood/serum metal ion measurements to assess implant wear and predict periprosthetic adverse reactions while Chapter 7 explores their potential role in the diagnosis of implant-related systemic toxicity.

### 4.4.1 Cobalt and chromium

Reference values for systemic cobalt and chromium levels in the general population (individuals without occupational exposure or metal implants) are 0.4 μg/L and 0.8 μg/L for serum and whole blood cobalt, respectively, and 0.5 μg/L for serum chromium (Delaunay et al., 2010). In the majority of individuals with MoM devices, blood cobalt and chromium concentrations range between 0.2 and 10 μg/L, with steady-state median concentrations between 1.5 and 2.3 μg/L (Prentice et al., 2013; Jantzen et al., 2013). In general, metal levels in patients with MoP arthroplasties (passive and galvanic corrosion mechanisms) are approximately 10-fold lower than in those with MoM bearings (Leikin et al., 2013). Previous studies of metal release from ceramic-on-polyethylene and ceramic-on-ceramic (CoC) implants reported that the associated cobalt and chromium concentrations were always below 1 μg/L, and in many cases undetectable (Beraudi et al., 2014; Nam et al., 2015).

It is believed that cobalt and chromium levels in the synovial fluid correlate with those in the serum, and that measurement of systemic cobalt and/or chromium levels can help detect accelerated implant wear before marked tissue/bone destruction has occurred, allowing for better outcomes in patients requiring implant revision (De Smet et al., 2008; De Smet et al., 2011; Grammatopoulos et al., 2009). Serum metal ion concentrations correlate well with the maximum depth of the wear scar and total volumetric wear rates of explanted femoral resurfacing prostheses (De Smet et al., 2008, Langton et al., 2011). Furthermore, patients with poorly performing MoM implants tend to have higher circulating levels of cobalt and/or chromium than those with well-functioning devices (Hart et al., 2014; Van Der Straeten et al., 2013a).

All major regulatory bodies around the world agree that revision should be considered in patients with abnormal cross-sectional imaging, especially if progressing, and/or rising blood cobalt/chromium levels; however, there is no international consensus on what level should mandate revision. The MHRA and Health Canada recommend 7 μg/L as a blood metal concentration of interest for further patient follow-up and cross-sectional imaging to assess the potential for local adverse reactions (Hart et al., 2011). The European Federation of National Associations of Orthopaedics and Traumatology proposes that blood cobalt levels between 2 and 7 μg/L should spark clinical concern, while those exceeding 20 μg/L indicate the need for revision.

The MHRA's 7 μg/L threshold represents the statistical extreme outlier of the range of measured blood metal concentrations from Hart et al. (2006, 2009), and has a reported sensitivity of 52% and specificity of 89% for differentiating failed from moderately functioning and well-functioning MoM HR (Hart et al., 2011, 2014). Reducing the cut-off to 4.97 μg/L marginally improved the sensitivity (63%) at the cost of reduced specificity (86%) (Hart et al., 2014). Other thresholds specific to HR implants have also been put forward. Van Der Straeten et al. (2013a,b) found that acceptable upper levels in patients with well-functioning unilateral or bilateral devices were 4.0 and 5.0 μg/L for cobalt and 4.6 and 7.4 μg/L for chromium, respectively. The specificity and sensitivity of these limits for predicting poor function were 96% and 22% for cobalt and 95% and 25% for chromium in unilateral implants, and 91% and 38.6% for cobalt and 93% and 43% for chromium in bilateral implants. Similar conclusions were reached by Sidaginamale et al. (2013), who suggested that a blood cobalt level of 4.5 μg/L is indicative of a poorly functioning HR implant, as defined by abnormal wear of $\geq 3\ mm^3$/year. Langton et al. (2011, 2013) proposed that blood cobalt values of $\geq 5$ μg/L are 'highly sensitive and specific for increased wear of explanted components', while those exceeding 10 μg/L signify 'unequivocally high rates of wear' in a healthy individual.

Although many authors support the use of serum/blood metal testing to assess MoM-bearing wear, even in asymptomatic patients, the relationship of systemic cobalt and chromium concentrations with the extent of soft-tissue damage, pain, and osteolysis is not straightforward (Langton et al., 2013). Griffin et al. (2012) and Langton et al. (2013) found no correlation between systemic metal load and the development of soft-tissue necrosis and bone loss in recipients of MoM THR and HR. Kiran et al. (2017), who performed serial annual metal ion testing in 256 patients with large-diameter MoM THR, reported that blood cobalt and chromium values had poor discriminatory ability to detect ARMD in asymptomatic patients at 7 μg/L (positive predictive value of 43.8% and 67.6%, respectively), and questioned the utility of blood metal measurements in this cohort beyond 7-year follow-up. The utility of blood cobalt/chromium ratio as a predictive marker for ALTR in MoM THR has also been challenged (Fehring et al., 2015).

The poor sensitivity of metal ion thresholds and lack of a clear association between blood cobalt or chromium levels and the development of ALTR means that patient management and the decision to revise should not be based on systemic metal values alone (Hart et al., 2014; Matharu et al., 2015b). Several risk stratification algorithms have been developed to help clinicians identify patients at risk of developing ARMD following MoM hip arthroplasty (Chalmers et al., 2016; Kwon et al., 2014; Van Der Straeten et al., 2013a). In these algorithms, systemic metal ion testing is combined with a thorough clinical evaluation involving patient symptomatology, implant track record, radiographs, cross-sectional imaging, histopathological analysis, infection work-up, and assessment of patient-specific factors and alternative diagnoses.

It should be kept in mind that poor renal function can impair the clearance of implant-derived metal ions, increasing their systemic levels and, potentially, biasing the results of trace metal analyses. This is especially important for cobalt, which is more soluble and reliant on urinary excretion than chromium and titanium. While mild or moderate renal insufficiency is unlikely to significantly impact the results, chronic renal failure can lead to marked increases in systemic metal levels in recipients of MoM hip arthroplasty. In Hur et al. (2008), the mean serum cobalt content in patients with chronic renal failure was two orders of magnitude higher than that in individuals with normal renal function matched for implant type and time *in situ*. The presence of additional metal implants, particularly bilateral MoM devices, can also complicate the interpretation of metal ion levels.

### 4.4.2 Titanium

Titanium is relatively slowly transported in the body so a modest increase in its systemic concentration likely reflects a massive release of titanium into the local joint space. Thus, monitoring blood titanium levels may be a useful tool to assess passive corrosion of titanium surfaces, abnormal contact between implant components, and the performance of modular junctions (Deny et al., 2018).

Due to the poor availability of reliable methods of detection and scarcity of studies that measure preoperative titanium levels in prospective hip implant recipients, it is unclear what blood titanium concentrations should be devoid of clinical concern. Past literature suggests that whole blood titanium levels in unexposed subjects are below 1 μg/L (Lavigne et al., 2011; Vendittoli et al., 2011, 2010). In patients with well-functioning modular implants, these can range from <1 to 9 μg/L, but are generally less than 3 μg/L (Vendittoli et al., 2011, 2010; Yao et al., 2020) (Table 4.3). Eichler et al. (2020) measured blood titanium concentration in 57 patients with well-functioning, unilateral CoC implants at a minimum of 5 years post-surgery (mean, 6.6 years) and reported a mean titanium level of 1.9 μg/L (range, 1.2–4.4 μg/L). Comparable results were reported by Swiatkowska et al. (2020), who evaluated 95 patients with titanium-stemmed CoC implants at a mean follow-up of 8.5 years. The median titanium values were 1.20 μg/L in whole blood and 1.70 μg/L in the plasma. Loose of otherwise failed devices have been associated with a wide range of values, from 1.5 to 620 μg/L (Table 4.4). Marked variations in the reported results are likely caused by a combination of factors, including inter-study differences in implant design and stability, number of titanium components implanted, severity of wear, follow-up time, sample type, analytical approach employed, and the patient's health status and environmental exposure to titanium.

Adverse local reactions to titanium have been reported but their incidence is much lower than in the case of cobalt and chromium. Due to little current concern about titanium toxicity and lack of a general agreement on what should constitute abnormal systemic titanium levels following THR, official guidelines for the follow-up of patients with titanium-based implants have not been issued.

**TABLE 4.3** Titanium levels associated with different types of well-functioning hip implants (Swiatkowska, 2019b).

| Reference | Blood Ti (μg/L)[a] | Serum Ti (μg/L)[a] | Assessment of implant fixation/ clinical function | Follow-up |
|---|---|---|---|---|
| Sarmiento-González et al. (2008) | 2.3 ($n$ = 11)<br>1.5 ($n$ = 11) | n/a<br>n/a | Not assessed | 14–22 m<br>70–106 m |
| Nuevo Ordóñez et al., (2009) | 3.0 ($n$ = 9),<br>2.2 ($n$ = 6) | n/a | Not assessed | n/a |
| Vendittoli et al. (2010) | 3.7; 1.4–8.8 ($n$ = 34)<br>2.8; 1.4–4.1 ($n$ = 33)<br>1.8; 0.9–4.6 ($n$ = 31)<br>1.3; 0.4–2.4 ($n$ = 24) | n/a<br>n/a<br>n/a<br>n/a | X-ray/WOMAC, Merle D'Aubigne and Postel scores | 3 m<br>6 m<br>1 y<br>2 y |
| Omlor et al. (2013) | n/a | 2.7; 1.1–7.0 ($n$ = 6) | Not assessed | 7–13 m |
| Levine et al. (2013) | n/a | 1.8; 1.7–1.9 ($n$ = 8) | X-ray/Harris Hip Score | 10 y |
| Gofton and Beaule (2015) | n/a<br>n/a | 2.5; 2.2–3.1 ($n$ = 23)<br>2.7; 2.1–3.3 ($n$ = 23) | Not assessed | 1 y<br>2 y |
| Nam et al. (2015) | 2.2 ($n$ = 15)<br>1.4 ($n$ = 15)<br>2.2 ($n$ = 15) | n/a<br>n/a<br>n/a | X-ray/Short-Form 12, WOMAC, Harris Hip Score | 1 y<br>2 y<br>5 y |
| Yi et al. (2016) | n/a | 2.28 ($n$ = 74) | X-ray/WOMAC, Harris Hip Score | 50 m |
| Swiatkowska et al. (2020) | 1.4; 0.6–4.4 ($n$ = 95) | n/a | X-ray/Oxford Hip Score | 64–143 m |
| Eichler et al. (2020) | 1.9; 1.2–4.4 ($n$ = 57) | n/a | X-ray/WOMAC, Patient's Joint Perception | 60–89 m |

*WOMAC*, Western Ontario and McMaster Universities Osteoarthritis Index.
[a]*Measured using high-resolution ICP-MS and reported as mean/median; range (sample size).*
*Source*: Adapted from Swiatkowska, I., 2019b. Toxicity of Metal Debris From Hip Implants (Ph.D. thesis). University College London, London/CC-BY-4.0.

**TABLE 4.4** Titanium levels associated with failed hip implants (Swiatkowska et al., 2019a).

| Failure mechanism | Analytical technique | Blood Ti (μg/L)[a] | Serum Ti (μg/L)[a] |
|---|---|---|---|
| Loosening of a titanium component | GFAAS | 38–602 ($n = 22$) Dorr et al. (1990) | n/a |
| | GFAAS | n/a | < LoD–17.2 ($n = 21$) Jacobs et al. (1991) |
| | HR ICP-MS | 1.5–1.8 ($n = 2$) Dunstan et al. (2005) | n/a |
| | ICP-OES | n/a | 9.1–61 ($n = 2$) Lazennec et al. (2009) |
| | Not specified | n/a | 15 ($n = 1$) McAlister and Abdel (2016) |
| | HR ICP-MS | n/a | 11.3[b] ($n = 1$) Omlor et al. (2013) |
| Manufacturing error resulting in modular neck breakage | HR ICP-MS | n/a | 4.5[b] ($n = 1$) Omlor et al. (2013) |
| Wear, osteolysis, and/or loosening | HR ICP-MS | LoD–140 ($n = 23$) Grosse et al. (2015) | n/a |
| Polyethylene wear-through leading to secondary wear of titanium acetabular shell | Not specified | n/a | 280[b] ($n = 1$) Quitmann et al. (2006) |
| | Not specified | n/a | 620 ($n = 1$) Malahias et al. (2019) |
| Corrosion at the neck/stem junction | Not specified | n/a | 1.6–5.8 ($n = 11$) Cooper et al. (2013) |
| Pseudotumour formation in local tissue | ICP-OES | n/a | 30 ($n = 1$) Sakamoto et al. (2016) |

*GFAAS*, graphite-furnace atomic absorption spectroscopy; *HR ICP-MS*, high-resolution inductively coupled plasma–mass spectrometry; *ICP-OES*, inductively coupled plasma–optical emission spectroscopy; *LoD*, limit of detection.
[a]*Expressed as range (sample size).*
[b]*Sample drawn post-revision of the failed implant.*
*Source*: Reprinted with permission of the Journal of Trace Elements in Medicine and Biology.

## 4.5 Summary and future directions

Measuring circulating cobalt and chromium concentrations is a valuable way to indirectly assess MoM-bearing wear and modular junction performance, but a clear relationship between systemic metal ion levels and the extent of ARMD has not been demonstrated. While certain regulatory agencies propose inflexible threshold values or ranges to guide the clinical management of patients with MoM hips, there is no international consensus on what systemic cobalt/chromium levels should mandate revision. High metal ion values should be analysed in view of the patient's symptoms, medical history, implant brand (some are known to be defective), cross-sectional imaging, and radiographs. The treating surgeon should also consider alternative causes of a painful hip, particularly periprosthetic infection and referred pain, and evaluate the risks and benefits of revision surgery as opposed to continued observation. Metal ion values within the normal range should not discourage further investigation of individuals with clinical or radiological signs and symptoms of implant dysfunction. Conversely, asymptomatic patients with highly elevated cobalt/chromium concentrations should be closely followed-up.

Several authors have called for the determination of blood/serum titanium content in patients with titanium-based hips on account of its potential as a marker of implant performance. Due to the differences in analytical approaches employed by different laboratories, meaningful inter-study comparisons and reliable conclusions regarding 'normal' and 'abnormal' levels are difficult to draw, however. The matter is further complicated by the ubiquity of titanium in the environment and the multifactorial nature of implant degradation, which implicates a number of surgeon, implant, and patient influences—some of which are impossible to control. Before reference ranges for titanium can be established, national standardisation of sampling and analytical protocols must be realised. In particular, spectroscopic interferences, which represent significant analytical barriers to obtaining accurate and clinically meaningful titanium measurements, are underappreciated by the orthopaedic community. ICP-MS instruments capable of resolving the interferences are largely limited to research institutions and their introduction into routine testing laboratories will require a significant investment of time and resources.

Blood samples must be collected using a technique that minimises contamination, and analysed using a suitable platform, preferably ICP-MS with interference-removal capabilities. Since imprecision increases near the lower end of the analytical measurement range, the instrument's LoD should always be disclosed and readings below the LoD ought to be clearly stated as such, rather than reported as 0 μg/L. To allow for statistical analysis, such values may also be reported as half of the instrument's LoD. To decrease the risk of clinical misinterpretation, the decision to revise a MoM hip replacement should not be based on a single elevated measurement, as absolute values taken out of context are unlikely to be meaningful, but metal ion trend over a period of time. Finally, serial measurements must, whenever possible, be sent to the same laboratory, as inter-laboratory variations can be a major source of bias.

## References


Affatato, S., Bersaglia, G., Junqiang, Y., Traina, F., Toni, A., Viceconti, M., 2006. The predictive power of surface profile parameters on the amount of wear measured in vitro on metal-on-polyethylene artificial hip joints. Proc. Inst. Mech. Eng. H. 220, 457–464.

Ashkanfar, A., Langton, D.J., Joyce, T.J., 2017. A large taper mismatch is one of the key factors behind high wear rates and failure at the taper junction of total hip replacements: a finite element wear analysis. J. Mech. Behav. Biomed. Mater. 69, 257–266.

Balcaen, L., Bolea-Fernandez, E., Resano, M., Vanhaecke, F., 2014. Accurate determination of ultra-trace levels of Ti in blood serum using ICP-MS/MS. Anal. Chim. Acta. 809, 1–8.

Barr, D.B., Wilder, L.C., Caudill, S.P., Gonzalez, A.J., Needham, L.L., Pirkle, J.L., 2005. Urinary creatinine concentrations in the U.S. population: implications for urinary biologic monitoring measurements. Environ. Health. Perspect. 113, 192–200.

Barry, J., Eichler, D., Robitaille, R., Vendittoli, P.A., 2020. Whole blood titanium metal ion measurement reproducibility of two laboratories. Pract. Lab. Med. 21, e00167.

Barry, J., Lavigne, M., Vendittoli, P.A., 2013. Evaluation of the method for analyzing chromium, cobalt and titanium ion levels in the blood following hip replacement with a metal-on-metal prosthesis. J. Anal. Toxicol. 37, 90–96.

Beraudi, A., Stea, S., De Pasquale, D., Bordini, B., Catalani, S., Apostoli, P., et al., 2014. Metal ion release: also a concern for ceramic-on-ceramic couplings? Hip Int. 24, 321–326.

Bishop, N., Witt, F., Pourzal, R., Fischer, A., Rutschi, M., Michel, M., et al., 2013. Wear patterns of taper connections in retrieved large diameter metal-on-metal bearings. J. Orthop. Res. 31, 1116–1122.

Butler, R.A., Barrack, R.L., 2004. Total hip wear debris presenting as lower extremity swelling. A report of two cases. J. Bone Joint Surg. Am. 86, 142–145.

Campbell, J.R., Estey, M.P., 2013. Metal release from hip prostheses: cobalt and chromium toxicity and the role of the clinical laboratory. Clin. Chem. Lab. Med. 51, 213–220.

Campbell, P., Takamura, K., 2020. Local and systemic consequences of metal-on-metal hip resurfacing implants. Ann. Joint 5, 1–8.

Chalmers, B.P., Perry, K.I., Taunton, M.J., Mabry, T.M., Abdel, M.P., 2016. Diagnosis of adverse local tissue reactions following metal-on-metal hip arthroplasty. Curr. Rev. Musculoskelet. Med. 9, 67–74.

Cheung, A.C., Banerjee, S., Cherian, J.J., Wong, F., Butany, J., Gilbert, C., et al., 2016. Systemic cobalt toxicity from total hip arthroplasties: review of a rare condition Part 1 - history, mechanism, measurements, and pathophysiology. Bone Joint J. 98-B, 6–13.

Clayton, R.A., Beggs, I., Salter, D.M., Grant, M.H., Patton, J.T., Porter, D.E., 2008. Inflammatory pseudotumor associated with femoral nerve palsy following metal-on-metal resurfacing of the hip. A case report. J. Bone Joint Surg. Am. 90, 1988–1993.

Collier, J.P., Surprenant, V.A., Jensen, R.E., Mayor, M.B., Surprenant, H.P., 1992. Corrosion between the components of modular femoral hip prostheses. J. Bone Joint Surg. Br. 74, 511–517.

Cooper, H.J., Urban, R.M., Wixson, R.L., Meneghini, R.M., Jacobs, J.J., 2013. Adverse local tissue reaction arising from corrosion at the femoral neck-body junction in a dual-taper stem with a cobalt-chromium modular neck. J. Bone Joint Surg. Am. 95, 865–872.

De Smet, K.A., De Haan, R., Calistri, A., Campbell, P.A., Ebramzadeh, E., Pattyn, C., et al., 2008. Metal ion measurement as a diagnostic tool to identify problems with metal-on-metal hip resurfacing. J. Bone Joint Surg. Am. 90, 202–208.

De Smet, K.A., Van Der Straeten, C., Van, O.M., Doubi, R., Backers, K., Grammatopoulos, G., 2011. Revisions of metal-on-metal hip resurfacing: lessons learned and improved outcome. Orthop. Clin. North. Am. 42, 259–269.

Delaunay, C., Petit, I., Learmonth, I.D., Oger, P., Vendittoli, P.A., 2010. Metal-on-metal bearings total hip arthroplasty: the cobalt and chromium ions release concern. Orthop. Traumatol. Surg. Res. 96, 894–904.

Deny, A., Barry, J., Hutt, J.R.B., Lavigne, M., Masse, V., Vendittoli, P.A., 2018. Effect of sleeved ceramic femoral heads on titanium ion release. Hip. Int. 28, 139–144.

Desy, N.M., Bergeron, S.G., Petit, A., Huk, O.L., Antoniou, J., 2011. Surgical variables influence metal ion levels after hip resurfacing. Clin. Orthop. Relat. Res. 469, 1635–1641.

Dorr, L., Bloebaum, R., Emmanual, J., Meldrum, R., 1990. Histologic, biochemical, and ion analysis of tissue and fluids retrieved during total hip arthroplasty. Clin. Orthop. Relat. Res. 1990, 82–95.

Dowson, D., Neville, A., 2014. Tribology and corrosion in hip joint replacements: materials and engineering. Joint Replacement Technology. Woodhead Publishing, pp. 401–442.

Dowson, D., Hardaker, C., Flett, M., Isaac, G.H., 2004. A hip joint simulator study of the performance of metal-on-metal joints: part I: the role of materials. J. Arthroplasty 19, 118–123.

Dunstan, E., Sanghrajka, A.P., Tilley, S., Unwin, P., Blunn, G., Cannon, S.R., et al., 2005. Metal ion levels after metal-on-metal proximal femoral replacements: a 30-year follow-up. J. Bone Joint Surg. Br. 87, 628–631.

Ebreo, D., Khan, A., El-Meligy, M., Armstrong, C., Peter, V., 2011. Metal ion levels decrease after revision for metallosis arising from large-diameter metal-on-metal hip arthroplasty. Acta Orthop. Belg. 77, 777–781.

Eichler, D., Barry, J., Lavigne, M., Masse, V., Vendittoli, P.A., 2020. No radiological and biological sign of trunnionosis with large diameter head ceramic bearing total hip arthroplasty after 5 years. Orthop. Traumatol. Surg. Res. 107, 102543.

Eltit, F., Wang, Q., Wang, R., 2019. Mechanisms of adverse local tissue reactions to hip implants. Front. Bioeng. Biotechnol. 7, 176.

Fehring, T.K., Carter, J.L., Fehring, K.A., Odum, S.M., Griffin, W.L., 2015. Cobalt to chromium ratio is not a key marker for ALTR in metal-on-metal hips. J. Arthroplasty 30, 107–109.

Finley, B.L., Unice, K.M., Kerger, B.D., Otani, J.M., Paustenbach, D.J., Galbraith, D.A., et al., 2013. 31-day study of cobalt(II) chloride ingestion in humans: pharmacokinetics and clinical effects. J. Toxicol. Environ. Health A 76, 1210–1224.

Gilbert, J.L., Mehta, M., Pinder, B., 2009. Fretting crevice corrosion of stainless steel stem-CoCr femoral head connections: comparisons of materials, initial moisture, and offset length. J. Biomed. Mater. Res. B Appl. Biomater. 88, 162–173.

Gofton, W., Beaule, P.E., 2015. Serum metal ions with a titanium modular neck total hip replacement system. J. Arthroplasty 30, 1781–1786.

Goldberg, J.R., Gilbert, J.L., Jacobs, J.J., Bauer, T.W., Paprosky, W., Leurgans, S., 2002. A multicenter retrieval study of the taper interfaces of modular hip prostheses. Clin. Orthop. Relat. Res. 401, 149–161.

Grammatopoulos, G., Pandit, H., Kwon, Y.M., Gundle, R., McLardy-Smith, P., Beard, D.J., et al., 2009. Hip resurfacings revised for inflammatory pseudotumour have a poor outcome. J. Bone Joint Surg. Br. 91, 1019–1024.

Griffin, W.L., Fehring, T.K., Kudrna, J.C., Schmidt, R.H., Christie, M.J., Odum, S.M., et al., 2012. Are metal ion levels a useful trigger for surgical intervention? J. Arthroplasty 27, 32–36.

Griffiths, H.J., Burke, J., Bonfiglio, T.A., 1987. Granulomatous pseudotumors in total joint replacement. Skeletal. Radiol. 16, 146–152.

Grosse, S., Haugland, H.K., Lilleng, P., Ellison, P., Hallan, G., Høl, P.J., 2015. Wear particles and ions from cemented and uncemented titanium-based hip prostheses—a histological and chemical analysis of retrieval material. J. Biomed. Mater. Res. B Appl. Biomater. 103, 709–717.

Harrington, C.F., McKibbin, C., Rahanu, M., Langton, D., Taylor, A., 2016. Measurement of titanium in hip-replacement patients by inductively coupled plasma optical emission spectroscopy. Ann. Clin. Biochem. 54(3), 362–369.

Hart, A.J., Buddhdev, P., Winship, P., Faria, N., Powell, J.J., Skinner, J.A., 2008. Cup inclination angle of greater than 50 degrees increases whole blood concentrations of cobalt and chromium ions after metal-on-metal hip resurfacing. Hip. Int. 18, 212–219.

Hart, A.J., Hester, T., Sinclair, K., Powell, J.J., Goodship, A.E., Pele, L., et al., 2006. The association between metal ions from hip resurfacing and reduced T-cell counts. J. Bone Joint Surg. 88, 449–454.

Hart, A.J., Sabah, S., Henckel, J., Lewis, A., Cobb, J., Sampson, B., et al., 2009. The painful metal-on-metal hip resurfacing. J. Bone Joint Surg. Br. 91, 738–744.

Hart, A.J., Sabah, S.A., Bandi, A.S., Maggiore, P., Tarassoli, P., Sampson, B., et al., 2011. Sensitivity and specificity of blood cobalt and chromium metal ions for predicting failure of metal-on-metal hip replacement. J. Bone Joint Surg. Br. 93, 1308–1313.

Hart, A.J., Sabah, S.A., Sampson, B., Skinner, J.A., Powell, J.J., Palla, L., et al., 2014. Surveillance of patients with metal-on-metal hip resurfacing and total hip prostheses: a prospective cohort study to investigate the relationship between blood metal ion levels and implant failure. J. Bone Joint Surg. Am. 96, 1091–1099.

Heisel, C., Silva, M., Skipor, A.K., Jacobs, J.J., Schmalzried, T.P., 2005. The relationship between activity and ions in patients with metal-on-metal bearing hip prostheses. J. Bone Joint Surg. Am. 87, 781–787.

Hur, C.I., Yoon, T.R., Cho, S.G., Song, E.K., Seon, J.K., 2008. Serum ion level after metal-on-metal THA in patients with renal failure. Clin. Orthop. Relat. Res. 466, 696–699.

Hutt, J., Lavigne, M., Lungu, E., Belzile, E., Morin, F., Vendittoli, P.A., 2016. Comparison of whole-blood metal ion levels among four types of large-head, metal-on-metal total hip arthroplasty implants: a concise follow-up, at five years, of a previous report. J. Bone Joint Surg. Am. 98, 257–266.

Jacobs, J., Skipor, A., Black, J., Urban, R., Galante, J., 1991. Release and excretion of metal in patients who have a total hip-replacement component made of titanium-base alloy. J. Bone Joint Surg. 73-A, 1475–1486.

Jacobs, J.J., Skipor, A.K., Patterson, L.M., Hallab, N.J., Paprosky, W.G., Black, J., et al., 1998. Metal release in patients who have had a primary total hip arthroplasty. A prospective, controlled, longitudinal study. J. Bone Joint. Surg. Am. 80, 1447–1458.

Jantzen, C., Jørgensen, H.L., Duus, B.R., Sporring, S.L., Lauritzen, J.B., 2013. Chromium and cobalt ion concentrations in blood and serum following various types of metal-on-metal hip arthroplasties: a literature overview. Acta Orthop. 84, 229–236.

Jelsma, J., Schotanus, M.G.M., van Kuijk, S.M.J., Buil, I.T.A.F., Heyligers, I.C., Grimm, B., 2020. Quality, but not quantity of physical activity is associated with metal ion concentrations in unilateral hip resurfacing. J. Orthop. Res. 38, 2206–2212.

Kerger, B.D., Gerads, R., Gurleyuk, H., Urban, A., Paustenbach, D.J., 2015. Total cobalt determination in human blood and synovial fluid using inductively coupled plasma-mass spectrometry: method validation and evaluation of performance variables affecting metal hip implant patient samples. Toxicol. Environ. Chem. 97, 1145–1163.

Khan, M., Takahashi, T., Kuiper, J.H., Sieniawska, C.E., Takagi, K., Richardson, J.B., 2006. Current in vivo wear of metal-on-metal bearings assessed by exercise-related rise in plasma cobalt level. J. Orthop. Res. 24, 2029–2035.

Kiran, M., Armstrong, C., Shivarathre, D., Peter, V.K., 2017. Blood metal ion levels have limited utility in the surveillance of asymptomatic large-head metal-on-metal total hip arthroplasties. J Arthroplasty 32(12), 3685–3688.

Koerten, H.K., Onderwater, J.J., Koerten, E.W., Bernoski, F.P., Nelissen, R.G., 2001. Observations at the articular surface of hip prostheses: an analytical electron microscopy study on wear and corrosion. J. Biomed. Mater. Res. 54, 591–596.

Koller, D., Bramhall, P., Devoy, J., Goenaga-Infante, H., Harrington, C.F., Leese, E., et al., 2018. Analysis of soluble or titanium dioxide derived titanium levels in human whole blood: consensus from an inter-laboratory comparison. Analyst 143, 5520–5529.

Kwon, Y.M., Lombardi, A.V., Jacobs, J.J., Fehring, T.K., Lewis, C.G., Cabanela, M.E., 2014. Risk stratification algorithm for management of patients with metal-on-metal hip arthroplasty: consensus statement of the American Association of Hip and Knee Surgeons, the American Academy of Orthopaedic Surgeons, and the Hip Society. J. Bone Joint Surg. Am. 96, e4.

Langton, D.J., Joyce, T.J., Mangat, N., Lord, J., Van Orsouw, M., De Smet, K., et al., 2011. Reducing metal ion release following hip resurfacing arthroplasty. Orthop. Clin. North Am. 42, 169–180.

Langton, D.J., Sidaginamale, R.P., Joyce, T.J., Natu, S., Blain, P., Jefferson, R.D., et al., 2013. The clinical implications of elevated blood metal ion concentrations in asymptomatic patients with MoM hip resurfacings: a cohort study. BMJ Open. 3, e001541.

Lavigne, M., Belzile, E.L., Roy, A., Morin, F., Amzica, T., Vendittoli, P.A., 2011. Comparison of whole-blood metal ion levels in four types of metal-on-metal large-diameter femoral head total hip arthroplasty: the potential influence of the adapter sleeve. J. Bone Joint Surg. Am. 93 (2), 128–136.

Lazennec, J.Y., Boyer, P., Poupon, J., Rousseau, M.A., Roy, C., Ravaud, P., et al., 2009. Outcome and serum ion determination up to 11 years after implantation of a cemented metal-on-metal hip prosthesis. Acta Orthop. 80, 168–173.

Leikin, J.B., Karydes, H.C., Whiteley, P.M., Wills, B.K., Cumpston, K.L., Jacobs, J.J., 2013. Outpatient toxicology clinic experience of patients with hip implants. Clin. Toxicol. 51, 230–236.

Levine, B.R., Hsu, A.R., Skipor, A.K., Hallab, N.J., Paprosky, W.G., Galante, J.O., et al., 2013. Ten-year outcome of serum metal ion levels after primary total hip arthroplasty. J. Bone Joint. 95, 512–518.

Levina, A., Pham, T.H., Lay, P.A., 2016. Binding of chromium(III) to transferrin could be involved in detoxification of dietary chromium(III) rather than transport of an essential trace element. Angew. Chem. Int. Ed. Engl. 55(28), 8104–8107.

Luetzner, J., Krummenauer, F., Lengel, A.M., Ziegler, J., Witzleb, W.C., 2007. Serum metal ion exposure after total knee arthroplasty. Clin. Orthop. Relat. Res. 461, 136–142.

Mabilleau, G., Kwon, Y.M., Pandit, H., Murray, D.W., Sabokbar, A., 2008. Metal-on-metal hip resurfacing arthroplasty: a review of periprosthetic biological reactions. Acta Orthop. 79, 734–747.

Malahias, M.A., De Martino, I., Gu, A., Baral, E., Wright, T.M., Sculco, P.K., 2019. Complete wear-through of a metal-backed acetabular cup in an ambulatory patient. Arthroplast. Today 2, 394–400.

Malek, I.A., Rogers, J., King, A.C., Clutton, J., Winson, D., John, A., 2015. The interchangeability of plasma and whole blood metal ion measurement in the monitoring of metal on metal hips. Arthritis 2015, 216785.

Matharu, G.S., Berryman, F., Brash, L., Pynsent, P.B., Treacy, R.B., Dunlop, D.J., 2015a. Influence of implant design on blood metal ion concentrations in metal-on-metal total hip replacement patients. Int. Orthop. 39, 1803–1811.

Matharu, G.S., Mellon, S.J., Murray, D.W., Pandit, H.G., 2015b. Follow-up of metal-on-metal hip arthroplasty patients is currently not evidence based or cost effective. J. Arthroplasty 30, 1317–1323.

May, T.W., Wiedmeyer, R.H., 1998. A table of polyatomic interferences in ICP-MS. At. Spectrosc. 19, 150–155.

McAlister, I.P., Abdel, M.P., 2016. Elevated serum titanium level as a marker for failure in a titanium modular fluted tapered stem. Orthopedics 39, e768–e770.

McTighe, T., Brazil, D., Keppler, L., Keggi, J., McPherson, E., 2015. Metallic modular taper junctions in total hip arthroplasty. Reconst. Rev. 5, 29–42.

Munro, J.T., Masri, B.A., Duncan, C.P., Garbuz, D.S., 2014. High complication rate after revision of large-head metal-on-metal total hip arthroplasty. Clin. Orthop. Relat. Res. 472, 523–528.

Nakano, N., Volpin, A., Bartlett, J., Khanduja, V., 2017. Management of metal-on-metal hips: a strategy for follow-up. Indian. J. Orthop. 51, 414–420.

Nam, D., Keeney, J.A., Nunley, R.M., Johnson, S.R., Clohisy, J.C., Barrack, R.L., 2015. Metal ion concentrations in young, active patients following total hip arthroplasty with the use of modern bearing couples. J. Arthroplasty 30, 2227–2232.

Nuevo Ordóñez, Y., Montes-Bayón, M., Blanco-González, E., Paz-Jiménez, J., Tejerina-Lobo, J.M., Peña-López, J.M., et al., 2009. Metal release in patients with total hip arthroplasty by DF-ICP-MS and their association to serum proteins. J. Anal. At. Spectrom. 24, 1037–1043.

Omlor, G.W., Kretzer, J.P., Reinders, J., Streit, M.R., Bruckner, T., Gotterbarm, T., et al., 2013. In vivo serum titanium ion levels following modular neck total hip arthroplasty–10 year results in 67 patients. Acta Biomater. 9, 6278–6282.

Osman, K., Panagiotidou, A.P., Khan, M., Blunn, G., Haddad, F.S., 2016. Corrosion at the head-neck interface of current designs of modular femoral components: essential questions and answers relating to corrosion in modular head-neck junctions. Bone Joint J. 98-B, 579–584.

Pandit, H., Glyn-Jones, S., McLardy-Smith, P., Gundle, R., Whitwell, D., Gibbons, C.L., et al., 2008. Pseudotumours associated with metal-on-metal hip resurfacings. J. Bone Joint Surg. Br. 90, 847–851.

Peacock, A., Say, J., Lawrence, T., 2008. Reactive synovitis following hip resurfacing: a case presentation. Hip. Int. 18, 224–227.

Prentice, J.R., Clark, M.J., Hoggard, N., Morton, A.C., Tooth, C., Paley, M.N., et al., 2013. Metal-on-metal hip prostheses and systemic health: a cross-sectional association study 8 years after implantation. PLoS One 8, 1–9.

Quitmann, H., Wedemeyer, C., Von Knoch, M., Russe, K., Saxler, G., 2006. Titanium serum levels may remain elevated despite hip revision surgery for wear-through of an acetabular component. Biomed. Tech. 51, 27–29.

Rahme, M., Lavigne, M., Barry, J., Cirtiu, C.M., Belanger, P., Vendittoli, P.A., 2014. Whole blood metal ion measurement reproducibility between different laboratories. J. Arthroplasty 29, 2214–2218.

Rambani, R., Kepecs, D.M., Mäkinen, T.J., Safir, O.A., Gross, A.E., Kuzyk, P.R., 2017. Revision total hip arthroplasty for fractured ceramic bearings: a review of best practices for revision cases. J. Arthroplasty 32, 1959–1964.

Rieker, C.B., Wahl, P., 2020. What the surgeon can do to reduce the risk of trunnionosis in hip arthroplasty: recommendations from the literature. Materials 13, 1950.

Rieker, C.B., Schon, R., Konrad, R., Liebentritt, G., Gnepf, P., Shen, M., et al., 2005. Influence of the clearance on in-vitro tribology of large diameter metal-on-metal articulations pertaining to resurfacing hip implants. Orthop. Clin. North Am. 36, 135–142.

Saini, R., Railton, P., Boyd, J., Sadrzadeh, H., Powell, J.N., 2019. Concordance between laboratories in metal ion testing in patients with metal-on-metal hip implants. Can. J. Surg. 62, 9–13.

Sakamoto, M., Watanabe, H., Higashi, H., Kubosawa, H., 2016. Pseudotumor caused by titanium particles from a total hip prosthesis. Orthopedics 39, e162–e165.

Sarmiento-González, A., Marchante-Gayón, J.M., Tejerina-Lobo, J.M., Paz-Jiménez, J., Sanz-Medel, A., 2005. ICP-MS multielemental determination of metals potentially released from dental implants and articular prostheses in human biological fluids. Anal. Bioanal. Chem. 382, 1001–1009.

Sarmiento-González, A., Marchante-Gayón, J.M., Tejerina-Lobo, J.M., Paz-Jiménez, J., Sanz-Medel, A., 2008. High-resolution ICP-MS determination of Ti, V, Cr, Co, Ni, and Mo in human blood and urine of patients implanted with a hip or knee prosthesis. Anal. Bioanal. Chem. 391, 2583–2589.

Savarino, L., Cadossi, M., Chiarello, E., Fotia, C., Greco, M., Baldini, N., et al., 2014. How do metal ion levels change over time in hip resurfacing patients? A cohort study. Sci. World J. 2014, 291925.

Sidaginamale, R.P., Joyce, T.J., Lord, J.K., Jefferson, R., Blain, P.G., Nargol, A.V., et al., 2013. Blood metal ion testing is an effective screening tool to identify poorly performing metal-on-metal bearing surfaces. Bone Joint Res. 2, 84–95.

Simonsen, L.O., Harbak, H., Bennekou, P., 2012. Cobalt metabolism and toxicology—a brief update. Sci. Total. Environ. 432, 210–215.

Skocaj, M., Filipic, M., Petkovic, J., Novak, S., 2011. Titanium dioxide in our everyday life; is it safe? Radiol. Oncol. 45, 227–247.

Sutphen, S.A., MacLaughlin, L.H., Madsen, A.A., Russell, J.H., McShane, M.A., 2016. Prevalence of pseudotumor in patients after metal-on-metal hip arthroplasty evaluated with metal ion analysis and MARS-MRI. J. Arthroplasty 31, 260–263.

Swiatkowska, I., Martin, N., Hart, A.J., 2019a. Blood titanium level as a biomarker of orthopaedic implant wear. J. Trace Elem. Med. Biol. 53, 120–128.

Swiatkowska, I., 2019b. Toxicity of Metal Debris From Hip Implants (Ph.D. thesis). University College London, London, UK.

Swiatkowska, I., Martin, N.G., Henckel, J., Apthorp, H., Hamshere, J., Hart, A.J., 2020. Blood and plasma titanium levels associated with well-functioning hip implants. J. Trace Elem. Med. Biol. 57, 9–17.

Urish, K.L., Giori, N.J., Lemons, J.E., Mihalko, W.M., Hallab, N., 2019. Trunnion corrosion in total hip arthroplasty—basic concepts. Orthop. Clin. North Am. 50, 281–288.

Van Der Straeten, C., Grammatopoulos, G., Gill, H.S., Calistri, A., Campbell, P., De Smet, K. A., 2013a. The 2012 Otto Aufranc Award: the interpretation of metal ion levels in unilateral and bilateral hip resurfacing. Clin. Orthop. Relat. Res. 471, 377–385.

Van Der Straeten, C., Van Quickenborne, D., De Roest, B., Calistri, A., Victor, J., De Smet, K., 2013b. Metal ion levels from well-functioning Birmingham hip resurfacings decline significantly at ten years. Bone Joint J. 95-B, 1332–1338.

Vendittoli, P.A., Amzica, T., Roy, A.G., Lusignan, D., Girard, J., Lavigne, M., 2011. Metal ion release with large-diameter metal-on-metal hip arthroplasty. J. Arthroplasty 26, 282–288.

Vendittoli, P.A., Masse, V., Kiss, M.O., Lusignan, D., Lavigne, M., 2019. Modular junction may be more problematic than bearing wear in metal-on-metal total hip arthroplasty. Hip. Int. 29, 262–269.

Vendittoli, P.A., Riviere, C., Roy, A.G., Barry, J., Lusignan, D., Lavigne, M., 2013. Metal-on-metal hip resurfacing compared with 28-mm diameter metal-on-metal total hip replacement: a randomised study with six to nine years' follow-up. Bone Joint J. 95-B, 1464–1473.

Vendittoli, P.A., Roy, A., Mottard, S., Girard, J., Lusignan, D., Lavigne, M., 2010. Metal ion release from bearing wear and corrosion with 28 mm and large-diameter metal-on-metal bearing articulations: a follow-up study. J. Bone Joint Surg. Br. 92, 12–19.

Walter, L.R., Marel, E., Harbury, R., Wearne, J., 2008. Distribution of chromium and cobalt ions in various blood fractions after resurfacing hip arthroplasty. J. Arthroplasty 23, 814–821.

Weir, A., Westerhoff, P., Fabricius, L., Hristovski, K., von Goetz, N., 2012. Titanium dioxide nanoparticles in food and personal care products. Environ. Sci. Technol. 46, 2242–2250.

Willert, H.G., Buchhorn, G.H., Fayyazi, A., Flury, R., Windler, M., Koster, G., et al., 2005. Metal-on-metal bearings and hypersensitivity in patients with artificial hip joints. A clinical and histomorphological study. J. Bone Joint Surg. Am. 87, 28–36.

Williams, D.H., Greidanus, N.V., Masri, B.A., Duncan, C.P., Garbuz, D.S., 2011. Prevalence of pseudotumor in asymptomatic patients after metal-on-metal hip arthroplasty. J. Bone Joint Surg. Am. 93, 2164–2171.

Wilschefski, S.C., Baxter, M.R., 2019. Inductively coupled plasma mass spectrometry: introduction to analytical aspects. Clin. Biochem. Rev. 40, 115–133.

Yan, Y., Neville, A., Dowson, D., 2006a. Biotribocorrosion–an appraisal of the time dependence of wear and corrosion interactions: I. The role of corrosion. J. Phys. D. Appl. Phys. 39, 3200–3205.

Yan, Y., Neville, A., Dowson, D., 2006b. Understanding the role of corrosion in the degradation of metal-on-metal implants. Proc. Inst. Mech. Eng. H. 220, 173–181.

Yao, J.J., Lewallen, E.A., Thaler, R., Dudakovic, A., Wermers, M., Day, P., et al., 2020. Challenges in the measurement and interpretation of serum titanium concentrations. Biol. Trace Elem. Res. 196, 20–26.

Yi, Z., Bo, Z., Bin, S., Jing, Y., Zongke, Z., Fuxing, P., 2016. Clinical results and metal ion levels after ceramic-on-metal total hip arthroplasty: a mean 50-month prospective single-center study. J. Arthroplasty 31, 438–441.

Chapter 5

# Biomarkers of compromised implant fixation

**Reshid Berber, Benjamin Bloch, Peter James and Andrew Manktelow**
*Nottingham University Hospitals, NHS Trust, Nottingham, United Kingdom*

## 5.1 Introduction

Total hip arthroplasty (THA) is one of the most common surgical interventions in medicine. The procedure is seen as highly successful for the treatment of end-stage arthritis of the hip, where implants have lasting durability (Evans et al., 2019). In the United Kingdom, over 200,000 THA and total knee arthroplasty (TKA) procedures were recorded in the National Joint Registry (NJR) in 2019 (NJR, 2020). Epidemiological studies have projected increased demands for THA up until 2030 and likely beyond (Kurtz et al., 2007).

As expected, increasing demand for joint replacement has led to a rise in the revision burden (Singh et al., 2019), translating into economic implications for health systems around the world. Compared with primary surgery, the costs associated with a re-do hip procedure are much greater (Vanhegan et al., 2012), averaging £11,897 for aseptic revisions. Besides the costs, revision surgery is technically more challenging, carries a higher complication rate, and can often result in higher re-revision rates for patients (Hoberg et al., 2016). As such, it is imperative to optimise primary outcomes in total joint arthroplasty (TJA).

One measure of the success of an implant is its revision rate, calculated as the number of revision procedures per 100 observed component-years, where one revision per 100 observed component-years corresponds to a revision rate of 1% at 1 year. National registries have reported excellent long-term implant survival across a variety of implant types, with an acceptable revision rate being less than 5% at 10 years (AOANJRR, 2020; NJR, 2020). There is, however, an expected variation in survival rates. In 2019, 6835 revision hip procedures were reported in the United Kingdom's NJR, and by far, the commonest indication was aseptic loosening (40%), followed by osteolysis (14%) (NJR, 2020). Notably, while the total annual number of hip revision procedures has been dropping since 2012 [most likely

Biomarkers of Hip Implant Function. DOI: https://doi.org/10.1016/B978-0-12-821596-8.00008-2

owing to the peak revision burden due to the metal-on-metal implant crises seen at that time, and the introduction of highly crosslinked polyethylene (Morlock and Jager, 2017)], the proportion of patients undergoing revision surgery for aseptic loosening has been consistent over the years (approximately 40%).

Aseptic loosening is defined as failure of integration between a prosthesis and the host bone in the absence of infection. Generation of excess wear particles at implant interfaces produces a pro-inflammatory state that stimulates osteoclast differentiation and macrophage production within the periprosthetic environment (Jiang et al., 2013). The ultimate pathway from progressive particle wear to aseptic loosening and construct failure is driven by inflammation-mediated bone loss (osteolysis) around the implant. Loosening reported in the first few years after implantation of a prosthesis is most likely due to failure of the implant to gain fixation to the bone, and loosening in later years often reflects loss of fixation secondary to osteolysis. Evaluation and accurate diagnosis are important because progressive osteolysis can lead to fracture if left untreated. The factors that can influence a person's susceptibility to aseptic loosening after TJA can be categorised into surgeon-, implant, and patient-related factors (Cherian et al., 2015).

Although serial radiography is traditionally employed to monitor implants for evidence of aseptic loosening over time, radiographs are not sensitive enough to detect abnormalities associated with the early stages of loosening and can appear unremarkable despite patient-reported symptoms (Mandalia et al., 2008, Hirschmann et al., 2011). Computed tomography (CT) and magnetic resonance imaging (MRI) are useful in cases of substantial osteolysis; however, like radiographs, they are compromised by artefacts arising from metal hardware, and early signs of osteolysis are likely to be misinterpreted. Periprosthetic osteolysis can therefore remain undiagnosed until substantial bone loss has occurred. Despite the development of metal artefact reduction techniques (Sneag et al., 2015), it is believed that traditional imaging techniques underestimate the true degree of periprosthetic osteolysis.

The inherent limitations of imaging diagnostic tools have driven the search for molecular biomarkers of compromised implant fixation to supplement the diagnosis. Biomarkers can serve as predictors of disease onset or as indicators of disease incidence or progression. This chapter explores the rationale for biomarker development in the area of impaired implant fixation and discusses the molecules that hold the most promise for early diagnosis and monitoring of aseptic loosening secondary to periprosthetic osteolysis.

## 5.2 Osseointegration of hip implants

Osseointegration is defined as the direct structural and functional connection between ordered, living bone and the surface of a load-bearing implant. The concept was first studied in the 1940s and the term was coined in 1952 by

Professor Per-Ingvar Brånemark, who performed both macroscopic and microscopic analyses of rabbit long bones treated with submerged titanium screws. He noted that following a period of immobilisation, the compact cortical bone grew directly onto the implant, without intervening fibrous tissue (Brånemark et al., 1969). Therefore, osseointegration can be compared to direct fracture healing, in which the fragment ends become united by bone, without intermediate fibrous tissue or fibrocartilage formation (Schenk and Buser, 1998).

As for fracture healing, the prerequisites for osseointegration include primary stability and adequate load. The tissue response to a freshly installed implant is largely dependent on the mechanical situation; absolute stability is required to encourage bone formation. In a fracture scenario, a bone union is achieved through anatomic reduction and compression of bone ends, with a small fracture gap. The implants used to aid fracture healing need to withstand functional loading to counteract forces that would create excessive movement at the fracture site.

Unlike fracture healing, osseointegration requires a union between bone and implant surfaces, with the implant material playing a crucial role in the achievement of this union. Upon exposure to body fluids, biomaterials become coated with various biomolecules, such as plasma proteins, lipids, and sugars, and the interactions between these adsorbed molecules and cells influence the host response. Therefore, implant parts whose fixation relies on bone integration need to be manufactured from materials whose surface configurations encourage bone formation. Titanium is considered to be more suitable than stainless steel for use in medical implant manufacture owing to its improved biocompatibility, strength-to-density ratio, corrosion resistance, and lower modulus of elasticity. These properties are further enhanced in titanium alloys, most notably in the β-alloy Ti-6Al-4V (Apostu et al., 2018). The biocompatibility of titanium can be explained by its low electrical conductivity, which facilitates the formation of a protective surface oxide layer that also encourages osseointegration via the adhesion of osteoblasts to the implant surface (Sidambe, 2014). However, damage to the oxide layer and subsequent corrosion can lead to the release of metal ions and adverse reactions such as inflammation and fibrosis around the implant. The latter can impede the formation of bone bridges by osteoblasts (Apostu et al., 2018). Bioactive coating materials can restrict this immune response and instigate a favourable tissue reaction by promoting cellular activities involved in bone formation, for example, through a chemical bond with tissue components, as seen with hydroxyapatite (Schenk and Buser, 1998).

When the stiffness of an implant exceeds that of adjacent bone, weight-bearing stresses are not distributed equally, leading to stress-shielding and, ultimately, bone resorption and implant loosening. It follows that the implant's elastic modulus (Young's modulus) should not be too dissimilar to that of bone, that is, 10–30 GPa. The values for cobalt-chromium alloys and stainless steel are around 210 and 180 GPa, respectively, while those for pure and alloyed titanium are generally lower, at approximately 110 GPa, making the latter materials more suitable for use as acetabular cups and femoral stems (Niinomi and Nakai, 2011). Besides alloy composition, Young's modulus can be affected by the implant's structure; specifically, increased porosity will make the material more bone-like. Importantly, while a low Young's modulus minimises stress-shielding, it produces more micromotion, leading to increased mechanical stress and growth of fibrous tissue at the implant–bone interface. Therefore, a compromise needs to be reached between the risk of stress-shielding and that of aseptic loosening (Apostu et al., 2018).

The notion that the implant surface properties influence its bonding to the host bone is an important consideration in orthopaedic implant design. It is assumed that rough surfaces improve the pull-out strength (the force required to remove an implant from its fixation) compared with smooth surfaces. However, the relationship between implant design and long-term stability also depends on the quality of the host bone, host factors, and surgeon factors, which all play a role in the individual susceptibility to aseptic loosening or failure to achieve osseointegration. Research into enhancing osseointegration, which centres around limiting fibrous tissue production and maintaining good osteoblast activity around the implant, could help alleviate the future revision burden due to aseptic loosening. Factors that have the potential to improve implant fixation, including the choice of implant, surgeon's technique, use of systemic drugs, rehabilitation protocol, and adjuvant therapies, are explored in the following sections.

## 5.2.1 Implant design

Histological analyses demonstrated that the formation of new bone along the surface of the native bone and deposition of bone along the implant surface occur in the immediate postoperative period. The bond matures over the first 3 months following surgery, and its formation is critical for the long-term survival of the implant. Successful bridging of gaps ($<2$ mm) between the bone and implant by *de novo* bone is dependent on the surface finish of the implant (Overmann et al., 2020). The most common implant types can be subdivided into those with roughened surfaces with a coating (e.g., plasma-sprayed or hydroxyapatite-coated titanium), roughened surfaces without a coating (e.g., sand-blasted, acid-etched, or anodically roughened), and those with machine-processed, uncoated titanium (e.g., machined or polished).

### *5.2.1.1 Bioactive coatings*

The major inorganic component of bone, hydroxyapatite, can be used as a surface coating on biomedical implants to promote osseointegration.

Hydroxyapatite coatings are superior to titanium alloy coatings for stimulating bone ingrowth from the native bone and, based on the NJR data, three of the top four (in terms of market share) cementless stems implanted utilise hydroxyapatite coatings (NJR, 2020). A meta-analysis showed that patients with hydroxyapatite-coated implants have a lower incidence of aseptic loosening than those with porous-coated prostheses (Chen et al., 2015). However, a study of 110 hips demonstrated no difference in implant survival between hydroxyapatite-coated and non-hydroxyapatite-coated titanium stems in patients who had received a simultaneous bilateral hip replacement, with one of each stem type implanted; none of the stems were revised for loosening at a mean follow-up of 15.6 years (Kim et al., 2012).

One perceived downside to hydroxyapatite is its highly osteophilic property, which makes it bind more strongly to the host bone than to the metal hardware, possibly resulting in separation between the coating and implant surface over time (Overmann et al., 2020; Buttaro et al., 2017).

Several growth and differentiation factors have been used experimentally as biocoatings of conventional implants to accelerate bone ingrowth. Bone morphogenetic protein (BMP)-2, BMP-7, and osteogenic protein-1 have all been shown to augment bone formation and osseointegration of implants. Other growth factors used to strengthen osseous fixation are platelet-derived growth factor, insulin-like growth factor, transforming growth factor-beta 1 (TGFβ-1), and TGFβ-2. Further research is required to establish the long-term efficacy of using such osteoinductive agents (Dimitriou and Babis, 2007).

### 5.2.1.2 *Surface properties*

Implant surface quality is considered in terms of its mechanical, topographical, and physicochemical properties, which work in concert, such that altering one will impact the other. With titanium, the mechanical properties are difficult to amend within the physiological range; however, the surface chemistry and topography can be readily manipulated. The primary interaction between the implant and host starts with a thin interface zone, where rapid protein adsorption and interaction with the connective tissues trigger a biological cascade that influences implant osseointegration. The formation of this zone is controlled by the physical and chemical properties of the material, such as chemical composition, wettability, roughness, and oxide layer thickness, and is critical for the long-term success of the implant.

#### Wettability

One factor affecting the interaction of implant biomaterials with biological fluids and tissues is surface wettability, that is, the ability of an aqueous solution to spread over a surface. Wettability is determined by the surface energy of the solid and surface tension of the liquid, and it is described by the contact angle between the liquid and solid interfaces: a low contact angle ($0^\circ < \theta < 90^\circ$) indicates a hydrophilic surface (liquid spreads across the surface) while a high contact angle ($90^\circ \leq \theta < 180^\circ$) indicates a hydrophobic surface (liquid forms droplets on the surface). Increasing the wettability may enhance fibrin adhesion and provide contact guidance for osteoblast migration along the surface of the implant, which is important in the early stages of osseointegration (Damiati et al., 2018; Apostu et al., 2018). Factors that affect a biomaterial's wettability include its microstructure (the rougher the surface, the poorer the wettability), chemical composition, sterilisation, and handling during implantation. However, the optimum range of wettability for proper osseointegration of hip implants has not been established.

#### Chemical composition

The chemical composition of a biomaterial's surface dictates the amount, type, and conformation of adsorbed proteins and the subsequent protein–cell interactions (Damiati et al., 2018). Osteoblasts are sensitive to subtle differences in surface chemistry, making it an important factor for improving implant osseointegration.

#### Oxide layer thickness

Titanium, in the presence of air or water, reacts with oxygen to form a protective and chemically stable oxide layer that reforms shortly after mechanical disturbance. This oxide layer strengthens the implant's corrosion resistance, lowers the rate of metal ion release, and bolsters biocompatibility. A thicker oxide layer leads to increased bone contact, which can accelerate the rate of bone formation (Sul et al., 2002).

#### Roughness

Surface roughness has an important role in enhancing osseointegration through its impact on the material's biomechanical properties, including stress distribution and implant interdigitation. The Ra value is a representation of surface roughness and is defined as the average deviation of the surface from a mean plane. The optimum Ra should be approximately 0.8 μm, although this was determined in combination with other factors, such as oxide layer thickness, porosity, and chemical composition (Sul et al., 2005). It is believed that surface roughness exceeding 1–1.5 μm can weaken primary implant fixation (Apostu et al., 2018).

### *5.2.1.3 Porous metals*

Cementless titanium hip implants have a porous surface to facilitate bone ingrowth and improve their fixation ability. The shape, dimension of pores, and pore throat size can affect the degree of implant osseointegration. It is accepted that a concave shape, wide pore throats, and pore dimensions between 150–600 μm facilitate bone ingrowth. In Taniguchi et al. (2016), a construct with 65% porosity and 600 μm pore size achieved better fixation than similar constructs with 300 or 900 μm pore size.

Sintered, bead-bonded, grit-blasted, and fibre mesh-bonded titanium surfaces are all in clinical use for hip prostheses. Besides conventional porous coatings on dense titanium, various methods have been developed to produce a porous titanium body, appropriate for acetabular shells and augments used in revision surgery. These include sintering with powders, solid-state foaming by an expansion of argon-filled pores, and compressing and sintering of titanium fibres. Notably, these manufacturing processes cannot precisely control the porosity, often leading to inconsistent throat size and poor interconnectivity, which are likely to result in poor bone ingrowth (Otsuki et al., 2006).

Selective laser melting is an additive manufacturing process that utilises a computer-controlled laser beam to selectively melt areas of powder layers, allowing for the production of complex 3D shapes (Taniguchi et al., 2016). Bartolomeu et al., (2021) demonstrated the accuracy of selective laser melting in reproducing implant porosity.

Another way to increase the porosity of hip implants is to manufacture them from porous metals, such as tantalum. A study comparing porous tantalum monobloc cups (porosity of 75%–80% and average pore size of 550 μm) and traditional porous-coated titanium cups (coated with three layers of 200–300 μm-diameter pure titanium beads with 30%–50% porosity) found that the former had better survivorship (100% vs 98%) and significantly less radiolucency (4% vs 33%) at 12 years post-implantation (Wegrzyn et al., 2015). Despite good results, the use of porous tantalum is currently limited to revision THA because of its high cost.

### 5.2.2 Patient-related factors

Age at the time of primary THA is an established risk factor for aseptic loosening. Several large, registry-based studies point to a linear inverse relationship between age and revision risk, with patients younger than 55 years being the highest-risk group (Malchau et al., 1993; Cnudde et al., 2018). Although these findings have been almost entirely attributed to greater activity levels and loading on the joints in younger patients, other age-related factors that affect bone quality may also be at play (Kremers et al., 2016).

A systematic review by Cherian et al., (2015) demonstrated a correlation of male sex [odds ratio (OR): 1.39; 95% confidence interval (CI): 1.22–1.58; $P = 0.1$] and higher activity levels (OR: 4.24; 95% CI: 2.46–7.31; $P = 0.1$) with increased risk of aseptic loosening after THA. The notion that males are at higher risk of aseptic loosening is not universally accepted, however, with data from certain studies pointing to a higher risk of aseptic THA revision in women (Inacio et al., 2013). Hormonal factors across a lifespan (including during menopause) may contribute to sex differences in implant osseointegration and require further study (Kremers et al., 2016).

There is little data on the risk of aseptic loosening and consequent implant revision across racial and ethnic groups. The aetiology of any differences is likely to be multifactorial and have outside influences, such as access to healthcare, patient preferences, environmental factors, and lifestyle choices (Kremers et al., 2016).

Although obesity is a risk factor for the development of osteoarthritis and adverse outcomes following THA, the evidence supporting an increased risk of aseptic loosening in individuals with obesity is conflicting (Cherian et al., 2015). In a meta-analysis by Haverkamp et al., (2011), pooled data from 5137 patients with a primary THA demonstrated that aseptic loosening occurs more often in those with a body mass index (BMI) of over 30 kg/m$^2$ (OR: 0.64; 95% CI: 0.43–0.96; $P = 0.3$). In contrast, a large, single-centre study in 17,774 patients found no statistically significant association between elevated BMI and implant failure due to aseptic loosening (Wagner et al., 2016).

Smoking and alcohol consumption can have a detrimental effect on bone mineral density, so by affiliation, they would be expected to increase the risk of aseptic loosening. However, evidence in this regard is also conflicting (Kremers et al., 2016).

The indication for surgery can influence the likelihood of aseptic implant failure. Some studies suggest that rheumatoid arthritis confers lower or similar risk for aseptic loosening than osteoarthritis, whereas hip dysplasia confers twice the risk compared with osteoarthritis (Kremers et al., 2016). Poor bone stock associated with developmental dysplasia of the hip is thought to contribute to the observed effect. Osteoporosis has been shown to impede the osseointegration of titanium implants (Aro et al., 2012). This may be related to the effects of oestrogen, whose circulatining levels are often inadequate in patients with osteoarthritis. Oestrogen increases the osteogenic activity of osteoblasts and indirectly inhibits osteoclasts via the RANKL/OPG signalling pathway, such that inadequate levels of the hormone translate into increased osteoclast activity and failure of mesenchymal stem cells (MSCs) to differentiate into osteoblasts (Apostu et al., 2018).

Although there is no concrete evidence to support a link between particular lifestyle factors and the risk of aseptic loosening, preoperative optimisation of patient weight, smoking cessation, diabetes control, and cardiovascular care might improve outcomes following TJA (Apostu et al., 2018).

### 5.2.3 Surgeon-related factors

Careful implant positioning is a key determinant of implant success. Incorrect placement may lead to a less than optimal biomechanical environment and accelerated wear at joint surfaces, which predisposes the patient to aseptic loosening. The surgeon's choice of surgical approach can impact the outcome, with the anterior and anterolateral approaches linked to a higher risk of revision for implant malposition (Lindgren et al., 2012; Eto et al., 2017). A study of the Dutch arthroplasty registry found that cementless femoral stems with a proximal shoulder were associated with early aseptic loosening when inserted through an anterior or anterolateral approach, compared with a posterior approach (Janssen et al., 2018).

Moreover, it is important that the bone is accurately prepared to accept the implant, with a minimal gap between the bone and implant surface; gaps larger than 150 μm are believed to impair bone ingrowth. A good fit between the bone and implant is a priority, and the surgeon ought to take care not to over-ream/broach for the risk of thermal damage to the bone and excessive micromotion at the implant–bone interface, which can compromise osseointegration (Apostu et al., 2018). Host bone coverage of the acetabular component greater than 60%, in combination with correct implant positioning, should create the optimal implant stability required for osseointegration.

The choice of femoral head size has been purported to impact the risk of osteolysis and subsequent aseptic loosening. Larger head sizes, which succumb to greater volumetric wear than their smaller counterparts, may stimulate the process; however, Lachiewicz et al. (2016) failed to demonstrate a link between head size and osteolysis at a 10-year follow-up. Until more long-term data are available, surgeons should take a balanced approach that considers the stability-promoting effects of larger head sizes and the potential risk of aseptic loosening.

## 5.3 Periprosthetic osteolysis and aseptic loosening

Bone is a living tissue that undergoes continuous remodelling to maintain elasticity and structural integrity. Removal of old bone by osteoclasts is countered by the deposition of new bone matrix by osteoblasts and its subsequent mineralisation by osteocytes, such that an equilibrium between bone formative and resorptive processes is reached. In individuals with periprosthetic osteolysis, this balance is disturbed in favour of bone resorption.

The convoluted molecular pathology of peri-implant osteolysis and aseptic loosening begins with the generation of implant wear debris and initiation of an inflammatory response within the joint. The strength and type of the inflammatory response depend on several factors, including the size, quantity, origin, and biochemical characteristics of the wear debris (Goodman and Gallo, 2019). Polyethylene and bone cement (polymethylmethacrylate) are the most common sources of inflammatory stimuli; however, corrosion products and metal wear debris can also lead to adverse reactions in the vicinity of the implant and induce hypersensitivity (Teo and Schalock, 2017). There is evidence to suggest that osteolysis is positively correlated with polyethylene wear (Yoon et al., 2022), although this is controversial. An X-ray- and CT-based study by Rames et al. (2021) assessed the effect of highly crosslinked polyethylene on osteolysis rates at a 16-year follow-up: 9 out of the 26 hips studied showed evidence of peri-implant osteolysis despite negligible linear and volumetric wear rates and zero revisions at the end of the study period.

Prosthetic by-products encounter innate immunity receptors located on the surface of immune cells and/or intracellularly. Activation of these receptors triggers an acute inflammatory response resulting in the upregulation and release of pro-inflammatory cytokines, chemokines, and reactive oxygen species (ROS). Low-grade inflammation can decrease oxygen and nutrient levels in the affected tissues, which leads to further tissue necrosis through the release of danger-associated molecular patterns (DAMPs) by stressed cells (Pearl et al., 2011). Stimulation of particular Toll-like receptors by DAMPs and activation of IL-1 signalling results in the activation of nuclear factor κB (NF-κB) and release of more pro-inflammatory cytokines. Tissue-resident macrophages detect wear particles and tissue damage within the periprosthetic environment and become activated as a result. Activated macrophages produce ROS, nitric oxide, and other inflammatory substances that mediate inflammatory bone resorption (Goodman and Gallo, 2019). The implant debris can also instigate an adaptive immune reaction, giving rise to the concept of implant-related metal sensitivity; however, the dominant form of the response is due to the innate reactivity of macrophages to implant debris danger signalling (DAMPs) and consequent release of pro-inflammatory cytokines and chemokines.

The interplay between the different bone cells is primarily regulated via the RANKL/RANK/OPG axis. RANKL, expressed by osteoblasts and osteocytes, stimulates RANK on the surface of osteoclasts and their precursors, which ultimately activates mitogen-activated protein kinases (MAPKs) and the transcription factors NF-κB and AP-1. Activated NF-κB influences osteoclast differentiation, activation, and survival via NFATc1 (nuclear factor of activated T-cells cytoplasmic 1) (Park et al., 2017; Goodman and Gallo, 2019). Resorptive processes are kept in check through activation of the osteogenic Wnt/β-catenin pathway and subsequent secretion of OPG by osteoblasts and osteogenic stromal stem cells. OPG competes with RANK for binding to RANKL to prevent excessive bone destruction.

## 5.4 Postoperative measures to stimulate osseointegration and inhibit osteolysis

### 5.4.1 Rehabilitation and postoperative drugs

The orthopaedic community has long contested the safety of immediate full weight-bearing following cementless THA because of concerns about compromised implant stability and reduced osseointegration. However, growing evidence suggests that the practice is safe. In Wolf et al. (2010), the periprosthetic bone mineral density or stability of the implanted stem did not differ between the immediate full weight-bearing and partial weight-bearing groups for 3 months. In a recent randomised controlled trial that compared the migration patterns of two cementless stem designs, all implants showed good osseointegration and secondary stability 2 years postoperatively despite all patients being allowed to fully weight-bear after the surgery (Reiner et al., 2020). Furthermore, in a meta-analysis of six randomised controlled trials and three cohort studies, early full weight-bearing was not associated with a demonstrable increase in the incidence of postoperative complications (Tian et al., 2017).

Non-steroidal anti-inflammatory drugs (NSAIDs) and opioids are a mainstay of pain management after THA, but several studies, mostly in patients with bone fractures, suggest that both drug classes interfere with bone growth. With regards to NSAIDs, the effect is mediated through the inhibition of cyclooxygenase-2 (COX-2) and decreased levels of prostaglandins, which play an important role in promoting inflammation and the subsequent migration of bone-forming cells. Yet, so far, studies in THA recipients are few and conflicting with respect to the effect of these drugs on implant osseointegration and the risk of aseptic loosening (Kremers et al., 2016). Other medications postulated to interfere with osseointegration include selective serotonin reuptake inhibitors and certain diuretics (Kremers et al., 2016). Limiting their use and looking for alternatives would be recommended in the postoperative period; however, there is currently a lack of guidance from orthopaedic societies concerning systemic drug use following TJA.

### 5.4.2 Pharmacological inhibition of periprosthetic osteolysis

There are no long-term studies that demonstrate the efficacy of pharmacological treatment for periprosthetic osteolysis, whether through enhancing bone ingrowth or stopping osteolysis at an early stage to prevent the need for revision surgery, but a number of promising treatments are being investigated.

One potentially beneficial class of drugs are bisphosphonates, which are synthetic analogues of pyrophosphate used for the treatment of osteoporosis and osteolysis associated with metabolic and metastatic bone disease. Bisphosphonates reduce osteoclast activity through induction of apoptosis, suppress the differentiation of osteoclast precursors, and can also cause cell death in macrophages (Couto et al., 2020). Meta-analyses of randomised clinical trials in which bisphosphonates were administered postoperatively to THA recipients highlight their antiresorptive effects on periprosthetic bone (Su et al., 2018).

RANKL is a protein released from osteoblasts that activates the transmembrane protein RANK on the surface of osteoclasts to promote their activity and subsequent bone resorption. Denosumab is a human monoclonal antibody to RANKL that blocks the binding of RANKL to RANK, thereby inhibiting osteoclast activity and function. In a proof-of-concept trial, where a single dose of denosumab was administered to THA recipients 8 weeks prior to revision surgery for symptomatic, radiographically proven osteolysis, the osteoclastic activity within the implant–bone interface was 83% lower in the denosumab group than in the control group (Mahatma et al., 2021).

Anabolic treatments have also been considered. The *N*-terminal 1–34 amino acid portion of parathyroid hormone is an anabolic substance that stimulates bone formation via protein kinase A and Wnt/β-catenin pathway signalling and increases the secretion of OPG, an antagonist of RANKL. It has been approved for the treatment of postmenopausal osteoporosis and might also be effective against particle-induced osteolysis (Russell, 2013).

Another drug approved for the treatment of postmenopausal osteoporosis that may have a role in particle-induced osteolysis is strontium ranelate—its bone-beneficial effects are mediated via stimulation of preosteoblast proliferation, suppression of osteoclastic differentiation, and facilitation of osteoclastic apoptosis (Goodman and Gallo, 2019).

The flavonoid icariin has been shown to protect against osteolysis by promoting osteogenic differentiation of MSCs and inhibiting the formation and activation of osteoclasts in a murine calvarial model (Wang et al., 2016).

More recently, a role for the hypoxia-inducible factor (HIF) family of transcription factors, in particular HIF-1α, has been established in bone formation. Within the skeletal system, hypoxia and HIF-α-driven signalling are involved in endochondral bone formation and angiogenesis through their regulation of osteoblast and osteoclast function, and the differentiation from MSC to osteoblast to osteocyte. Disruption of blood vessels in response to bone fracture creates a localised hypoxic environment. Haematoma formation isolates the site of injury from perfusion, further augmenting localised hypoxia. Since osseointegration is similar to fracture repair, and since this process closely mimics stages of embryonic bone development (in which hypoxia and HIF-α signalling play a crucial role), it is unsurprising that trauma-induced hypoxia promotes skeletal repair involving HIF-α signalling. Therefore, pharmacological manipulation of neoangiogenesis may promote bone formation and bone repair, as seen with the HIF-1α-stabiliser desferrioxamine, which increases vascularity and callus size in femoral fractures and hastens bone bridging in femoral segmental defects (Yellowley and Genetos, 2019).

### 5.4.3 Biophysical stimulation

Biophysical stimulation represents a non-invasive and locally applied strategy to enhance bone healing. Two methods of biophysical stimulation have been used in orthopaedic practice: pulsed electromagnetic fields (PEMFs) and low-intensity pulsed ultrasound (LIPUS). In the case of PEMF therapy, the positive effect on bone ingrowth is primarily linked to improved vascular function secondary to the release of angiogenic factors, such as IL-8, βFGF, and VEGF (Dimitriou and Babis, 2007). LIPUS, on the other hand, is a form of mechanical energy that is transmitted through living tissue as acoustic pressure waves and absorbed at a rate proportional to the density of the tissues it passes through. It has been hypothesised that the micromechanical strains produced by LIPUS in biological tissues result in biochemical events that stimulate fracture healing (Dimitriou and Babis, 2007). Although both techniques were initially developed, and are currently employed, to stimulate bone regeneration during fracture healing, their positive clinical outcomes and safety highlight their potential as adjunct therapies to enhance implant osseointegration. There is a paucity of data on the impact of biophysical stimulation on osseointegration of prosthetic implants, but some advantages in terms of early recovery have been described in patients treated with these procedures, suggesting that biophysical stimulation could reduce bone oedema, pain, and bone reabsorption around femoral stems following THA (Massari et al., 2015).

## 5.5 Monitoring patients for signs of periprosthetic osteolysis and aseptic loosening

The assessment of suspected periprosthetic osteolysis or aseptic loosening is largely driven by clinical judgement. Pain felt on the first few steps after arising—termed 'start-up pain'—is almost pathognomonic of aseptic loosening in both hip and knee replacements, but there is no single specific clinical finding that would signal early pathological osteolysis. Often, substantial bone loss will have occurred before a patient becomes symptomatic. Clinically, symptoms of pain or lack of function might indicate structural failure, implant loosening, periprosthetic fracture, or infection. Differential diagnosis is not straightforward and requires consideration of factors such as time since surgery, type of implant, peri-operative wound concerns, falls/trauma, and the patient's activity levels. Physical examination by the orthopaedic surgeon should include an assessment of the soft tissues, surgical site, and the joint's range of motion and stability. Individuals with periprosthetic osteolysis often present with a relatively normal clinical picture, though aseptic loosening or failure of osseointegration are likely to be detected on examination through antalgic gait and limited range of motion at the joint.

The primary imaging tool used to detect periprosthetic osteolysis or loosening is standard radiography of the joint. The British Orthopaedic Association blue book on good practice relating to THR recommends 'radiographic follow-up in the form of anteroposterior and lateral X-rays at 1 year, 7 years, and each subsequent 5 years following surgery' (Bannister et al., 2012). Serial radiographs offer an assessment of implant position and penetration of the femoral head into the polyethylene acetabular liner over time, providing clues as to the quality of implant fixation and the amount of particulate debris generated. The sensitivity and specificity of plain radiography for the detection of aseptic loosening are fairly high, at 82% (95% CI: 76–87) and 81% (95% CI: 73–87), respectively (Temmerman et al., 2005). However, X-rays often underestimate the degree of periprosthetic osteolysis and lack the precision necessary to yield useful information at short-term follow-up and in patients with low-wearing hip implants (Malchau et al., 2008). Typical findings on radiographs that are suggestive of compromised implant fixation are radiolucent lines at the implant–bone interface and areas of radiolucency within bone that lies in close proximity to the implant (Fig. 5.1).

CT and MRI are more useful than plain radiography for detecting and evaluating the extent of osteolysis, and can additionally reveal local soft-tissue reactions to wear debris (e.g., oedema, tendinitis, and pseudotumours). Multidetector CT can alert to soft-tissue, bone, and metal hardware complications, and it is frequently used in addition to radiographs in the imaging evaluation of a symptomatic joint replacement. The downside of CT and MRI is that they are both prone to image degradation by susceptibility artefacts from metallic implants, although these can be minimised through the use of a metal artefact reduction sequence (MARS) (Blum et al., 2016). In MARS-MRI, low T1 and high T2 signals around the implant components may support the diagnosis of loosening in suspected cases (Lohmann et al., 2017).

Nuclear medicine imaging techniques such as single-photon emission computed tomography (SPECT) have been used in the second line of a painful hip work-up, particularly when assessing for loosening or infection in patients with equivocal radiographs. The sensitivity and specificity of nuclear bone scans are 85% (95% CI: 79–89) and 72% (95% CI: 64–79), respectively (Temmerman et al., 2005). Coupling SPECT with CT provides combined functional and anatomical information, with better spatial localisation than plain radiography and higher tolerance to metal artefacts than MRI. Focal periprosthetic radiolucencies exceeding 2 mm with corresponding increased tracer uptake at the implant–bone interface is a common radiological manifestation of loosening (Tam et al., 2014). Despite their obvious strengths, nuclear medicine studies have certain drawbacks, including false-positive findings and the inability to differentiate competing diagnoses, such as periprosthetic infection or fracture (Berber et al., 2015).

Digital tomosynthesis (DTS) is an emerging imaging technique derived from radiographic tomography with various clinical applications, including the imaging of orthopaedic implants. It provides multiple-section planes obtained from several projections of various angles (Blum et al., 2018). DTS was previously shown to offer better diagnostic accuracy than radiographs or CT for the detection of periprosthetic loosening (Guo et al., 2018); however, sensitivity was inferior to that of MARS-CT or plain radiography (Gillet et al., 2019).

Radiostereometric analysis (RSA) is currently the most accurate method to examine component migration and predict implant survival with regard to

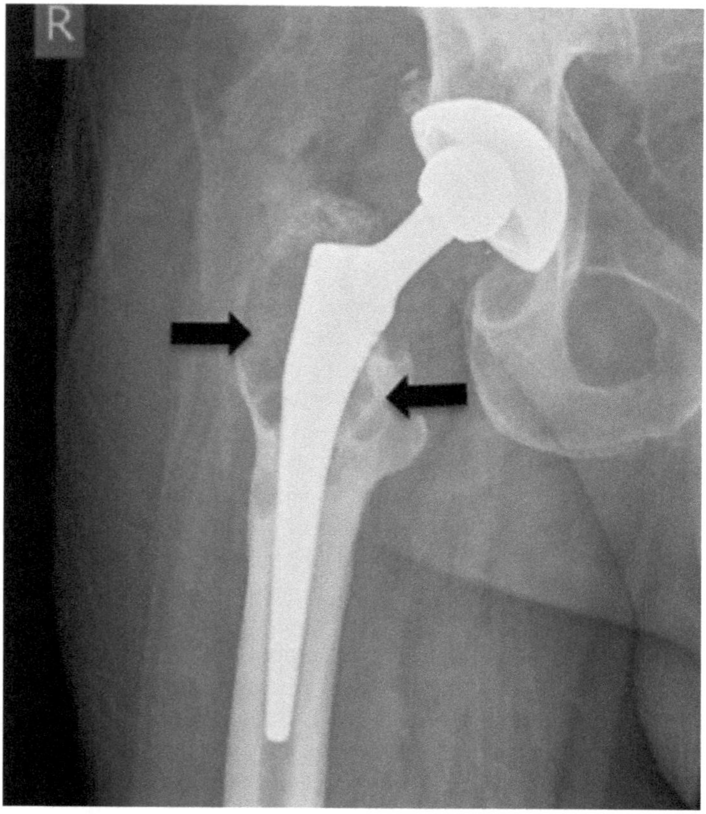


**FIGURE 5.1** Radiographic appearance of peri-implant osteolysis. Arrows point to radiolucent areas seen at and around the implant–bone interface.

aseptic loosening. It involves simultaneously taking X-rays from two different angles to create a 'stereo' image. To enable this, small tantalum beads are embedded into the host bone at the time of hip replacement surgery—these act as references that allow the position of the prosthesis to be precisely mapped. Stereoradiographs taken during regular follow-up visits are used to quantify any shifting of implant components over time. Unfortunately, the requirement for specialised software, marker beads, and expertise to interpret the results restricts the use of RSA in prospective research settings. Computer-assisted methods that rely on edge detection are less accurate than RSA because they use conventional anteroposterior and lateral radiographs of the hip. Their advantage, however, is that they can be applied to larger groups of patients (Malchau et al., 2008).

While laboratory tests for suspected osteolysis are currently limited, any patient presenting with a painful joint replacement should undergo measurement of C-reactive protein (CRP) level, erythrocyte sedimentation rate, and a full blood count. If any of these first-line markers are elevated, then the joint should be aspirated to exclude periprosthetic infection. Currently, there are no specific blood tests in use for the assessment of periprosthetic osteolysis or loosening; however, there are several experimental markers that might be used to support the diagnosis.

## 5.6 Molecular biomarkers of periprosthetic osteolysis and aseptic loosening

Periprosthetic osteolysis often progresses ‘silently’ until mechanical failure of the implant–bone interface gives rise to symptoms, such as patient-reported pain. Conventional radiography can be used to document the progression of implant wear; however, it often underestimates the extent of bone loss and the only treatment for aseptic loosening caused by osteolysis is revision surgery. Revision procedures are associated with higher failure rates, higher patient mortality, and worse pain and functional outcomes than primary THA. Substantial bone loss prior to revision surgery is a contributing factor to these poorer outcomes, highlighting the importance of timely diagnosis and early intervention while the patient still maintains adequate bone mass. Preclinical data highlight the potential of biological and pharmacological treatments to curtail osteolysis progression if initiated in its early stages (Section 5.4.2), but clinical trial evidence is scant. One of the reasons that so few clinical trials have been initiated is the inability to identify patients at risk for osteolysis prior to radiographic diagnosis or emergence of pain.

A biomarker is a measurable characteristic that can be used to indicate the presence or severity of a disease state and/or to measure the response to intervention for such a state. Systemic biomarkers have shown considerable utility in diagnosing or monitoring the response to treatment in musculoskeletal diseases such as osteoporosis. Using biomarkers to monitor patients post-THA could provide information on periprosthetic inflammation or bone loss which, in turn, could help surgeons identify patients at risk for osteolysis. Although there are no validated biomarkers to diagnose or monitor periprosthetic osteolysis, the following molecules show promise.

### 5.6.1 Inflammatory markers

The implant–bone interface is essentially a series of bone multicellular units, which are composed of osteoclasts, osteoblasts, and other cells of the MSC-osteoblast lineage. Chronic low-grade inflammation associated with products of implant wear suppresses bone formation by inhibiting the proliferation, differentiation, maturation, and function of progenitor cells and their downstream lineage cells. Ongoing inflammation also favours areas of cell necrosis and fibrosis, undermining the stability of the prosthesis and its ability to withstand physiological loads (Goodman and Gallo, 2019).

Inflammatory molecules involved in this process include TNF-α, IL-1, IL-6, IL-8, CRP, and CD cell surface molecules. Central chemokines (such as IL-8) are associated with implant debris reactivity as related to the innate immune system activation/cytokine expression through danger signalling (IL-1β), TLR activation (IL-6, TNF-α), and hypoxia responses (HIF-1α) (Hallab and Jacobs, 2017).

A systematic review by Sumner et al. (2014) summarised the findings of 22 studies that compared the levels of various inflammatory markers in patients with aseptic loosening and those with stable implants. In the case of TNF-α, only 3 out of 11 studies reported significantly higher levels in association with loose implants. Similarly, IL-1β was only elevated in 2 out of 11 studies, whereas IL-11 and sIL-2r failed to show an effect in any of the comparisons performed. Only a handful of studies assessed IL-6 and IL-8 levels: IL-6 was elevated in 3 out of 5 studies and IL-8 was elevated in both studies that assessed it. CRP was elevated in 1 of 2 studies, whereas the CD molecules were elevated in most comparisons (CD2, CD14, CD16, CD22, and CD25 elevated in 100% of the studies, CD4 and CD8 elevated in 50% of the studies). The authors concluded that the apparent insensitivity of these markers to detect periprosthetic osteolysis may reflect their early involvement in the disease course, with relatively minor roles in the end stage of the disease when most of the reviewed studies were conducted.

Ross et al. (2018) used a repository of 24-hour urine samples collected prior to surgery and annually thereafter in 26 patients, of whom 16 developed osteolysis and 10 did not. Seven candidate biomarkers were assessed for their ability to predict osteolysis prior to radiographic diagnosis. Levels of IL-6 ($P = 0.20$) were significantly lower in the osteolysis group preoperatively, but subsequently elevated at 6 ($P = 0.31$) and 4 ($P = 0.48$) years prior to radiographic diagnosis, and at the time of diagnosis ($P = 0.14$). IL-8 levels were comparable between the two groups at all postoperative timepoints. Of all the biomarkers evaluated in this study, IL-6 had the highest diagnostic accuracy preoperatively and at 1, 4, and 5 years prior to diagnosis (AUC: 0.775, 0.822, 0.768, and 0.741, respectively). These results demonstrate the potential of using non-invasive biomarkers to identify patients at risk for peri-implant osteolysis long before the emergence of radiographic signs.

### 5.6.2 Markers of bone turnover

Type I collagen accounts for 90% of organic bone mass, and using molecular biomarkers to assess its synthesis and degradation before and after THA might help identify those who are at risk of unbalanced bone remodelling after the procedure and monitor implant fixation in the long term (Savarino et al., 2005).

Measuring crosslinked N-terminal telopeptide (NTx) levels can shed light on the extent of the bone formation while C-terminal propeptide of type I procollagen (PICP), crosslinked C-terminal telopeptide of type I collagen (CTX-I), and deoxypyridinoline (DPD) are indicators of bone resorption. Markers of osteoblast differentiation (e.g., OPG) and osteoclast differentiation/activity (e.g., RANKL and TRAP 5b) have also been studied.

Osteocalcin is the most abundant non-collagenous protein in a mineralised bone matrix; its levels are often elevated in pathologies such as primary hyperparathyroidism, Paget's disease, and chronic renal failure. Plasma osteocalcin concentration generally correlates well with the histological measures of bone formation while that of osteocalcin fragments (released during bone resorption) can inform on the extent of bone loss.

Serum OPG provided good diagnostic accuracy in detecting implant failure in a study of 128 individuals (33 patients with stable implants, 36 patients with aseptic loosening of implant components, 39 controls with osteoarthritis, and 20 healthy volunteers) (Granchi et al., 2006). Positive and negative likelihood ratios were 7.1 (95% CI: 2.5–20.8) and 0.2 (95% CI: 0.1–0.4), respectively, suggesting that osteolysis was seven-fold more likely if serum OPG levels exceeded 2296 pg/mL. Additionally, the sum of osteolytic areas seen radiographically around the femoral stem was found to be moderately correlated with serum RANKL level ($r = 0.38$, $P = 0.2$) and the OPG/RANKL ratio ($r = -0.29$, $P = 0.4$).

Landgraeber et al. (2010) compared the serum levels of CTX-I and TRAP 5b in patients with established aseptic loosening versus patients with stable implants and a group of healthy volunteers. CTX-I values were comparable between the aseptic loosening group and the control group. In contrast, TRAP 5b values were significantly higher in patients with aseptic loosening, and a cut-off of 3.365 U/L differentiated a loose implant from an intact arthroplasty with a sensitivity of 83.3% and specificity of 91.7% (AUC: 0.958). In Lawrence et al. (2015), serum TRAP 5b >2.46 U/L showed a sensitivity of 100% and specificity of 31% to detect aseptic loosening (AUC: 0.67), resulting in a positive predictive value (PPV) of 57% and a negative predictive value (NPV) of 100%. Corresponding results for serum CTX-I >5.5 ng/L were sensitivity 91%, specificity 69% (AUC: 0.77), PPV 73%, and NPV 90%. The high NPVs point to the potential use of serum TRAP 5b and CTX-I as screening biomarkers for excluding aseptic loosening.

Sumner et al. (2014), who summarised the relevant biomarker literature published through December 2013, reported that TRAP 5b had the highest diagnostic accuracy (0.96; although a separate comparison for this biomarker yielded an accuracy of 0.76). Accuracy of >0.80 was also found for NTx (sensitivity 82%; specificity 87%), OPG (sensitivity 92%; specificity 75%), and DPD (sensitivity 83%; specificity 83%). Osteocalcin was considerably less accurate (0.67), and its sensitivity and specificity were also poorer (69% and 65%, respectively).

In the urine study by Ross et al. (2018), a combination of α-CTX and IL-6 was highly predictive of osteolysis at all postoperative timepoints

(AUC: $\geq$0.941) as well as preoperatively (AUC: 1.000). In terms of individual biomarkers, DPD had the highest ability to identify at-risk patients, with an AUC of 0.844 at 6 years prior to radiographic diagnosis.

### 5.6.3 Markers of oxidative stress

ROS are highly reactive by-products of mitochondrial respiration; additional ROS derive from reactions catalysed by redox-active metals (e.g., Fe, Co, and Cr) and NADPH oxidases. They are broadly classed into free radicals, such as hydroxyl radical ($\cdot OH$) and superoxide ($\cdot O_2^-$), and molecules that can generate free radicals, primarily hydrogen peroxide ($H_2O_2$). Depending on their concentrations, ROS can exert beneficial or deleterious effects: at low levels, they function as cellular messengers that help regulate gene expression, receptor activation, and other vital cellular processes, but can cause oxidative injury to proteins (carbonylation), lipids (peroxidation), and DNA (mutations, double-strand breaks) when present in excess. ROS levels are normally kept in check by specialised enzymes and other antioxidants. However, in certain situations, the rate of ROS generation can overwhelm the antioxidant defences of the body, resulting in a state of oxidative stress (Galliera et al., 2021). Several patient-specific factors related to the indication for THA, lifestyle choices, and comorbidities such as osteoarthritis, hypertension, and diabetes mellitus are associated with heightened oxidative stress. Pre-existing oxidative stress may influence the surgical outcome of THA and compromise implant longevity (Hameister et al., 2020).

ROS are considered the major common and final pathway of tissue damage in different organ systems. In bone metabolism, ROS are involved in both physiological and pathological bone processes. Under normal conditions, they accelerate the destruction of calcified tissue, assisting in physiological bone remodelling (Savvidis et al., 2020). In the presence of oxidative stress, they stimulate osteoclast production while inhibiting osteoblast differentiation, inducing TNF-$\alpha$ and other pro-inflammatory cytokines, and causing cell death in osteoblasts and osteocytes, with a net effect of bone erosion (Sheweita and Khoshhal, 2007).

After a bone fracture, the bone ends become ischaemic before gradually re-establishing the blood supply through fracture healing. Paradoxically, restoration of blood flow causes further damage to the tissues (referred to as 'ischaemia-reperfusion injury'), which is mediated via the generation of ROS by polymorphonuclear neutrophils. The ROS stimulate osteoclastic activity and inhibit fracture healing by initiating a chain reaction that causes lipid peroxidation and, eventually, cell membrane damage and cell lysis (Sheweita and Khoshhal, 2007).

Implantation of an orthopaedic device leads to the same ischemia-reperfusion pattern seen after a fracture, at least during the initial host response. Electrochemical reactions taking place at metallic implant surfaces contribute to the generation of free radicals and ROS, with material properties such as surface chemical composition, porosity, surface roughness, surface energy, surface modifications, and wettability influencing the amount and type of reactive species generated and their interaction with inflammatory cells (Hameister et al., 2020). Thus, localised oxidative stress is likely to occur after TJA—the ROS produced amplify chronic inflammation and promote both RANKL-induced osteoclastogenesis and osteoblast apoptosis, interfering with osseointegration and promoting periprosthetic tissue fibrosis and bone destruction.

Biomarkers of oxidative stress include various antioxidants as well as end products of protein carbonylation and lipid peroxidation. Glutathione is the master antioxidant responsible for removing free radicals and other toxins from the body. Besides key roles in the antioxidant defence and free radical scavenging, glutathione regenerates other important antioxidants, such as vitamins C and E, and maintains the redox state of critical protein sulfhydryl groups that are necessary for DNA repair. Under oxidative stress conditions, glutathione becomes oxidised and the ratio of reduced to oxidised glutathione drops. Assessing the levels of oxidised glutathione (GSSG) and reduced glutathione (GSH), as well as their ratio, can lend clues as to the amount of oxidative stress in the body (Savvidis et al., 2020). Another frequently used indicator of oxidative stress is malondialdehyde, an endogenous genotoxic product of lipid peroxidation and a potentially important contributor to protein damage and DNA mutations. GSSG and malondialdehyde were both shown to be more abundant in periprosthetic tissues from patients with aseptic loosening than those with stable implants, and the GSH/GSSG ratio was lower in the former group (Kinov et al., 2006).

In terms of antioxidant enzymes, superoxide dismutase-2 (SOD2) and peroxiredoxin-2 (PRDX2) were significantly upregulated in the capsular tissue from patients undergoing revision for radiologically confirmed aseptic loosening compared with control specimens from patients with polyethylene wear alone or individuals undergoing primary THA (SOD2: $P = 0.11$ and $P = 0.12$; PRDX2: $P = 0.12$ and $P = 0.15$, respectively), pointing to their potential role as biomarkers of aseptic loosening (Peng et al., 2011).

Other oxidative stress-related molecules with increased expression in aseptic loosening include COX-2, its oxidised product (4-HNE), and intercellular nitric oxide synthase isoform (iNOS) (Steinbeck et al., 2014).

In one study, the danger signalling protein and osteoclast differentiation factor HMGB1 (high mobility group box protein 1) was more abundant in tissues surrounding implants revised for osteolysis than those revised for non-osteolytic reasons, and its levels directly correlated with the severity of osteolysis (Steinbeck et al., 2014). The authors proposed that assessing serum HMGB1, along with plasma biomarkers of bone remodelling, could be a sensitive way to detect and monitor the progression of osteolysis.

Although there is ample evidence to support the role of oxidative stress in particle-induced periprosthetic osteolysis, further studies are needed to evaluate the sensitivity and specificity of serum oxidative stress biomarkers for predicting and diagnosing the condition in patients.

### 5.6.4 Single-nucleotide polymorphisms

Genetic polymorphisms are normal genetic variations that can be found in more than 1% of the population. Depending on where in the DNA sequence they are located (i.e., coding, regulatory, or non-coding regions), polymorphisms may influence protein transcription and can be used to predict an individual's response to a drug or susceptibility to a pathological condition.

Single-nucleotide polymorphism (SNP) is a type of genetic polymorphism that occurs when a single DNA base (G, C, A, or T) is substituted by another at a specific genomic position, giving rise to alternative versions (alleles) of the same gene. For each individual SNP, three genotypes are possible: XX (homozygous), YY (homozygous), or XY (heterozygous).

SNPs have been described within genes encoding various cytokines, cell receptors, intracellular mediators, and enzymes involved in bone metabolism, and some have been associated with susceptibility to osteolysis or time to aseptic failure after THA.

#### *5.6.4.1 Cytokines*

Pro-inflammatory cytokines such as TNF-a and IL-1 are heavily implicated in the initiation and propagation of the immune response to wear particles and recruitment and maturation of osteoclast precursors at the implant–bone interface. Genetic variations in genes encoding these proteins may be able to explain the variability in osteolysis severity and implant survival in patients with similar wear rates.

In Gallo et al. (2009), carriers of the rare TNF-238 A allele were six times more likely to develop severe acetabular osteolysis ($P = 0.5$) and had an elevated cumulative hazard of THA failure ($P = 0.24$) than GG homozygotes. The findings were in line with those of Wilkinson et al. (2003), who reported that presence of the TNF-238 A allele was associated with an approximately two-fold higher incidence of femoral and pelvic osteolysis, which was independent of other risk factors. Carriers of the IL6-17ca4 G allele were 2.5 times more likely to develop severe osteolysis that non-carriers ($P = 0.7$), albeit without a corresponding increase in risk of implant failure. Interestingly, the *IL2*-330 G allele seemed to confer protection from severe osteolysis (OR: 0.55; $P = 0.43$) and implant failure ($P = 0.19$) (Gallo et al., 2009).

In a preliminary study, Kolundzić et al. (2006) demonstrated a markedly increased risk of aseptic loosening associated with both the IL6-597 G→A and IL6-572 G→C transitions in combination (HR: 5.7; $P = 0.16$), but without a demonstrable influence on prosthesis survival. The authors also assessed the TGFβ1-27 T→C signal sequence transitions, pointing to a higher risk of developing aseptic loosening in TT homozygotes compared with carriers of the C allele (HR: 8.23; $P = 0.17$).

### 5.6.4.2 *Proteins, receptors, and intracellular mediators*

The RANKL/RANK/OPG system has been implicated in the biological cascade of events initiated by particulate wear debris around joint replacements. Malik et al. (2006) conducted a case-control study of 91 patients with early aseptic loosening and 150 patients with well-fixed stems to study the possible association of the OPG and RANK genes with aseptic failure. Significant associations were identified for the OPG-163 A allele ($P < 0.1$) and AA genotype ($P < 0.1$), and RANK +575 SNP T allele ($P = 0.4$) and TT genotype ($P = 0.8$).

The same authors reported that certain polymorphic positions within the gene encoding mannose-binding lectin (MBL) were also associated with aseptic failure; specifically, the C allele (OR: 2.23; $P = 0.1$) and CC genotype ($P = 0.4$) for the promoter −550 SNP, and the codon 54 SNP G allele (OR: 2.17; $P = 0.12$) and GG genotype ($P = 0.27$). MBL is a liver-derived serum protein whose production increases during the acute phase of inflammatory response (Malik et al., 2007a).

BCL-2 is a crucial controller of cell death that is also important in bone metabolism, where it may influence the process of aseptic loosening. While no significant influence of the BCL2−938 polymorphism on time to aseptic loosening was found in one study (Wedemeyer et al., 2009), a later analysis reported that the CC genotype was associated with a higher risk of aseptic loosening (OR: 1.93; 95% CI: 1.15−3.25; $P = 0.13$ vs AA), and that carriers of the low-risk AA genotype whose implants succumbed to aseptic failure showed a significantly shorter time to loosening than carriers of the C allele ($P = 0.4$) (Stelmach et al., 2016).

### 5.6.4.3 *Enzymes*

Matrix metalloproteinases (MMPs) are the enzymes responsible for bone cleavage during bone resorption. Chronic inflammation promotes an imbalance between bone metalloproteinases and their tissue inhibitors, leading to the overexpression of MMPs at the implant−bone interface, which fuels bone erosion (Jonitz-Heincke et al., 2016).

Godoy-Santos et al. (2009) assessed polymorphisms within the MMP-1 gene promoter among patients with early aseptic loosening. The allele 2 G was observed with 20.97% frequency in the control group, compared with 83.33% in the aseptic loosening group. The genotype 2 G/2 G was observed in 66.66% of patients with aseptic loosening, whereas the genotype 1 G/1 G was found in 67.74% of the control subjects. Therefore, individuals with the allele 2 G seemed to be more prone to implant loosening ($P < 0.1$).

Yan et al. (2014) demonstrated that MMP-1 rs5854 C/T polymorphism was associated with an increased risk of early aseptic loosening, particularly for carriers of the T allele (OR: 2.72; $P < 0.1$) and TT homozygotes (OR: 5.13; $P = 0.113$ vs CC genotype).

Malik et al. (2007b), who studied a number of different SNPs in 91 patients with aseptic loosening and 150 controls with stable, asymptomatic implants, reported that the MMP-1 C allele (T/C SNP in the 39 untranslated regions; OR: 3.27) and CC genotype ($P = 0.1$) were highly associated with aseptic failure, while none of the studied MMP2 or MMP4 SNPs yielded statistically significant results. The authors concluded that MMP-1 SNPs could serve as predictors of implant survival and aid in the pharmacological prevention of implant failure.

## 5.7 Summary and future directions

The long-term success of orthopaedic hip implants relies on a lasting bond between the implant and the host bone, which requires careful attention to implant choice, design, implantation technique, and host acceptability. Failure to establish this union through osseointegration or late osteolysis that culminates in aseptic loosening can cause pain and loss of function. Despite advances in the field, aseptic loosening remains the most common reason for revision surgery and carries a heavy morbidity and financial cost burden for the patient and health services.

Within clinical practice, we rely on patient-reported symptoms and examination findings to highlight a concern with a poorly performing implant. Imaging techniques have traditionally been preferred since the results provide spatial information that helps the surgeon both make a diagnosis and plan surgical intervention. However, traditional radiography tends to only demonstrate aseptic loosening once significant periprosthetic osteolysis has occurred. At that point, it is usually too late to prevent the need for revision surgery. Modalities such as nuclear imaging are more sensitive but lack the required specificity. Therefore, a need exists for highly sensitive and specific tests that would compliment routine radiographic surveillance to assist in the early detection of an implant at risk of aseptic failure. Timely diagnosis would afford the opportunity to prolong the life of the implant through the promotion of osseointegration, be it through biophysical stimulation or pharmacological means.

The feasibility of molecular biomarkers for the identification of aseptic loosening is an important consideration, with non-invasive testing of blood or urine being most likely to be accepted by clinicians and patients. Notably, the majority of clinical studies to date have focused on markers of end-stage disease, but what is really needed is a way to detect osteolysis early, before any radiographic signs appear and implant fixation becomes compromised. This aim is considered achievable but not yet realised, with a growing list of emerging local and circulating biomarkers, including pro-inflammatory cytokines, markers of bone turnover, and indicators of oxidative stress. Owing to the multifactorial nature of implant failure and the plethora of different cell types and signalling pathways involved in the inflammatory and bone-resorptive processes at the implant–bone interface, biomarker panels that include molecules from several of these groups are likely to be more valuable than any one of them in isolation (Sumner et al., 2014). This was highlighted by He et al., (2013), who only observed a significant difference between patients with aseptic loosening and those with stable implants when plasma levels of six biomarkers (TNF-α, IL-1β, OPG, RANKL, NTx, and PICP) were analysed together. Notably, serum levels of osteoclast activity markers may be increased by other bone-resorptive processes, such as advanced osteoporosis or bone metastasis, limiting their utility for evaluating aseptic loosening in certain patient populations (unless reference values obtained pre-THA or when the implant was stable are available).

While a large number of studies evaluated the role of molecular biomarkers, their findings rarely corroborate each other, possibly due to the variability in the assays used, imperfect classification of patients into study groups, and/or failure to select well-matched controls. In addition, prospective analyses, in which the patient's biomarker levels are monitored over time and related to disease progression are extremely rare. Thus, the main obstacle that remains is the identification of a biomarker, or their combination, that is sensitive, specific, and reproducible with regard to clinical utility. Various techniques have been used to measure target biomarkers, including immunoassays and, more recently, proteomic-based approaches that include mass spectrometric techniques. These newer techniques are the most likely source for the identification of novel circulating biomarkers of periprosthetic osteolysis (Wilson, 2019).

Considerable efforts have also been directed towards the development of genetic prognostic markers for aseptic loosening, which could help identify at-risk individuals prior to primary THA, allowing them to benefit from a personalised strategy aimed at maximising implant survivorship. Research in this area has focused on candidate gene-driven analyses but the importance of genome-wide association studies cannot be understated (Brüggemann et al., 2022). Despite significant associations of several SNPs with the risk for and time to aseptic loosening, none are assessed routinely.

The biomarkers highlighted in this chapter represent promising leads, but more well-designed, prospective, longitudinal studies will be needed to validate their clinical utility to detect early osteolysis or impending aseptic loosening.

## References

Apostu, D., Lucaciu, O., Berce, C., Lucaciu, D., Cosma, D., 2018. Current methods of preventing aseptic loosening and improving osseointegration of titanium implants in cementless total hip arthroplasty: a review. J. Int. Med. Res. 46, 2104–2119.

Aro, H.T., Alm, J.J., Moritz, N., Makinen, T.J., Lankinen, P., 2012. Low BMD affects initial stability and delays stem osseointegration in cementless total hip arthroplasty in women: a 2-year RSA study of 39 patients. Acta Orthop. 83, 107–114.

AOANJRR, 2020. Australian Orthopaedic Association National Joint Replacement Registry: Annual Report 2020.

Bannister, G. H. F., Nolan, J., Skinner, J., Hodgkinson, J., Timperley, J., et al., 2012. Primary Total Hip Replacement: A Guide to Good Practice.

Bartolomeu, F., Costa, M.M., Alves, N., Miranda, G., Silva, F.S., 2021. Selective laser melting of Ti6Al4V sub-millimetric cellular structures: prediction of dimensional deviations and mechanical performance. J. Mech. Behav. Biomed. Mater. 113, 104123.

Berber, R., Henckel, J., Khoo, M., Wan, S., Hua, J., Skinner, J., et al., 2015. Clinical usefulness of SPECT-CT in patients with an unexplained pain in metal on metal (MOM) total hip arthroplasty. J. Arthroplasty 30, 687–694.

Blum, A., Meyer, J.B., Raymond, A., Louis, M., Bakour, O., Kechidi, R., et al., 2016. CT of hip prosthesis: new techniques and new paradigms. Diagn. Interv. Imaging 97, 725–733.

Blum, A., Noel, A., Regent, D., Villani, N., Gillet, R., Teixeira, P.G., 2018. Tomosynthesis in musculoskeletal pathology. Diagn. Interv. Imaging. 99, 423–441.

Brånemark, P.I., Adell, R., Breine, U., Hansson, B.O., Lindstrom, J., Ohlsson, A., 1969. Intra-osseous anchorage of dental prostheses. I. experimental studies. Scand. J. Plast. Reconstr. Surg. 3, 81–100.

Brüggemann, A., Eriksson, N., Michaelsson, K., Hailer, N.P., 2022. Risk of revision after arthroplasty associated with specific gene loci: a genomewide association study of single-nucleotide polymorphisms in 1,130 twins treated with arthroplasty. J. Bone Jt. Surg. Am.

Buttaro, M.A., Onativia, J.I., Slullitel, P.A., Andreoli, M., Comba, F., Zanotti, G., et al., 2017. Metaphyseal debonding of the corail collarless cementless stem: report of 18 cases and case-control study. Bone Jt. J. 99-B, 1435–1441.

Chen, Y.L., Lin, T., Liu, A., Shi, M.M., Hu, B., Shi, Z.L., et al., 2015. Does hydroxyapatite coating have no advantage over porous coating in primary total hip arthroplasty? A meta-analysis. J. Orthop. Surg. Res. 10, 21.

Cherian, J.J., Jauregui, J.J., Banerjee, S., Pierce, T., Mont, M.A., 2015. What host factors affect aseptic loosening after THA and TKA? Clin. Orthop. Relat. Res. 473, 2700–2709.

Cnudde, P.H.J., Nemes, S., Bulow, E., Timperley, A.J., Whitehouse, S.L., Karrholm, J., et al., 2018. Risk of further surgery on the same or opposite side and mortality after primary total hip arthroplasty: a multi-state analysis of 133,654 patients from the Swedish hip arthroplasty register. Acta Orthop. 89, 386–393.

Couto, M., Vasconcelos, D.P., Sousa, D.M., Sousa, B., Conceição, F., Neto, E., et al., 2020. The mechanisms underlying the biological response to wear debris in periprosthetic inflammation. Front. Mater. 7, 274.

Damiati, L., Eales, M.G., Nobbs, A.H., Su, B., Tsimbouri, P.M., Salmeron-Sanchez, M., et al., 2018. Impact of surface topography and coating on osteogenesis and bacterial attachment on titanium implants. J. Tissue Eng. 9, 2041731418790694.

Dimitriou, R., Babis, G.C., 2007. Biomaterial osseointegration enhancement with biophysical stimulation. J. Musculoskelet. Neuronal Interact. 7, 253–265.

Eto, S., Hwang, K., Huddleston, J.I., Amanatullah, D.F., Maloney, W.J., Goodman, S.B., 2017. The direct anterior approach is associated with early revision total hip arthroplasty. J. Arthroplasty 32, 1001–1005.

Evans, J.T., Evans, J.P., Walker, R.W., Blom, A.W., Whitehouse, M.R., Sayers, A., 2019. How long does a hip replacement last? A systematic review and meta-analysis of case series and national registry reports with more than 15 years of follow-up. Lancet 393, 647–654.

Galliera, E., Massaccesi, L., Banfi, G., De Vecchi, E., Ragone, V., Corsi Romanelli, M.M., 2021. Effect of oxidative stress on bone remodeling in periprosthetic osteolysis. Clin. Rev. Bone Miner. Metab. 19, 14–23.

Gallo, J., Mrazek, F., Petrek, M., 2009. Variation in cytokine genes can contribute to severity of acetabular osteolysis and risk for revision in patients with ABG 1 total hip arthroplasty: a genetic association study. BMC Med. Genet. 10, 109.

Gillet, R., Teixeira, P., Bonarelli, C., Coudane, H., Sirveaux, F., Louis, M., et al., 2019. Comparison of radiographs, tomosynthesis and CT with metal artifact reduction for the detection of hip prosthetic loosening. Eur. Radiol. 29, 1258–1266.

Godoy-Santos, A.L., D'elia, C.O., Teixeira, W.J., Cabrita, H.B., Camanho, G.L., 2009. Aseptic loosening of total hip arthroplasty: preliminary genetic investigation. J. Arthroplasty 24, 297–302.

Goodman, S.B., Gallo, J., 2019. Periprosthetic osteolysis: mechanisms, prevention and treatment. J. Clin. Med. 8, 2091.

Granchi, D., Pellacani, A., Spina, M., Cenni, E., Savarino, L.M., Baldini, N., et al., 2006. Serum levels of osteoprotegerin and receptor activator of nuclear factor-kappaB ligand as markers of periprosthetic osteolysis. J. Bone Jt. Surg. Am. 88, 1501–1509.

Guo, S., Tang, H., Zhou, Y., Huang, Y., Shao, H., Yang, D., 2018. Accuracy of digital tomosynthesis with metal artifact reduction for detecting osteointegration in cementless hip arthroplasty. J. Arthroplasty 33, 1579–1587.

Hallab, N.J., Jacobs, J.J., 2017. Chemokines associated with pathologic responses to orthopedic implant debris. Front. Endocrinol. 8, 5.

Hameister, R., Kaur, C., Dheen, S.T., Lohmann, C.H., Singh, G., 2020. Reactive oxygen/nitrogen species (ROS/RNS) and oxidative stress in arthroplasty. J. Biomed. Mater. Res. B Appl. Biomater. 108, 2073–2087.

Haverkamp, D., Klinkenbijl, M.N., Somford, M.P., Albers, G.H., Van Der Vis, H.M., 2011. Obesity in total hip arthroplasty—does it really matter? A meta-analysis. Acta Orthop. 82, 417–422.

He, T., Wu, W., Huang, Y., Zhang, X., Tang, T., Dai, K., 2013. Multiple biomarkers analysis for the early detection of prosthetic aseptic loosening of hip arthroplasty. Int. Orthop. 37, 1025–1031.

Hirschmann, M.T., Konala, P., Iranpour, F., Kerner, A., Rasch, H., Friederich, N.F., 2011. Clinical value of SPECT/CT for evaluation of patients with painful knees after total knee arthroplasty—a new dimension of diagnostics? BMC Musculoskelet. Disord. 12, 36.

Hoberg, M., Konrads, C., Engelien, J., Oschmann, D., Holder, M., Walcher, M., et al., 2016. Similar outcomes between two-stage revisions for infection and aseptic hip revisions. Int. Orthop. 40, 459–464.

Inacio, M.C., Ake, C.F., Paxton, E.W., Khatod, M., Wang, C., Gross, T.P., et al., 2013. Sex and risk of hip implant failure: assessing total hip arthroplasty outcomes in the United States. JAMA Intern. Med. 173, 435–441.

Janssen, L., Wijnands, K.A.P., Janssen, D., Janssen, M.W.H.E., Morrenhof, J.W., 2018. Do stem design and surgical approach influence early aseptic loosening in cementless THA? Clin. Orthop. Relat. Res. 476, 1212–1220.

Jiang, Y., Jia, T., Wooley, P.H., Yang, S.Y., 2013. Current research in the pathogenesis of aseptic implant loosening associated with particulate wear debris. Acta Orthop. Belg. 79, 1–9.

Jonitz-Heincke, A., Lochner, K., Schulze, C., Pohle, D., Pustlauk, W., Hansmann, D., et al., 2016. Contribution of human osteoblasts and macrophages to bone matrix degradation and proinflammatory cytokine release after exposure to abrasive endoprosthetic wear particles. Mol. Med. Rep. 14, 1491–1500.

Kim, Y.H., Kim, J.S., Joo, J.H., Park, J.W., 2012. Is hydroxyapatite coating necessary to improve survivorship of porous-coated titanium femoral stem? J. Arthroplasty 27, 559–563.

Kinov, P., Leithner, A., Radl, R., Bodo, K., Khoschsorur, G.A., Schauenstein, K., et al., 2006. Role of free radicals in aseptic loosening of hip arthroplasty. J. Orthop. Res. 24, 55–62.

Kolundzić, R., Orlić, D., Trkulja, V., Pavelić, K., Troselj, K.G., 2006. Single nucleotide polymorphisms in the interleukin-6 gene promoter, tumor necrosis factor-alpha gene promoter, and transforming growth factor-beta1 gene signal sequence as predictors of time to onset of aseptic loosening after total hip arthroplasty: preliminary study. J. Orthop. Sci. 11, 592–600.

Kremers, H.M., Lewallen, E.A., Van Wijnen, A.J., Lewallen, D.G., 2016. Clinical factors, disease parameters, and molecular therapies affecting osseointegration of orthopedic implants. Curr. Mol. Biol. Rep. 2, 123–132.

Kurtz, S., Ong, K., Lau, E., Mowat, F., Halpern, M., 2007. Projections of primary and revision hip and knee arthroplasty in the United States from 2005 to 2030. J. Bone Jt. Surg. Am. 89, 780–785.

Lachiewicz, P.F., Soileau, E.S., Martell, J.M., 2016. Wear and osteolysis of highly crosslinked polyethylene at 10 to 14 years: the effect of femoral head size. Clin. Orthop. Relat. Res. 474, 365–371.

Landgraeber, S., Loer, F., Heep, H., Classen, T., Grabellus, F., Totsch, M., et al., 2010. Tartrate-resistant acid phosphatase 5b and C-terminal telopeptides of type I collagen as markers for diagnosis of aseptic loosening after total hip replacement. Arch. Orthop. Trauma Surg. 130, 441–445.

Lawrence, N.R., Jayasuriya, R.L., Gossiel, F., Wilkinson, J.M., 2015. Diagnostic accuracy of bone turnover markers as a screening tool for aseptic loosening after total hip arthroplasty. Hip Int. 25, 525–530.

Lindgren, V., Garellick, G., Karrholm, J., Wretenberg, P., 2012. The type of surgical approach influences the risk of revision in total hip arthroplasty: a study from the Swedish Hip Arthroplasty Register of 90,662 total hipreplacements with 3 different cemented prostheses. Acta Orthop. 83, 559–565.

Lohmann, C.H., Rampal, S., Lohrengel, M., Singh, G., 2017. Imaging in peri-prosthetic assessment: an orthopaedic perspective. EFORT Open. Rev. 2, 117–125.

Mahatma, M.M., Jayasuriya, R.L., Hughes, D., Hoggard, N., Buckley, S.C, Gordon, A, et al., 2021. Effect of denosumab on osteolytic lesion activity after total hip arthroplasty: a single-centre, randomised, double-blind, placebo-controlled, proof of concept trial. Lancet Rheumatol. 3, E195–E203.

Malchau, H., Herberts, P., Ahnfelt, L., 1993. Prognosis of total hip replacement in Sweden. Follow-up of 92,675 operations performed 1978–1990. Acta Orthop. Scand. 64, 497–506.

Malchau, H., Potter, H.G., Implant Wear Symposium Clinical Work Group, 2008. How are wear-related problems diagnosed and what forms of surveillance are necessary? J. Am. Acad. Orthop. Surg. 16 (Suppl 1), S14–S19.

Malik, M.H.A., Bayat, A., Jury, F., Ollier, W.E.R., Kay, P.R., 2006. Genetic susceptibility to hip arthroplasty failure association with the RANK/OPG pathway. Int. Orthop. 30, 177–181.

Malik, M.H.A., Bayat, A., Jury, F., Kay, P.R., Ollier, W.E.R, 2007a. Genetic susceptibility to total hip arthroplasty failure–positive association with mannose-binding lectin. J. Arthroplasty 22, 265–270.

Malik, M.H.A., Jury, F., Bayat, A., Ollier, W.E.R., Kay, P.R., 2007b. Genetic susceptibility to total hip arthroplasty failure: a preliminary study on the influence of matrix metalloproteinase 1, interleukin 6 polymorphisms and vitamin D receptor. Ann. Rheumatic Dis. 66, 1116–1120.

Mandalia, V., Eyres, K., Schranz, P., Toms, A.D., 2008. Evaluation of patients with a painful total knee replacement. J. Bone Jt. Surg. Br. 90, 265–271.

Massari, L., Osti, R., Lorusso, V., Setti, S., Caruso, G., 2015. Biophysical stimulation and the periprosthetic bone: is there a rationale in the use of pulsed electromagnetic fields after a hip or knee implant? J. Biol. Regul. Homeost. Agents 29, 1013–1015.

Morlock, M.M., Jager, M., 2017. Endoprostheses in the elderly: biomaterials, implant selection and fixation technique. Orthopade. 46, 4–17.

Niinomi, M., Nakai, M., 2011. Titanium-based biomaterials for preventing stress shielding between implant devices and bone. Int. J. Biomater. 2011, 836587.

NJR, 2020. National Joint Registry: 17th Annual Report.

Otsuki, B., Takemoto, M., Fujibayashi, S., Neo, M., Kokubo, T., Nakamura, T., 2006. Pore throat size and connectivity determine bone and tissue ingrowth into porous implants: three-dimensional micro-CT based structural analyses of porous bioactive titanium implants. Biomaterials 27, 5892–5900.

Overmann, A.L., Aparicio, C., Richards, J.T., Mutreja, I., Fischer, N.G., Wade, S.M., et al., 2020. Orthopaedic osseointegration: implantology and future directions. J. Orthop. Res. 38, 1445–1454.

Park, J.H., Lee, N.K., Lee, S.Y., 2017. Current understanding of RANK signaling in osteoclast differentiation and maturation. Mol. Cell 40, 706–713.

Pearl, J.I., Ma, T., Irani, A.R., Huang, Z., Robinson, W.H., Smith, R.L., et al., 2011. Role of the Toll-like receptor pathway in the recognition of orthopedic implant wear-debris particles. Biomaterials 32, 5535–5542.

Peng, K.T., Hsu, W.H., Shih, H.N., Hsieh, C.W., Huang, T.W., Hsu, R.W., et al., 2011. The role of reactive oxygen species scavenging enzymes in the development of septic loosening after total hip replacement. J. Bone Jt. Surg. Br. 93, 1201–1209.

Rames, R.D., Hillen, T.J., Pashos, G.E., Maloney, W.J., Clohisy, J.C., 2021. Incidence and characteristics of osteolysis in HXLPE THA at 16-year follow up in patients 50 years and less. J. Arthroplasty 36, 641–646.

Reiner, T., Sonntag, R., Kretzer, J.P., Clarius, M., Jakubowitz, E., Weiss, S., et al., 2020. The migration pattern of a cementless hydroxyapatite-coated titanium stem under immediate full weight-bearing – a randomized controlled trial using model-based RSA. J. Clin. Med. 9, 2077.

Ross, R.D., Deng, Y., Fang, R., Frisch, N.B., Jacobs, J.J., Sumner, D.R., 2018. Discovery of biomarkers to identify peri-implant osteolysis before radiographic diagnosis. J. Orthop. Res. 36, 2754–2761.

Russell, L.A., 2013. Osteoporosis and orthopedic surgery: effect of bone health on total joint arthroplasty outcome. Curr. Rheumatol. Rep. 15, 371.

Savarino, L., Granchi, D., Cenni, E., Baldini, N., Greco, M., Giunti, A., 2005. Systemic cross-linked N-terminal telopeptide and procollagen I C-terminal extension peptide as markers of bone turnover after total hip arthroplasty. J. Bone Jt. Surg. Br. 87, 571–576.

Savvidis, M., Papavasiliou, K., Taitzoglou, I., Giannakopoulou, A., Kitridis, D., Galanis, N., et al., 2020. Postoperative administration of alpha-tocopherol enhances osseointegration of stainless steel implants: an in vivo rat model. Clin. Orthop. Relat. Res. 478, 406–419.

Schenk, R.K., Buser, D., 1998. Osseointegration: a reality. Periodontol. 2000 17, 22–35.

Sheweita, S.A., Khoshhal, K.I., 2007. Calcium metabolism and oxidative stress in bone fractures: role of antioxidants. Curr. Drug Metab. 8, 519–525.

Sidambe, A.T., 2014. Biocompatibility of advanced manufactured titanium implants – a review. Mater. 7, 8168–8188.

Singh, J.A., Yu, S., Chen, L., Cleveland, J.D., 2019. Rates of total joint replacement in the united states: future projections to 2020–2040 using the national inpatient sample. J. Rheumatol. 46, 1134–1140.

Sneag, D.B., Bogner, E.A., Potter, H.G., 2015. Magnetic resonance imaging evaluation of the painful total knee arthroplasty. Semin. Musculoskelet. Radiol. 19, 40–48.

Steinbeck, M.J., Jablonowski, L.J., Parvizi, J., Freeman, T.A., 2014. The role of oxidative stress in aseptic loosening of total hip arthroplasties. J. Arthroplasty 29, 843–849.

Stelmach, P., Wedemeyer, C., Fuest, L., Kurscheid, G., Gehrke, T., Klenke, S., et al., 2016. The BCL2 -938C > A promoter polymorphism is associated with risk for and time to aseptic loosening of total hip arthroplasty. PLoS One 11, e0149528.

Su, J., Wei, Y., Li, X.M., Diao, Y.P., Liu, H.G., Zhang, L., 2018. Efficacy of risedronate in improving bone mineral density in patients undergoing total hip arthroplasty: a meta-analysis of randomized controlled trials. Medicine 97, e13346.

Sul, Y.T., Johansson, C.B., Kang, Y., Jeon, D.G., Albrektsson, T., 2002. Bone reactions to oxidized titanium implants with electrochemical anion sulphuric acid and phosphoric acid incorporation. Clin. Implant Dent. Relat. Res. 4, 78–87.

Sul, Y.T., Johansson, C., Wennerberg, A., Cho, L.R., Chang, B.S., Albrektsson, T., 2005. Optimum surface properties of oxidized implants for reinforcement of osseointegration: surface chemistry, oxide thickness, porosity, roughness, and crystal structure. Int. J. Oral Maxillofac. Implant. 20, 349–359.

Sumner, D.R., Ross, R., Purdue, E., 2014. Are there biological markers for wear or corrosion? A systematic review. Clin. Orthop. Relat. Res. 472, 3728–3739.

Tam, H.H., Bhaludin, B., Rahman, F., Weller, A., Ejindu, V., Parthipun, A., 2014. SPECT-CT in total hip arthroplasty. Clin. Radiol. 69, 82–95.

Taniguchi, N., Fujibayashi, S., Takemoto, M., Sasaki, K., Otsuki, B., Nakamura, T., et al., 2016. Effect of pore size on bone ingrowth into porous titanium implants fabricated by additive manufacturing: an in vivo experiment. Mater. Sci. Eng. C Mater. Biol. Appl. 59, 690–701.

Temmerman, O.P., Raijmakers, P.G., Berkhof, J., Hoekstra, O.S., Teule, G.J., Heyligers, I.C., 2005. Accuracy of diagnostic imaging techniques in the diagnosis of aseptic loosening of the femoral component of a hip prosthesis: a meta-analysis. J. Bone Jt. Surg. Br. 87, 781–785.

Teo, W.Z.W., Schalock, P.C., 2017. Metal hypersensitivity reactions to orthopedic implants. Dermatol. Ther. 7, 53–64.

Tian, P., Li, Z.J., Xu, G.J., Sun, X.L., Ma, X.L., 2017. Partial versus early full weight bearing after uncemented total hip arthroplasty: a meta-analysis. J. Orthop. Surg. Res. 12, 31.

Vanhegan, I.S., Malik, A.K., Jayakumar, P., Ul Islam, S., Haddad, F.S., 2012. A financial analysis of revision hip arthroplasty: the economic burden in relation to the national tariff. J. Bone Jt. Surg. Br. 94, 619–623.

Wagner, E.R., Kamath, A.F., Fruth, K.M., Harmsen, W.S., Berry, D.J., 2016. Effect of body mass index on complications and reoperations after total hip arthroplasty. J. Bone Jt. Surg. Am. 98, 169–179.

Wang, J., Tao, Y., Ping, Z., Zhang, W., Hu, X., Wang, Y., et al., 2016. Icariin attenuates titanium-particle inhibition of bone formation by activating the Wnt/beta-catenin signaling pathway in vivo and in vitro. Sci. Rep. 6, 23827.

Wedemeyer, C., Kauther, M.D., Hanenkamp, S., Nuckel, H., Bau, M., Siffert, W., et al., 2009. BCL2−938C > A and CALCA-1786T > C polymorphisms in aseptic loosened total hip arthroplasty. Eur. J. Med. Res. 14, 250−255.

Wegrzyn, J., Kaufman, K.R., Hanssen, A.D., Lewallen, D.G., 2015. Performance of porous tantalum vs. titanium cup in total hip arthroplasty: randomized trial with minimum 10-year follow-up. J. Arthroplasty 30, 1008−1013.

Wilkinson, J.M., Wilson, A.G., Stockley, I., Scott, I.R., Macdonald, D.A., Hamer, A.J., et al., 2003. Variation in the TNF gene promoter and risk of osteolysis after total hip arthroplasty. J. Bone Min. Res. 18, 1995−2001.

Wilson, B., 2019. Biomarkers of Compromised Implant Fixation (Ph.D. thesis). Rush University, Chicago, IL, USA.

Wolf, O., Mattsson, P., Milbrink, J., Larsson, S., Mallmin, H., 2010. Periprosthetic bone mineral density and fixation of the uncemented CLS stem related to different weight bearing regimes: a randomized study using DXA and RSA in 38 patients followed for 5 years. Acta Orthop. 81, 286−291.

Yan, Y., Hu, J., Lu, H., Wang, W., 2014. Genetic susceptibility to total hip arthroplasty failure: a case-control study on the influence of MMP 1 gene polymorphism. Diagn. Pathol. 9, 177.

Yellowley, C.E., Genetos, D.C., 2019. Hypoxia signaling in the skeleton: implications for bone health. Curr. Osteoporos. Rep. 17, 26−35.

Yoon, B.H., Park, J.W., Lee, Y.K., Koo, K.H., Chang, C.B., 2022. Long-term wear-related complications of cross-linked versus conventional polyethylene after total hip arthroplasty: a meta-analysis. J. Arthroplasty 37, 2308−2315.e2.

Chapter 6

# Biomarkers of periprosthetic joint infection

**Marc-Olivier Kiss and Vincent Massé**
*Department of Surgery, Hospital Maisonneuve-Rosemont, University of Montreal, Montreal, QC, Canada*

## 6.1 Introduction

Periprosthetic joint infection (PJI) is a devastating complication of joint replacement surgery and the third most common indication for revision total hip arthroplasty (THA) (Bozic et al., 2009). It is characterised by progressive inflammation in the joints, leading to persistent pain and stiffness at the arthroplasty site, impaired mobility, discharge from the surgical wound, fever, malaise, and substantially reduced quality of life. If left untreated, PJI can cause disability and/or death.

Although the reported incidence of PJI may vary depending on the definition used to classify the condition, in general, it is estimated that 0.4%–1% of primary THA recipients and 2%–3% of patients who have undergone revision THA develop deep PJI severe enough to warrant surgical revision (Carroll et al., 2020). When compared with primary THA and aseptic revision, septic revision procedures are associated with longer operating times, higher blood loss, increased likelihood of postoperative complications, and higher costs (Garfield et al., 2020). Given the rise in life expectancy and growing expectations for mobility among older patients, the demand for hip arthroplasty and implant indwelling time are on the rise. With a higher number of primary THAs being performed and longer implant time in situ, the incidence of PJI and the associated socio-economic, healthcare, and financial burden will also increase.

The management of PJI is technically demanding and requires collaboration between orthopaedic surgeons, infectious disease physicians, and microbiologists. Typically, treatment strategies are complex and involve debridement, one- or two-stage surgical revision of the implant, and long-term antimicrobial therapy. Timely and accurate diagnosis of PJI, with the identification of the infectious agent and its drug susceptibility, is crucial for

**Biomarkers of Hip Implant Function. DOI: https://doi.org/10.1016/B978-0-12-821596-8.00002-1**

selecting the most appropriate treatment. However, this is challenging because of the varied and non-specific nature of the presenting symptoms (joint dysfunction accompanied by a variable degree of pain), the lack of universally accepted criteria by which PJI is diagnosed, and the multitude of factors that can alter the diagnostic performance of the currently available assays and tests, including the patient's age, comorbidities, time from index arthroplasty, duration of symptoms, microorganism's virulence, culture-negative infection, and prior antibiotic use (Carli et al., 2019). Thus, treating surgeons not only have to select the appropriate investigations but also correctly interpret the results, taking into consideration their specific caveats and all aspects of the patient's health. If PJI is missed or undertreated, the patient may develop persistent infection and require multiple surgical revisions, resulting in poor function, disability, and higher short-term mortality risk. On the other hand, overdiagnosis risks inappropriate invasive treatment, which is associated with high morbidity, prolonged hospitalisation, extended periods of immobility, and significant healthcare costs.

A combination of imaging studies, laboratory testing, histopathology, and microbiology is required for the most accurate diagnosis of PJI. In this chapter, we summarise the most recent diagnostic recommendations for PJI, with a focus on the existing and emerging biomarkers and their application to clinical practice.

## 6.2 Periprosthetic joint infection

### 6.2.1 Pathogenesis and bacterial aetiology

The majority of PJI cases that occur in the first year after joint replacement are caused by the introduction of microorganisms onto the implant during surgery, either through direct contact between the implant and the environment or the patient's skin. The second most common pathogenesis of PJI is bacterial seeding from an existing infection elsewhere in the body via the bloodstream. Haematogenic infection typically originates from a respiratory, cardiovascular, cutaneous, urinary, dental, or gastrointestinal focus, and can occur at any point throughout the life of the implant. In the early postoperative period, a superficial wound infection may spread to the prosthesis. Contiguous spread may also occur later if implant-adjacent tissue becomes damaged due to trauma or surgery.

Once the invading bacteria have adhered to the implant, they begin to multiply and form microcolonies of increasing complexity. Within weeks, a bacterial biofilm is formed on the implant surface, which contains sessile colonies surrounded by a layer of self-produced extracellular polysaccharide matrix. The biofilm is a highly organised structure that supports bacterial communication, as well as providing protection from the host's immune system and antimicrobial agents. As it matures, the bacteria embedded within become more difficult to eradicate.

Depending on the time from index arthroplasty to the onset of first symptoms, PJI is classified as early ($\leq$3 months), delayed (3–24 months), or late ($>$24 months). Early PJI occurring within the first 4 weeks following implantation is commonly caused by high-virulent microorganisms, such as *Staphylococcus aureus*, streptococci, and enterococci, while delayed/late PJI occurring between 3 months and 3 years following implantation is commonly due to low-virulent microorganisms, such as coagulase-negative staphylococci (CNS) or *Cutibacterium acnes*. Gram-negative bacteria, anaerobes, fungi, and mycobacteria have also been implicated in PJI, albeit more rarely (Izakovicova et al., 2019).

Causative organisms are typically identified through microbiological testing of synovial fluid or periprosthetic tissue. Whilst a key element of PJI diagnosis, microbiological culture has poor sensitivity and yields false-negative results in 7%–50% of patients (Goswami and Parvizi, 2020). Cases of PJI in which no microorganisms are isolated from microbiological cultures are referred to as culture-negative infection. Factors related to culture-negative infection include premature antibiotic use, inability to recover biofilm-encapsulated bacteria, and presence of indolent, slow-growing, or fastidious pathogens. It is estimated that fungi and mycobacteria are responsible for 85% of culture-negative infections, with the remainder attributed to fastidious bacteria such as *Brucella* or *Coxiella burnetti* (Palan et al., 2019). Successful identification of fungi and atypical bacteria may necessitate the use of molecular microbiological techniques, such as polymerase chain reaction (PCR)-based assays or next-generation sequencing (NGS).

### 6.2.2 Clinical presentation

The clinical presentation of PJI varies widely, both in terms of symptom onset and clinical features, and can be summarised as shown in Table 6.1. Many of the symptoms experienced by patients with PJI are also seen in aseptic loosening, introducing the possibility of misdiagnosis and inappropriate treatment, which can have profound consequences for the patient.

Patient-related factors consistently associated with an increased risk of developing PJI after primary THA include obesity, diabetes mellitus, rheumatoid arthritis, congestive heart failure, male sex, and smoking (Lenguerrand et al., 2018).

### 6.2.3 Treatment

In early PJI occurring within 4 weeks of surgery, or acute haematogenous infections with symptoms lasting less than 3 weeks, DAIR (debridement, antimicrobial therapy, irrigation, and implant retention) has a fair chance of

**TABLE 6.1 The spectrum of clinical presentation of periprosthetic joint infection (Barrett and Atkins, 2014).**

| Classification | | Clinical features |
|---|---|---|
| Early<br><3 months from index surgery | Acute<br><3 weeks symptom duration | Acute pain, warmth, red/swollen joint, often with features of systemic sepsis (e.g., fever) |
| | Chronic<br>≥3 weeks symptom duration | Prolonged discharge from postoperative wound |
| Delayed/late<br>3 months from index surgery | Acute<br><3 weeks symptom duration | Acute pain, warmth, red/swollen joint, often with features of systemic sepsis (e.g., fever) |
| | Chronic<br>≥3 weeks symptom duration | Chronic pain ± sinus tract; prosthesis loosening may be apparent on X-rays |

*Source*: Based on Barrett, L., Atkins, B., 2014. The clinical presentation of prosthetic joint infection. J. Antimicrob. Chemother. 69, i25–i27.

success since the biofilm is not yet fully developed and antibiotics are more effective. In cases of delayed or late PJI, the treatment approach takes into account that mature biofilm is likely to be present on the prosthesis—most of the time, prosthesis removal, radical debridement of the infected tissue, and fitting of a new prosthesis with subsequent long-term antibiotic treatment is performed. The use of antibiofilm-active agents (antibiotics that are able to penetrate the biofilm and destroy the bacteria within) is often necessary. In carefully selected cases, the treating surgeon may opt for lifelong antibiotic suppression without surgery, or prosthesis removal with no reimplantation. The best course of action in any given case will be determined based on the characteristics of the infecting pathogen, its drug susceptibility, the patient's general condition (age, frailty, comorbidities, extent of tissue damage, and quality of bone stock), and surgeon experience.

## 6.3 Clinical definition of periprosthetic joint infection

The criteria by which PJI is diagnosed determine the treatment strategy and, thus, have a strong impact on epidemiological, socio-economic, and medico-legal evaluations. Despite the collaborative efforts of highly regarded scientific institutions and societies, a single, universally accepted definition of PJI remains elusive. The diagnostic algorithms and the associated biomarker thresholds have continued to evolve over the past decade, causing considerable confusion in the orthopaedic community and making meaningful comparisons of the published literature difficult.

The first attempt to standardise the PJI definition was made in 2011 by the Musculoskeletal Infection Society (MSIS), who produced a set of major and minor diagnostic criteria featuring the classical blood and synovial biomarkers of inflammation in addition to histological and microbiological assessment of the periprosthetic tissue and synovial fluid (Parvizi et al., 2011). In 2013, the MSIS definition was endorsed at the International Consensus Meeting on Musculoskeletal Infection (ICM) and released with slight modifications. The modified recommendations provided separate diagnostic thresholds for identifying acute and chronic infection (occuring less or more than 90 days from the index surgery, respectively), and added a positive synovial leukocyte esterase test to the minor criteria (Parvizi et al., 2013).

In 2013, the Infectious Diseases Society of America (IDSA) published its own evidence- and expert opinion-based clinical practice guidelines for the diagnosis and management of PJI (Osmon et al., 2013). Even though this definition has been cited a number of times in the literature, its clinical efficacy has not been evaluated.

In 2018, the MSIS definition was further reworked to reflect newly accrued evidence and incorporate promising biomarkers such as serum D-dimer, synovial alpha-defensin, and synovial C-reactive protein (CRP). The new system comprised a set of major and minor criteria, the latter of which was divided into preoperative tests and postoperative investigations to be considered when the preoperative diagnosis is inconclusive. Based on their relative weights, each test within the minor criteria was assigned a score, and the aggregate score determined the likelihood of PJI. The scoring system was validated in an external patient cohort, revealing an overall sensitivity and specificity of 97.7% and 99.5%, respectively. The criteria were more sensitive than the 2013 ICM definition (86.9%) and the 2011 MSIS definition (79.3%) while retaining excellent specificity (Parvizi et al., 2018). The validated system was modified several months after the conclusion of the clinical trial. The modified version, which excluded synovial CRP, scored synovial fluid biomarkers differently, and combined the preoperative and postoperative investigations into one minor criteria set only reached a 'weak consensus' at the reconvened ICM and was not endorsed by the MSIS (Parvizi and Gehrke, 2018).

To overcome some of the limitations of existing PJI definitions, the World Association against Infection in Orthopaedics and Trauma (WAIOT) developed an alternative approach to diagnosing the condition, based on the patient's clinical symptoms and the relative ability of currently available tests to confirm or exclude PJI (Romanò et al., 2019). The balance between positive 'rule in' tests and negative 'rule out' tests allowed to classify patients with suspected infection into one of the five groups: high-grade PJI, low-grade PJI, biofilm-related implant malfunction, contamination, or no infection. The definition included several new biomarkers and was the first to recommend nuclear imaging studies to inform the diagnosis. A subsequent multi-centre, retrospective validation study in 210 consecutive patients concluded that the 2019 WAIOT definition was a reliable tool to distinguish patients with and without PJI at a minimum of 90 days from index THA or total knee arthroplasty (TKA) (Bozhkova et al., 2020).

In 2018, the European Bone and Joint Infection Society (EBJIS) proposed yet another diagnostic definition of PJI, whose modified version was recently endorsed by the MSIS and the European Society of Clinical Microbiology and Infectious Diseases Study Group for Implant-Associated Infections (ESGIAI) (McNally et al., 2021; Renz et al., 2018). The criteria adopt a 'traffic-light' system that classifies infection as confirmed (red), likely (orange), or unlikely (green), based on the significance of each test and, unlike previous definitions, does not differentiate PJI based on the duration of the infection or its severity. The emphasis on simple, inexpensive, and widely available tests aims to facilitate clinical application of the criteria while increasing the sensitivity of diagnosis and limiting false-positive results.

As the EBJIS definition awaits validation studies, the 2013 ICM and 2018 MSIS criteria remain the most commonly used for clinical and research purposes.

The major published definitions of PJI are summarised in Table 6.2 and Table 6.3, and the different diagnostic tests featured within them are explained in Section 6.4.

## 6.4 Diagnostic categories

### 6.4.1 Clinical symptoms

Pain and reduced range of movement are the most sensitive clinical findings in infected cases, but they can be easily misinterpreted as aseptic implant failure. The only definitive clinical signs confirming infection are the presence of a sinus tract, exposed prosthesis, or visible purulence around the prosthesis (if other likely causes, such as adverse tissue reaction to implant debris, inflammatory arthritis, and crystal arthropathy, are excluded). Other variables that should raise suspicion of infection include history of impaired wound healing after primary implantation, multiple surgeries in the index joint, recent bacteraemia, and presence of erythema and/or tachycardia (McNally et al., 2021).

### 6.4.2 Imaging studies

Conventional imaging studies rarely have a definitive role in the diagnosis of PJI and are mainly used to rule out complications that may produce

**TABLE 6.2** Summary of the diagnostic criteria featured in the five major periprosthetic joint infection (PJI) definitions.

| | Criteria | | Diagnosis |
|---|---|---|---|
| MSIS (2011) | *Major:*<br>1. Sinus tract communicating with the prosthesis<br>2. Pathogen isolated by culture from ≥2 separate tissue/fluid samples from the affected joint | *Minor:*<br>1. Elevated ESR (>30 mm/h) AND serum CRP (>10 mg/L)<br>2. Elevated synovial leukocyte count (no specific threshold given)<br>3. Elevated PMN% (no specific threshold given)<br>4. Purulence in the affected joint<br>5. Microorganism isolated in one culture of periprosthetic tissue/synovial fluid<br>6. Positive histology (>5 neutrophils/HPF in 5 HPFs on periprosthetic tissue at 400× magnification) | PJI confirmed if 1 of 2 major criteria met OR ≥4 of 6 minor criteria met<br>PJI possible if <4 minor criteria met |
| IDSA (2013) | 1. Sinus tract communicating with the prosthesis<br>2. Purulence without other aetiology surrounding the prosthesis<br>3. Acute inflammation on histopathological examination of periprosthetic tissue<br>4. ≥2 intraoperative cultures or combination of preoperative aspiration and intraoperative cultures yielding an indistinguishable organism<br>5. Growth of a virulent organism, such as *S. aureus*, in a single specimen of a tissue biopsy or synovial fluid | | PJI confirmed if ≥1 criteria met<br>PJI possible if no criteria met |
| ICM (2013) | *Major:*<br>1. Sinus tract communicating with the prosthesis<br>2. Two positive periprosthetic cultures with phenotypically identical organisms | *Minor:*<br>1. Elevated ESR (not helpful in acute infections; >30 mm/h for chronic PJI) AND serum CRP (>100 mg/L for acute PJI; >10 mg/L for chronic PJI)<br>2. Elevated synovial fluid WBC count (>10,000 cells/μL for acute PJI; >3000 cells/μL for chronic PJI) OR + or ++ result on LE test strip (for both acute and chronic PJI)<br>3. Elevated PMN% (>90% for acute PJI; >80% for chronic PJI)<br>4. Positive histology (>5 neutrophils/HPF in 5 HPFs on periprosthetic tissue at 400× magnification in both acute and chronic PJI)<br>5. Single positive culture (both acute and chronic PJI) | PJI confirmed if ≥1 major criteria OR ≥3 minor criteria met<br>PJI possible without meeting the above criteria, especially in case of less virulent organisms (e.g., *C. acnes*) |

(*Continued*)

**TABLE 6.2 (Continued)**

| | Criteria | | | Diagnosis |
|---|---|---|---|---|
| MSIS (2018)[a] | *Major:*<br>1. Sinus tract with evidence of communication to the joint or visualisation of the prosthesis<br>2. Two positive growths of the same organism using standard culture methods | *Minor preoperative:*<br>1. Elevated serum CRP (> 100 mg/L for acute PJI; >10 mg/L for chronic PJI) OR D-dimer (unknown threshold for acute PJI; >860 ng/mL FEU for chronic PJI) (*score: 2*)<br>2. Elevated ESR (no role in acute PJI; >30 mm/h for chronic PJI) (*score: 1*)<br>3. Elevated synovial WBC count (>10,000 cells/μL for acute PJI; >3000 cells/μL for chronic PJI) OR positive LE (+ + for acute and chronic PJI) (*score: 3*)<br>4. Elevated PMN% (>90% for acute PJI; >80% for chronic PJI) (*score: 2*)<br>5. Positive alpha-defensin test (signal-to-cut-off ratio >1.0 in both acute and chronic PJI) (*score: 3*)<br>6. Elevated synovial CRP (>6.9 mg/L for acute PJI; unknown threshold for chronic PJI) (*score: 1*) | *Minor intraoperative*[b]*:*<br>1. Preoperative score (score: ?)<br>2. Positive histology (>5 neutrophils/HPF in 5 HPFs on periprosthetic tissue at 400 × magnification) (*score: 3*)<br>3. Positive intraoperative purulence (*score: 3*)<br>4. Single positive culture (*score: 2*) | PJI confirmed if ≥1 major criteria met OR total preoperative score ≥6<br>Possibly infected if total preoperative score 2–5[b]<br>Not infected if total preoperative score ≤1 |

| | | | | |
|---|---|---|---|---|
| EBJIS (2020) | *Red:*<br>1. Sinus tract with evidence of communication to the joint or visualisation of the prosthesis<br>2. Increased synovial leukocyte count (>3000 cells/μL)[d,e]<br>3. Elevated PMN% (>80%)[d,e]<br>4. Positive alpha-defensin immunoassay or lateral flow assay[i]<br>5. ≥2 positive samples with the same microorganism in intraoperative fluid/tissue[f]<br>6. >50 CFU/mL of any organism in sonication fluid[h,f] | *Orange:*<br>1. Radiological signs of loosening within the first 5 years after implantation OR previous wound healing problems OR history of recent fever or bacteraemia OR purulence around the prosthesis[c]<br>2. Elevated serum CRP (>10 mg/L)[d]<br>3. Elevated synovial WBC count (>1500 cells/μL)[d,e]<br>4. Elevated PMN% (>65%)[d,e]<br>5. Positive aspiration fluid culture[f]<br>6. Single positive culture[f,g]<br>7. >1 CFU/mL of any organism in sonication fluid[f,g,h]<br>8. Positive histology (≥5 neutrophils in a single HPF)[d,j]<br>9. Positive WBC scintigraphy[k] | *Green:*<br>1. Clear alternative reason for implant dysfunction (e.g., fracture, implant breakage, malposition, tumour)<br>2. Synovial WBC count ≤1500 and PMN% ≤65%[d,e]<br>3. All intraoperative cultures negative[f]<br>4. No growth in sonication fluid culture[f,h]<br>5. Negative histology[d,j]<br>6. Negative three-phase isotope scan[d] | PJI confirmed if ≥1 red tests positive<br>PJI likely if there is a positive clinical feature or elevated serum CRP, together with another orange test; consider further comprehensive investigation<br>PJI unlikely if all green criteria met |

(*Continued*)

**TABLE 6.2 (Continued)**

| | Criteria | | | Diagnosis |
|---|---|---|---|---|
| | 7. Positive histology (≥5 neutrophils/HPF in ≥5 HPFs)[d,j]<br>8. Presence of visible microorganisms in periprosthetic tissue[d,j] | | | |

*ALTR*, adverse local tissue reaction; *CFU*, colony-forming units; *CRP*, C-reactive protein; *EBJIS*, European Bone and Joint Infection Society; *EDTA*, ethylenediaminetetraacetic acid; *ESR*, erythrocyte sedimentation rate; *FEU*, fibrinogen-equivalent units; *ICM*, International Consensus Meeting on Musculoskeletal Infection; *IDSA*, Infectious Diseases Society of America; *LE*, leukocyte esterase; *MSIS*, Musculoskeletal Infection Society; *NGS*, next-generation sequencing; *PMN%*, synovial polymorphonuclear cell percentage; *WBC*, white blood cell.

[a] *Proceed with caution in cases of ALTR, crystal deposition disease, or slow-growing organisms.*

[b] *In ambiguous cases, intraoperative criteria can be used to refute or confirm PJI: if total intraoperative score ≥6, infected; 4–5, inconclusive (consider further molecular diagnostics, such as NGS); ≤3, not infected.*

[c] *Except in ALTR or crystal arthropathy cases.*

[d] *Interpret with caution when other possible causes of inflammation are present, such as crystal deposition disease, metallosis, periprosthetic fracture, active inflammatory joint disease (e.g., rheumatoid arthritis), or early postoperative period.*

[e] *Parameters are only valid when clear fluid is obtained and no lavage has been performed. Volume for the analysis should be >250 μL, ideally 1 mL, collected in an EDTA-containing tube and analysed in <1 hour, preferentially using automated techniques. For viscous samples, pretreatment with hyaluronidase improves the accuracy of optical or automated techniques. In case of bloody samples, the adjusted synovial WBC = synovial WBC observed − [WBC blood/RBC blood × RBC synovial fluid] should be used.*

[f] *Results of microbiological analysis may be compromised by prior antibiotic treatment (not simple prophylaxis). In these cases, molecular techniques may have a place. Results of culture may be obtained from preoperative synovial aspiration, preoperative synovial biopsies, or (preferred) from intraoperative tissue samples.*

[g] *Interpretation of single positive culture (or <50 CFU/mL in sonication fluid) must be cautious and taken together with other evidence. If a preoperative aspiration identified the same microorganism, they should be considered as two positive confirmatory samples. Uncommon contaminants or virulent organisms (e.g., S. aureus or gram-negative rods) are more likely to represent infection than common contaminants (such as coagulase-negative staphylococci, micrococci, or C. acnes).*

[h] *If centrifugation is applied then the suggested cut-off is 200 CFU/mL to confirm infection—if other variations to the protocol are used then the published cut-offs for each protocol must be applied.*

[i] *Not valid in cases of ALTR, haematomas, acute inflammatory disease, or gout.*

[j] *Histological analysis can be from preoperative biopsy or intraoperative tissue samples, with either paraffin or frozen section preparation.*

[k] *Uptake at the 20-hour scan is increased compared with the earlier scans (especially when combined with complementary bone marrow scan).*

*Source*: Based on McNally et al. (2021), Parvizi et al. (2018, 2013, 2011), and Osmon et al. (2013).

**TABLE 6.3** The WAIOT-proposed definition of periprosthetic joint infection (PJI).

| | Preoperative diagnosis | | | | |
|---|---|---|---|---|---|
| | **No infection** | **Contamination** | **Biofilm-related implant malfunction** | **Low-grade PJI** | **High-grade PJI** |
| Clinical presentation | One or more condition(s) other than infection can cause the symptoms or be the reason for reoperation (e.g., wear debris, metallosis, recurrent dislocation or joint instability, fracture, malposition, neuropathic pain) | | One or more of the following: otherwise 'unexplained' pain, swelling, stiffness | | Two or more of the following: pain, swelling, redness, warmth, *functio laesa* |
| Total 'rule in' score minus total 'rule out' score | <0 | <0 | <0 | ≥0 | ≥1 |
| Postoperatively confirmed if: | Negative cultural examination | One preoperative or intraoperative positive culture, with negative histology | Positive culture examination (preferably with antibiofilm techniques) and/or positive histology | | |
| | **Preoperative and intraoperative tests[a]** | | | | |
| | ***'Rule out' tests (sensitivity >90%): each negative test scores -1; positive test scores 0*** | | ***'Rule in' tests (specificity >90%): each positive test scores +1; negative test scores 0*** | | |
| Clinical examination | – | | Purulence or draining sinus or exposed joint prosthesis[b] | | |
| Serum | CRP (>10 mg/L); ESR (>30 mm/h) | | IL-6 (>10 pg/mL); procalcitonin (>0.5 ng/mL); D-dimer (>850 ng/mL) | | |
| Synovial fluid | WBC (>1500 cells/μL); leukocyte esterase (+ +); alpha-defensin immunoassay (>5.2 mg/L) | | Culture examination; WBC count (>3000 cells/μL); leukocyte esterase (+ +); alpha-defensin immunoassay (>5.2 mg/L) or lateral flow test | | |
| Imaging | $^{99m}$Tc bone scintigraphy | | $^{99m}$Tc WBC scintigraphy combined with bone marrow scintigraphy | | |
| Histology | – | | Five neutrophils in ≥3 HPFs in a frozen tissue section | | |

*CRP*, C-reactive protein; *ESR*, erythrocyte sedimentation rate; HPF, high-power field (400×); *IL-6*, interleukin-6; *WBC*, white blood cell.
Preoperative and intraoperative tests are classified according to their sensitivity and specificity and, hence, their ability to exclude (rule out) or to confirm (rule in) PJI. Reference cut-off values are shown in parentheses (Romanò et al., 2019).
[a]*A minimum of two 'rule in' tests and two 'rule out' tests should be performed in a given patient to define the presence of PJI. If concurrent local or systemic inflammatory conditions are present, more than two 'rule in' and two 'rule out' tests should be performed, and the tests should be repeated to evaluate their pattern over time or once the confounding inflammatory condition has settled.*
[b]*If present, consider as infected.*
*Source*: Adapted with permission from The Journal of Clinical Medicine/CC-BY-4.0.

infection-like symptoms. Radiographs are often normal despite the presence of infection and are more useful in assessing for periprosthetic fracture, dislocation, and osteolysis. While soft-tissue swelling and periprosthetic lucency are commonly seen in aseptic loosening as well as infection, gas formation and presence of active, immature periostitis may reveal PJI with high specificity (Signore et al., 2019) (Fig. 6.1). Ultrasound can also detect abnormalities around the prosthesis, but its capacity for detecting PJI is debated. Advanced imaging studies such as computed tomography and magnetic resonance imaging (MRI) have a high spatial resolution, allowing for better evaluation of bone and periprosthetic soft tissue. Metal artefact reduction sequence MRI is particularly valuable in the differential diagnosis of metallosis-related adverse local tissue reactions (ALTR).

The emerging role of nuclear imaging in the diagnosis of PJI is reflected in the inclusion of bone scintigraphy and white blood cell (WBC) scintigraphy within both the 2019 WAIOT and 2020 EBJIS criteria (Romanò et al., 2020). In bone scintigraphy, patients are injected with a $^{99m}$Tc-labelled tracer

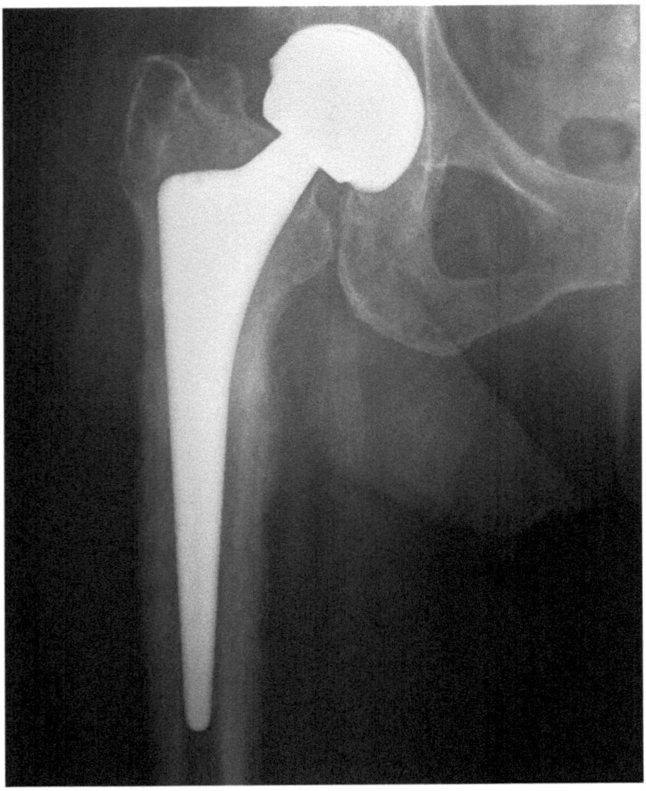

**FIGURE 6.1** Conventional radiography showing signs of PJI: soft-tissue swelling, active periostitis, and implant loosening with distal migration of the femoral stem.

before undergoing a three-phase scan to assess perfusion, blood pool, and uptake of the tracer by bone. Increased activity in the periprosthetic area reflects bone remodelling around the implant, which can be indicative of conditions such as aseptic loosening or infection. Although the specificity of a positive result is low for PJI—particularly during the first 2 years after hip arthroplasty when physiological bone remodelling typically takes place—a negative result rules out infection with high certainty. In radiolabelled WBC scintigraphy, the patient's leukocytes are extracted, tagged with radionuclides, then injected back into the patient. The radiolabelled cells migrate to the sites of inflammation and their distribution helps to differentiate between aseptic loosening and infection in a painful THA. The technique has an estimated diagnostic accuracy of >90% when combined with bone marrow scintigraphy, and is more reliable in the early postoperative period than three-phase bone scintigraphy.

### 6.4.3 Blood biomarkers

Thanks to its wide availability, repeatability, and relatively low invasiveness, blood testing remains a fundamental part of the evaluation for many disease states, including PJI. The following paragraphs describe the traditional and emerging blood biomarkers used to inform the diagnosis of PJI.

#### *6.4.3.1 C-reactive protein and erythrocyte sedimentation rate*

CRP is a positive acute-phase reactant synthesised in the liver in response to cytokine stimulation. It acts as an opsonin for invading microorganisms, activates the classical complement pathway, and, as such, forms an integral part of the innate immune system. Serum CRP levels tend to rise sharply in response to acute inflammatory stimuli, with the magnitude of the increase proportional to the intensity of the insult, then drop quickly once inflammation subsides. Aside from infection, factors that stimulate CRP synthesis include systemic autoimmune diseases, trauma, neoplasms, smoking, and obesity. In Kim et al. (2016), only 24% of cases of elevated CRP levels after primary TKA were due to postoperative infection, while 20% were idiopathic, and the remainder were caused by urologic, gastrointestinal, vascular, or respiratory disorders.

Serum CRP is the most accurate serum biomarker for PJI, with reported sensitivity and specificity of 74%–94% and 20%–100%, respectively (Saleh et al., 2018). Following THA, CRP levels typically peak on postoperative day 2 but remain elevated for another 2 weeks, reflecting postinterventional inflammation. This means that measurements taken in the early postoperative period should be interpreted with caution (Parvizi et al., 2010). It should also be kept in mind that concurrent use of systemic antibiotics or glucocorticoids can decrease serum CRP levels, potentially compromising its diagnostic value in PJI.

Despite the poor specificity, serum CRP >10 mg/L has prominence in all previous PJI definitions.

Erythrocyte sedimentation rate (ESR) is the rate at which red blood cells settle to the bottom of the test tube in an anticoagulated blood specimen, and is usually reported in millimetres per hour. Abnormally high values reflect active inflammation, proving useful in the diagnosis and monitoring of conditions such as rheumatoid arthritis and infection. The ESR is higher in females than in males, and increases with age in adults. The parameter can also be influenced by various metabolic and lifestyle factors, such as obesity, alcoholism, smoking, and drug use, and tends to be elevated in anaemia (Alende-Castro et al., 2019). As with CRP, previous administration of systemic antibiotics or anti-inflammatory treatment may affect the ESR and lower its ability to detect PJI in those who have it.

The diagnostic performance of the ESR in PJI is similar to that of serum CRP, with reported sensitivity and specificity of 42%–94% and 33%–87%, respectively (Saleh et al., 2018). However, unlike CRP, the ESR has no role in suspected acute PJI because it can remain elevated for up to 6 weeks after THA (Parvizi et al., 2010).

ESR >30 mm/h features in all published definitions of PJI except for the 2020 EBJIS criteria, where it was judged to offer no additional value over serum CRP (McNally et al., 2021).

The ESR has shown improved sensitivity and specificity for PJI when combined with serum CRP, with a reported sensitivity and specificity of 89%–96% and 56%–72%, respectively (Austin et al., 2008; Bingham et al., 2020). The high diagnostic performance of the combined test was demonstrated in a study of 414 THA recipients undergoing surgical treatment for PJI, in which preoperative ESR and serum CRP concentration were both normal in only 4% of the cases (McArthur et al., 2015). Individuals testing negative for both biomarkers had a high prevalence of slow-growing pathogens, such as CNS and *C. acnes*, which may elicit a relatively mild inflammatory response that fails to raise the inflammatory biomarkers enough to satisfy PJI criteria. Lowering the diagnostic threshold from 10 mg/L to 5 mg/L for serum CRP and from 30 mm/h to 10 mm/h for the ESR maximises the sensitivity of the combined test to help avoid missed PJI (Bingham et al., 2020).

Because of their familiarity, low cost, wide availability, and rapid turnaround time, diagnostic guidelines from the MSIS and ICM recommend serum CRP and ESR as first-line tests for suspected PJI. However, both markers only reflect general inflammation and have limited utility in the early postoperative period and low-grade infection (Pérez-Prieto et al., 2017). As a result, they cannot be used in isolation to confirm or exclude PJI and must be supplemented by synovial and/or microbiological testing.

#### 6.4.3.2 D-dimer

D-dimer is a fibrin degradation product generated after a blood clot is dissolved by plasmin. High D-dimer concentration reflects hypercoagulation and thrombus formation, making it a good initial screening test for patients with suspected venous thromboembolism (VTE). More recently, D-dimer has garnered attention for its role in detecting increased fibrinolytic activity that accompanies systemic inflammation and infection, and an increasing number of studies evaluate its diagnostic value in PJI.

Following THA, serum D-dimer levels peak on postoperative day 1, return to baseline the following day, then slowly rise again until 2 weeks postoperatively and normalise by week 6. Because D-dimer levels fluctuate more rapidly than CRP levels and the ESR in the early postoperative period, serial measurement of D-dimer concentration may be more helpful in the detection of early acute PJI. The lack of a rapid decrease in D-dimer levels by postoperative day 2–3 is considered abnormal after primary THA. Conversely, low D-dimer levels in the 6 weeks after the surgery may be reassuring in cases of suspected PJI (Lee et al., 2018).

In Shahi et al. (2017a), serum D-dimer outperformed serum CRP and ESR in predicting PJI, with a sensitivity of 89% and specificity of 93% at a 850 ng/mL cut-off. Encouraging early results prompted the incorporation of D-dimer into the 2018 MSIS and 2019 WAIOT definitions of PJI to maximise the sensitivity of diagnosis (Parvizi et al., 2018; Romanò et al., 2019). However, a recent meta-analysis of eight studies involving 1587 patients with PJI concluded that the marker had limited performance for the diagnosis of PJI (Lu et al., 2020). The thresholds used in the reviewed studies ranged from 760 to 1250 ng/mL, and the pooled sensitivity and specificity was 82% and 70%, respectively—poorer than those of serum CRP (85% and 81%, respectively) and ESR (82% and 79%, respectively) (Carli et al., 2019). In the 2020 EBJIS guidelines for PJI, D-dimer was determined to offer no additional value over the traditional blood inflammatory biomarkers and was excluded from the diagnostic criteria (McNally et al., 2021).

Aside from controversial diagnostic accuracy, there are other issues that hinder the clinical adoption of D-dimer in the setting of PJI (Moser et al., 2020). Firstly, the official diagnostic thresholds for D-dimer apply to the serum, but the commonly used laboratory assays measure D-dimer levels in the plasma. While D-dimer levels in the two blood fractions are thought to be strongly correlated, their diagnostic accuracy is not comparable (Lu et al., 2020). Further, D-dimer levels can be quantified via enzyme-linked immunosorbent assay (ELISA), immunoturbidimetric automated assay, or latex-based immunoassay. The three assay types vary in their analytical performance, with ELISA displaying the highest sensitivity. Although studies in patients with suspected VTE suggest that the choice of D-dimer assay does not affect the utility of the diagnostic cut-off, it is unclear whether this is also the case in patients with suspected PJI (Shohat and Parvizi, 2020). Compounding the issue is the fact that D-dimer levels are seldom reported using unit type [fibrinogen-equivalent units (FEU) or D-dimer units (DDU)] as well as unit magnitude (ng/mL, ug/L, etc.). Given that 2 ng/mL FEU is approximately equal to 1 ng/mL DDU, it is essential for the units to be fully disclosed, because a two-fold difference may influence clinical interpretation of the results and hinder meaningful literature comparisons.

Another limitation of D-dimer is that its production can be stimulated by a large number of conditions, including thrombosis, malignancy, autoimmune disease, and polyethylene wear particle-induced local tissue reaction (Lu et al., 2020). While the available evidence suggests that D-dimer levels are markedly higher in septic than in aseptic implant failure, further studies are warranted to corroborate these findings (Shahi et al., 2017a).

D-dimer is a promising biomarker for PJI, but future work is needed to verify its superiority over serum CRP and ESR. The lack of assay standardisation and ambiguities in the reporting of D-dimer levels must also be addressed if the test is to be successfully integrated into the diagnostic algorithm for PJI.

#### 6.4.3.3 *Interleukin-6*

IL-6 is a cytokine produced by monocytes, fibroblasts, and macrophages in response to infection and inflammation. IL-6 stimulates the synthesis of acute-phase reactants, including CRP, and plays a role in the regulation of various metabolic, regenerative, and neural processes. Elevated serum IL-6 concentration has been described in numerous pathologies, such as autoimmune diseases, inflammatory diseases, Paget's disease, and immunodeficiency syndromes, and its utility in the setting of PJI is an active area of research.

The baseline serum IL-6 levels are approximately 1 pg/mL. Following THA, they may increase to up to 430 pg/mL for as long as 3 days, with peak values on postoperative day 2 (Di Cesare et al., 2005). Since IL-6 response to infection is much more rapid than that of serum CRP, IL-6 may be the better indicator of acute PJI. Notably, IL-6 can be raised in cases of polyethylene wear without evidence of infection; however, the available evidence suggests that the magnitude of the increase is markedly higher in infection than in aseptic loosening (Bottner et al., 2007; Randau et al., 2014). In a meta-analysis of ten studies, the pooled sensitivity and specificity of serum IL-6 for distinguishing PJI from other causes of implant failure was 88% and 82%, respectively (Yoon et al., 2018).

The optimal diagnostic cut-off for serum IL-6 is unknown. In Randau et al. (2014), 2.6 pg/mL yielded a sensitivity of 80% and specificity of 58%; when the threshold was raised to 6.6 pg/mL, the specificity increased to 88% while the sensitivity dropped to 48%. Elgeidi et al. (2014), who used

10.4 pg/mL as the cut-off, reported a sensitivity of 100% and specificity of 91%, whereas Bottner et al. (2007) reported a sensitivity of 95% and specificity of 87% with 12 pg/mL.

Based on the excellent specificity, the 2019 WAIOT diagnostic criteria include serum IL-6 >10 ng/mL as one of the 'rule in' tests for PJI (Romanò et al., 2019).

#### 6.4.3.4 *Procalcitonin*

Procalcitonin is a peptide precursor of the hormone calcitonin, which is involved in calcium homoeostasis. Under normal conditions, the molecule is mainly produced by parafollicular cells of the thyroid and its serum levels are negligible. In response to pro-inflammatory cytokines and bacterial products, procalcitonin is released from nearly all tissues and cell types in the body, resulting in a sharp increase in its circulating concentration. Typically, procalcitonin levels rise rapidly within 3–4 hours of bacterial infection, peak at approximately 24 hours, and promptly return to baseline once the infection is resolved; the magnitude of the increase is correlated with the severity of infection (Samsudin and Vasikaran, 2017). The procalcitonin assay can detect sepsis and severe bacterial infection at a relatively early stage, and is more widely available than the IL-6 assay. Moreover, unlike serum CRP and ESR, procalcitonin is unaffected by the administration of anti-inflammatory therapy.

Although serum procalcitonin has shown high diagnostic accuracy for the identification of systemic infections such as septic arthritis and diabetic foot infections, its diagnostic value in suspected PJI and the optimal diagnostic threshold remain debated. One study in 78 patients found that serum procalcitonin levels exceeding 0.3 ng/mL had a low sensitivity (33%) but high specificity (98%) for diagnosing PJI (Bottner et al., 2007). In a recent meta-analysis of 12 studies, threshold values between 0.025 and 46 ng/mL had a pooled sensitivity and specificity of 53% and 92%, respectively (Xie et al., 2017). Combining procalcitoninwith other biomarkers seems to improve its ability to distinguish aseptic loosening from PJI (Glehr et al., 2013).

Owing to the high specificity, the 2019 WAIOT criteria feature procalcitonin >0.5 ng/mL as one of the 'rule in' tests for PJI (Romanò et al., 2019).

#### 6.4.3.5 *Fibrinogen*

Fibrinogen is an acute-phase glycoprotein and precursor of the insoluble fibrin mesh that forms blood clots. In addition to its role in the haemostatic response, it promotes the synthesis of proinflammatory cytokines (IL-6 and TNF-$\alpha$) in mononuclear cells and is an emerging biomarker for PJI.

In a study of patients undergoing revision THA or TKA, a serum fibrinogen threshold of 574 mg/dL had a sensitivity of 81% and specificity of 25% for PJI, suggesting that a low fibrinogen level could help rule out PJI (Klim et al., 2018). These conclusions were supported by Li et al. (2019), who reported a sensitivity of 76% and specificity of 86% for excluding PJI (Li et al., 2019). More recent studies suggest that the performance of plasma fibrinogen is comparable to that of serum CRP and ESR, and that the test can be especially useful for assessing infection outcomes after first-stage surgery (Bin et al., 2020; Zhang et al., 2021).

### 6.4.4 Synovial biomarkers

Joint aspiration and synovial fluid tests are usually the second step in the diagnosis of a painful THA. Measuring biomarker levels in the fluid surrounding the affected joint leads to a more reliable and accurate diagnosis than determining their circulating concentrations, particularly in the case of PJI caused by slow-growing microorganisms capable of forming a biofilm. The existing and emerging synovial biomarkers for PJI are described in the following sections.

#### *6.4.4.1 White blood cell count and polymorphonuclear leukocyte percentage*

Changes in the relative number of different WBC types in the blood can aid in the diagnosis of various inflammatory health conditions, infections, and blood disorders. In the setting of suspected PJI, two parameters are typically measured: the total WBC count, frequently expressed as a number of cells per microlitre of sample, and its proportion of polymorphonuclear leukocytes (PMN%). WBC differential may be generated automatically, using an automated cell counter, or manually, by microscopic examination of a blood smear.

While serum WBC count and PMN% have no role in PJI diagnosis (Toossi et al., 2012), their synovial fluid counterparts are the cornerstone of PJI evaluation and feature in all of the major PJI definitions. Typical sensitivity and specificity are 84%–93% and 51%–100% for synovial WBC count, and 81%–93% and 69%–83% for synovial PMN%, respectively, with the exact result dependent on the thresholds applied (Wyatt et al., 2016). The ICM diagnostic cut-offs for the two biomarkers were set depending on the timing of symptoms. During the immediate postoperative period, particularly within the first 6 weeks, a degree of postsurgical inflammatory response is expected, resulting in relatively high cut-off values for diagnosing early PJI: 10,000 cells/μL for synovial WBC count and 90% for PMN%. For delayed-onset PJI occurring >90 days from index arthroplasty, the values are lowered to 3,000 cells/μL and 80%, respectively (Parvizi et al., 2018). The 2020 EBJIS criteria, which do not differentiate between acute and chronic PJI, dictate that synovial WBC count exceeding 1500 cells/μL and PMN% of over 65% are suggestive of infection while WBC count exceeding 3000 cells/μL and PMN% of over 80% confirm it (McNally et al., 2021).

Lee et al. (2017) found that the ICM thresholds have a sensitivity of 89% and specificity of 86% for both synovial WBC count and PMN%. However, the diagnostic accuracy of both tests can be influenced by a number of factors, including virulence of the causative microorganism, prior antimicrobial therapy, concurrent inflammatory arthritis, presence of metal-on-metal (MoM) arthroplasty, viscosity of the synovial fluid, time between aspiration and analysis, and blood contamination of the sample (Ottink et al., 2019). For these reasons, synovial WBC count and PMN% should be interpreted in view of the patient's overall condition and the sample preparation protocol used. To preserve diagnostic sensitivity, it is advised to withhold systemic antibiotics in suspected chronic PJI until a proper diagnostic work-up has been carried out.

### *6.4.4.2 Leukocyte esterase*

Leukocyte esterase is an enzyme released by activated neutrophils after they have been recruited to sites of infection. Detection of leukocyte esterase in the urine has been traditionally used to diagnose lower urinary tract infections. More recently, the test has been applied to synovial fluid samples in the work-up of a suspected PJI, where it demonstrated a sensitivity of 66%–100% and specificity of 77%–100%, and outperformed the traditional serum and synovial inflammatory biomarkers (Shahi et al., 2017b; Signore et al., 2019).

The leukocyte esterase test uses a colorimetric strip to measure the esterase activity in a sample, providing a qualitative estimation of the leukocyte count: the more neutrophils, the higher the esterase activity and the more intense the colour change. Leukocyte esterase levels are most commonly categorised as 'negative', 'trace', '+', or '++'. In some countries, a '+++' reading may also be encountered. A recent meta-analysis demonstrated that the '++' reading is highly accurate for PJI, with a pooled sensitivity and specificity of 81% and 97%, respectively (Wyatt et al., 2016).

The assay is inexpensive, quick, easy to perform, and is not affected by prior antibiotic use (Shahi et al., 2019). It also offers the benefit of immediate results: if the synovial fluid is clear, the reading can be obtained in a few minutes (Aggarwal et al., 2013). Optimal timing for reading the strip in the diagnosis of PJI has not been established, however, and while manufacturers of the urine strip recommend 3 minutes, it is unclear if the same timepoint can be reliably applied to synovial fluid, which is more viscous than urine (Zheng et al., 2021). Another limitation of the test is that the reagent strip cannot be adequately read in the presence of blood or other debris. Bloody samples must first be centrifuged, which not only delays the results but may also decrease the test's sensitivity (Li et al., 2018a, b).

Leukocyte esterase features as a minor criterion in both the 2013 ICM and 2018 MSIS definitions of PJI, but because of its subjective nature and colour interference by erythrocytes, it was excluded from the 2020 EBJIS criteria for PJI (McNally et al., 2021).

### 6.4.4.3 *Alpha-defensin*

Alpha-defensins are small antimicrobial peptides released from activated neutrophils in response to infection. They are active against many gram-positive and gram-negative bacteria, certain fungi, and enveloped viruses, and exert their lethal effect by permeabilising the microbial membrane. The presence of alpha-defensin in the synovial fluid has unparalleled ability to distinguish septic from aseptic implant failure after THA (Wyatt et al., 2016).

Synovial alpha-defensin can be detected using an ELISA-based test or lateral flow device. Lateral flow tests are qualitative, with the alpha-defensin concentration expressed as signal-to-cut-off ratio, and can be performed on-site, even intraoperatively, providing results in as little as 10 minutes. ELISA can quantify alpha-defensin levels but requires more analytical steps and has to be performed by specialised personnel in a certified laboratory, resulting in higher costs and longer turnaround time (24–48 hours). Both methods have good sensitivity and excellent specificity for PJI. In Han et al. (2019), the pooled sensitivity, specificity, and AUC of synovial alpha-defensin were 96%, 97%, and 0.99, respectively, when measured by laboratory-based immunoassay, and 86%, 96%, and 0.95, respectively, when a lateral flow device was used. A recent meta-analysis of all prospective studies comparing the diagnostic performance of the two test types in PJI diagnosis found no differences in terms of sensitivity and specificity, supporting the implementation of either test in diagnostic routines (Kuiper et al., 2020).

Aside from exceptional diagnostic accuracy, synovial alpha-defensin performs well in challenging situations such as blood-contaminated samples and culture-negative PJI, and is not impacted by prior antibiotic administration unlike the ESR, serum CRP, and synovial WBC count and PMN%. Moreover, its release is stimulated to a similar extent by a wide spectrum of organisms, providing consistent results regardless of pathogen characteristics (Deirmengian et al., 2014b, 2015; Shahi et al., 2016). However, the presence of a draining sinus or crystal deposition disease may predispose the test to false-negative results, while metallosis-associated ALTR may increase the likelihood of a false-positive result (Bonanzinga et al., 2017; Okroj et al., 2018; Plate et al., 2018).

Alpha-defensin immunoassay forms part of the minor criteria in the 2018 MSIS definition of PJI, and serves as a ‘rule out’ test in the 2019 WAIOT definition. The 2020 EBJIS criteria stipulate that a positive alpha-defensin ELISA or lateral flow test confirms infection, but not in cases of ALTR, haematomas, acute inflammatory disease, or gout (McNally et al., 2021). While the optimal cut-off value for ELISA-based alpha-defensin test is unclear, 4.8, 5.2, and 7.72 mg/L were all associated with a sensitivity and specificity of >95% in previous studies (Chen et al., 2019).

#### 6.4.4.4 Calprotectin

Calprotectin is an abundant metal-binding protein released by activated neutrophils in response to increased mucosal permeability, tissue damage, and infection. Faecal calprotectin is a well-established marker of intestinal inflammation and a key part of the diagnostic work-up for inflammatory bowel disease. In addition to being pro-inflammatory, calprotectin exerts antimicrobial and antifungal effects thanks to its ability to sequester zinc and manganese from invading pathogens. Recent evidence suggests that calprotectin measured in the synovial fluid is more specific for infection than synovial WBC count, hinting at its potential utility in the preoperative evaluation of PJI (Wouthuyzen-Bakker et al., 2017). As with alpha-defensin, calprotectin concentration can be measured using ELISA or lateral flow immunoassay.

To date, the use of synovial calprotectin in PJI has only been evaluated in a limited number of small studies; however, the results are promising. In two recent reports, synovial calprotectin concentration determined using lateral flow assay showed a sensitivity of 87%–89%, specificity of 90%–92%, AUC of >0.9, positive predictive value of 81%, and negative predictive value of 94%–95% at a cut-off of 50 mg/L, suggesting that the test could be particularly useful for ruling out PJI in patients with chronic pain (Wouthuyzen-Bakker et al., 2018, 2017). When measured using an ELISA-based approach, synovial calprotectin displayed a sensitivity of 100%, specificity of 95%, and AUC approaching 1 for the diagnosis of PJI (Salari et al., 2020).

Excellent diagnostic performance combined with the relatively low cost, wide availability, and point-of-care nature of lateral flow devices make synovial calprotectin an attractive screening test for suspected PJI. However, larger prospective studies are needed to assess the effect of immunosuppressive drugs, antibiotics, and different causative organisms on synovial calprotectin levels, and determine the optimal diagnostic threshold for PJI, particularly in patients with concurrent inflammatory conditions.

#### 6.4.4.5 Synovial C-reactive protein

Accumulating evidence suggests that synovial CRP is more accurate than its serum counterpart in the diagnosis of PJI. Omar et al. (2015) measured CRP levels in 80 patients undergoing septic or aseptic revision THA and reported that synovial CRP concentration was markedly higher in the former cohort. The synovial test was associated with a sensitivity of 95.5%, specificity of

93.3%, and AUC of 0.96 at a threshold of 2.5 mg/L, while the serum test yielded a sensitivity of 78.3%, specificity of 86.4%, and AUC of 0.87 at a threshold of 9.5 mg/L. A meta-analysis of six studies including 456 patients with painful THAs or TKAs reported a pooled sensitivity of 92%, specificity of 90%, and AUC of 0.97 associated with synovial CRP thresholds between 2.8 and 12.2 mg/L (Wang et al., 2016).

In Sharma et al. (2020), synovial CRP ranked among the best-performing biomarkers for hip/knee PJI (AUC, 0.921), behind only synovial WBC count and PMN% (AUC, 0.952 and 0.935, respectively), and outperformed alpha-defensin (AUC, 0.916). In contrast, Lee et al. (2017) reported that synovial CRP had a lower diagnostic accuracy for PJI than other synovial biomarkers, such as alpha-defensin, leukocyte esterase, IL-6, and IL-8. An exhaustive meta-analysis evaluating the diagnostic accuracy of all serologic, synovial, and tissue tests used in the diagnosis of PJI concluded that synovial CRP, PMN%, and WBC count performed similarly (Carli et al., 2019).

Since CRP assays are commonplace in medical laboratories, the measurement of synovial CRP should be easily transferred into clinical practice. The test features in the 2018 MSIS definition of PJI but was removed from the 2020 EBJIS criteria due to the lack of a clear advantage over other inflammatory biomarkers (McNally et al., 2021).

#### 6.4.4.6 *Synovial interleukin-6*

As with CRP measurements, IL-6 assay appears to be more sensitive when performed in synovial fluid than serum samples. A meta-analysis reported a sensitivity of 91% and 72%, and a specificity of 90% and 89% for synovial and serum IL-6, respectively (Xie et al., 2017). It may be possible to improve the diagnostic accuracy of synovial IL-6 by combining it with serum IL-6; however, since synovial IL-6 alone is already fairly specific for PJI, combining the two tests may not add significant value to the diagnosis of PJI.

The 2019 WAIOT definition of PJI is the first to include synovial IL-6 in its diagnostic criteria, with the recommended threshold of 10 pg/mL (Romanò et al., 2019).

#### 6.4.4.7 *Synovial interleukin-8*

IL-8 is a cytokine produced by phagocytes and mesenchymal cells exposed to infectious, inflammatory, ischaemic, or traumatic stimuli, and functions to promote the influx of neutrophils to the sites of injury. Several studies have noted marked increases in IL-8 levels in the synovial fluid of patients with PJI, which generated considerable interest in its potential role as a biomarker for PJI (Deirmengian et al., 2010; Erdemli et al., 2018; Mirza et al., 2019; Prince et al., 2020; Sharma et al., 2020).

A meta-analysis of 33 studies investigating various synovial biomarkers of PJI reported a pooled sensitivity of 87%, pooled specificity of 94%, and AUC of 0.96 for IL-8 (Lee et al., 2017). An analysis of synovial fluid samples from 95 patients, including 29 with an established infection based on the MSIS criteria, found that IL-8 had a sensitivity of 100% and specificity of 95% for diagnosing PJI when a cut-off of 6.5 ng/mL was used (Deirmengian et al., 2014a). In another study, a threshold of 8.79 ng/mL yielded a sensitivity of 90.3% and specificity of 97.7% (Jacovides et al., 2011).

Despite encouraging results, the clinical use of synovial IL-8 in suspected PJI cases is hindered by the high cost of the assay and lack of a clear diagnostic threshold.

### 6.4.5 Microbiology

When synovial fluid biomarkers are unable to confirm PJI, preoperative and intraoperative cultures should be obtained to gather information about the causative microorganisms.

#### *6.4.5.1 Joint aspiration culture*

Synovial fluid culture is the most accurate preoperative examination for the detection of PJI, with a sensitivity ranging from 45% to 80%, and specificity of up to 95% (Kapadia et al., 2016; Lee et al., 2017; Trampuz et al., 2004). The low sensitivity means that a negative result cannot be used to rule out PJI. A positive result is, however, strongly suggestive of infection. The samples should be inoculated into paediatric blood culture bottles and incubated for at least 14 days to allow enough time to detect low-virulent and fastidious pathogens (Hughes et al., 2001; Izakovicova et al., 2019).

#### *6.4.5.2 Preoperative periprosthetic biopsy culture*

Preoperative periprosthetic tissue biopsy has high accuracy for the diagnosis of PJI, with the reported sensitivity and specificity of 79%–88% and 90%–100%, respectively (Sadiq et al., 2005; Meermans and Haddad, 2010; Fink et al., 2013). When combined with a histologic examination of the specimen, microbiologic culture can identify the causative pathogen and its drug susceptibility to guide the surgical strategy and aid in the choice of appropriate antibiotic regimen. However, in cases of polyinfection, it may be difficult to identify all causative microorganisms preoperatively.

Preoperative periprosthetic biopsy culture has several limitations, including additional medical expenses and risk of complications related to the biopsy procedure itself. Moreover, the test is influenced by prior antibiotic use and is only slightly more sensitive than the more easily obtained synovial fluid culture. For these reasons, preoperative synovial biopsy culture is not commonly performed in suspected PJI cases.

#### 6.4.5.3 *Intraoperative periprosthetic tissue culture*

Microbiologic cultures of intraoperative tissue samples have high diagnostic utility in PJI. In a study of 132 patients, including 40 with a definite diagnosis of PJI, intraoperative cultures had a sensitivity of 80% and specificity of 97% for the identification of PJI (Marín et al., 2012). When intraoperative culture results were combined with PCR analysis, the sensitivity and specificity of the test reached 94% and 100%, respectively.

To improve the culture yield and decrease the likelihood of misdiagnosis, it is recommended that at least five reliable periprosthetic tissue samples are obtained from visibly inflamed areas, and that each sample is collected using separate, sterile instruments to avoid contamination (Parvizi et al., 2011; McNally et al., 2021). Samples should be cultured for both aerobic and anaerobic microorganisms to increase the chances of successful identification of the causative pathogen. While it is not necessary to routinely culture for fungi and mycobacteria, these cultures may also be included. The presence of phenotypically indistinguishable microorganisms with identical antibiotic susceptibility pattern in at least two different samples clearly defines infection. If a common contaminant such as CNS or *C. acnes* is detected in a single sample only, infection cannot be confirmed and prompt further investigation should be initiated (McNally et al., 2021).

Although the use of antibiotic prophylaxis does not appear to affect microbiologic culture results, prior long-term antibiotic therapy is a major risk for culture-negative PJI (Bedenčič et al., 2016; Tetreault et al., 2014; Yoon et al., 2017). In a study of patients undergoing revision for aseptic failure or PJI, tissue culture sensitivity decreased from 76.9% to 47.8% and then to 41.2% as the preoperative antibiotic-free interval was reduced from >14 days to 4–14 days to 0–3 days (Trampuz et al., 2007).

#### 6.4.5.4 *Sonication fluid culture*

Sonication of explanted prostheses helps to disrupt the bacterial biofilm and increase the number of bacterial cells available for culture, resulting in improved diagnostic performance, with reported sensitivity and specificity of 60%–97% and 90%–99%, respectively (Palan et al., 2019). In Trampuz et al. (2007), sonication fluid cultures were more sensitive than periprosthetic tissue cultures (78% vs 61%), even in patients receiving systemic antimicrobial therapy in the 14 days before the surgery (75% vs 45%).

The EBJIS criteria suggest a cut-off of 50 colony-forming units per millilitre (CFU/mL) to confirm infection if the sonication fluid had not been concentrated by centrifugation, and 200 CFU/mL if a concentration technique had been applied.

### 6.4.6 Histology

Periprosthetic tissue samples collected preoperatively or during revision surgery may also be evaluated for the presence of visible microorganisms and acute inflammation. Since the inflammatory infiltrate may not be uniformly distributed throughout the joint, it is essential to take at least three deep samples, favouring the bone–implant interface membrane, synovium/pseudocapsule, or other abnormal tissue, and to examine 5–10 high-power fields (HPFs) in each case. Positive histology result is defined as the presence of at least five PMNs per HPF in at least 5 HPFs at 400 × magnification (Parvizi et al., 2018, 2011; McNally et al., 2021). Additionally, the finding of at least five neutrophils in at least 1 HPF is suggestive of infection (McNally et al., 2021). While both fixed and frozen intraoperative tissue samples are suitable for histologic analysis, assessment of frozen sections is considered less sensitive for detecting PJI (Ko et al., 2005; Tsaras et al., 2012). Despite lower sensitivity, frozen tissue samples can be prepared and analysed more readily than fixed specimens, providing the orthopaedic surgeon with results while revision surgery is still underway.

An important limitation of histological assessment is that it may not be sensitive enough in cases of low-virulent pathogens that trigger few inflammatory changes in the periprosthetic tissue. Furthermore, because histologic examination is carried out by musculoskeletal pathologists, the results are dependent on the examiner's expertise and their understanding of what constitutes a positive histology test. To minimise false-positive results, care must be taken to avoid counting PMNs found in superficial fibrin and those adhering to endothelial tissue or small veins (Parvizi et al., 2011). Caution must also be exercised when interpreting the results for patients with conditions or inflammatory disorders known to increase PMN numbers, such as recent periprosthetic fracture or inflammatory arthropathy.

#### *6.4.6.1 Gram stain*

Out of the many bacterial strains implicated in PJI, the most frequently found are gram-positive cocci such as *S. aureus* and CNS (Beam and Osmon, 2018). A retrospective review of 704 joint arthroplasties revised for aseptic reasons and 299 arthroplasties revised for deep infection reported that while Gram stains had excellent specificity (98%–100%) and high positive predictive values (89%–100%) for diagnosing PJI, their sensitivity (30%–50%) and negative predictive values (70%–79%) were insufficient (Ghanem et al., 2009). Another limitation of Gram stains is that they can produce false-positive results if the reagents used are contaminated by bacteria (Oethinger et al., 2011).

Although intraoperative Gram stains may be the deciding factor when the preoperative diagnosis is unclear, they are not an adequate solitary test for PJI. According to the 2020 EBJIS criteria, Gram staining is useful to confirm the presence of microorganisms, particularly atypical ones that may not be cultured on routine microbiological protocols (McNally et al., 2021).

## 6.4.7 Molecular techniques

### *6.4.7.1 Polymerase chain reaction*

PCR techniques used in the diagnosis of PJI include 16S rRNA PCR, which detects bacterial 16S rRNA to establish the presence of bacteria in general, and multiplex PCR panels, which use specific primers to identify specific bacterial strains. Reported sensitivity and specificity of 16S rRNA PCR carried out on intraoperative periprosthetic tissue samples range from 67.1% to 73.3% and from 95.5% to 97.8%, respectively (Marín et al., 2012; Bémer et al., 2014). Multiplex PCR performed on the same sample type has comparable specificity, but markedly lower sensitivity (16%–41.7%) (Ryu et al., 2014; Villa et al., 2017). A recent meta-analysis evaluating the use of PCR assays on sonication fluid reported a pooled sensitivity and specificity of 75% and 96%, respectively (Liu et al., 2018). Thus, neither 16S rRNA PCR or multiplex PCR have the required sensitivity for diagnosing PJI, whether being performed on periprosthetic tissue samples or sonication fluid. Although they cannot be relied on exclusively to make a diagnosis, the high specificity means that they can be useful in excluding PJI.

PCR methods can also be used to determine the mRNA expression of Toll-like receptors (TLRs) in the periprosthetic tissue, which may be particularly helpful to detect culture-negative PJI. TLRs are transmembrane receptors that play an integral role in the activation of host inflammatory response against microbial infections. TLR-1 and TLR-6 are activated by bacterial lipoproteins and are emerging biomarkers in the diagnosis of PJI, with a reported sensitivity of 95.2% and 85.7%, and specificity of 100% and 82.8%, respectively (Cipriano et al., 2014).

PCR assays can be performed on fresh tissue, without the need for any microbiologic culturing prior to DNA extraction, and provide results within a few hours of sample collection (Arvieux and Common, 2019). In addition to timely results, PCR analysis of periprosthetic tissues is able to identify difficult-to-culture bacteria and is not influenced by prior antibiotic use. Limitations of PCR techniques include relatively high rate of false-positive results due to exogenous bacterial contamination, and limited choice of primers (Beam and Osmon, 2018; Li et al., 2018a, b; Wasterlain et al., 2020).

### *6.4.7.2 Next-generation sequencing*

Recent evidence supports the use of NGS to identify pathogenic microorganisms in orthopaedic infections, particularly among patients with culture-negative PJI (Goswami and Parvizi, 2020; Tarabichi et al., 2018). Unlike PCR, which relies on the use of predetermined primers, NGS sequences the entire genome of all DNA in a sample, enabling identification of all strains and genomic variants of a given microorganism. The sequenced DNA can not only be compared to known microbial sequences but also to viral, fungal, and parasitic sequences to help identify the causative pathogen. Moreover, NGS is faster, less expensive, more sensitive, and requires less manual manipulation than other sequencing methods. The concern remains, however, that the increased sensitivity may lead to overdiagnosis and consequent overtreatment of patients who do not have PJI. Careful handling of synovial fluid from patient to laboratory to avoid sample contamination is particularly important with NGS methods.

## 6.5 Confounding factors

The diagnosis of chronic PJI following joint replacement can be confounded by conditions with similar presenting symptoms, e.g., pain, joint inflammation, and/or elevated serum inflammatory markers. The most notable confounders and ways to distinguish them from PJI are discussed below.

### 6.5.1 Adverse reaction to metal debris

Adverse reaction to metal debris (ARMD) refers to the chronic inflammatory response and periprosthetic soft-tissue destruction resulting from the release of metal debris from MoM bearings or from corrosion at the head/neck junction in modular THAs. Local tissue effects range from asymptomatic soft-tissue lesions to striking periprosthetic effusions, tissue necrosis, and osteolysis, which can be accompanied by cystic and/or solid masses (Pisanu et al., 2019).

When assessing for possible PJI in patients with MoM implants or suspected ARMD, several factors should be considered. Firstly, the intraoperative finding of periprosthetic purulence cannot be used to confirm PJI, as purulence has also been documented in cases of failed MoM arthroplasty (Earll et al., 2011; Mikhael et al., 2009). Secondly, the presence of metal debris, amorphous material, fragmented cells, or clots in the synovial fluid of patients with ARMD may produce false-positive results in automated WBC counts (Yi et al., 2015). ARMD may also have a detrimental effect on the accuracy of alpha-defensin and leukocyte esterase tests (Okroj et al., 2018; Wetters et al., 2012).

Keeping in mind that ARMD and PJI can coexist, extensive patient work-up is warranted to accumulate as much information as possible prior to revision surgery. Manual synovial WBC count, PMN%, and adjunctive tests, such as alpha-defensin, should be accompanied by preoperative synovial fluid culture to increase the likelihood of an appropriate evaluation for PJI.

### 6.5.2 Inflammatory arthritis

Inflammatory arthritis encompasses a group of systemic disorders characterised by progressive destruction of articular cartilage due to an overactive immune system. Biomarkers of inflammation are often higher in patients with inflammatory arthritis than in the general population, which can potentially alter the accuracy of common diagnostic tests for PJI and lead to the identification of a large number of false-positive cases (Barrack et al., 2019).

Cipriano et al. (2012) found that the ESR, serum CRP, synovial WBC count, and PMN% had similar optimal cut-off values and overall testing performance for chronic PJI in individuals with inflammatory and non-inflammatory arthritis. A larger, multi-institutional analysis of patients undergoing septic or aseptic joint revision surgery reported comparable results, reinforcing the MSIS recommendation that the same diagnostic thresholds should be used for suspected chronic PJI regardless of the underlying inflammatory arthropathy (Shohat et al., 2018). However, it must be kept in mind that the levels of inflammatory biomarkers can be affected by many factors, including concurrent presence of other inflammatory events (e.g., ARMD, dislocation, periprosthetic fracture), action of disease-modifying antirheumatic drugs, and the level of inflammation (flare vs controlled disease) (Barrack et al., 2019). For these reasons, it may be prudent to apply alternative diagnostic thresholds in patients with active disease or flares.

### 6.5.3 Crystal-induced arthritis

Crystal-induced arthritis is a type of inflammatory arthritis. The most common forms of crystal-induced arthritis are gout and pseudo-gout, which are caused by the build-up of monosodium urate (MSU) or calcium pyrophosphate (CPP) crystals in the joints, respectively. Although many cases of crystal deposition are asymptomatic, some patients go on to develop joint trauma and inflammation with a clinical presentation similar to that of PJI, i.e., rapid-onset joint pain, swelling, erythema, and constitutional symptoms such as fever and general malaise. The prevalence of crystal-induced arthritis after THA is not well described, as only a few case reports have been published (Chernoff et al., 2020).

Laboratory findings commonly noted in both crystalline arthropathy and PJI include elevated serum CRP levels, ESR, synovial WBC count, and PMN% (Archibeck et al., 2001). An increased synovial WBC count may, in turn, influence alpha-defensin and leukocyte esterase levels, as both of these markers are released by activated neutrophils (Chernoff et al., 2020). Indeed, false-positive alpha-defensin results have been reported in patients with gout and pseudo-gout (Partridge et al., 2017; Plate et al., 2018). Polarised light microscopy, which can detect MSU and CPP crystals in the synovial fluid, may be useful to differentiate PJI from crystalline arthropathy (Archibeck et al., 2001). Notably, findings of cloudy or creamy yellow fluid have been reported during revision surgery in patients with gout who did not have PJI (Archibeck et al., 2001). While such observations are suggestive of purulence, further tests must be conducted to rule out infection.

## 6.6 Summary and future directions

The differentiation between septic and aseptic implant failure is crucial to ensure appropriate medical and surgical treatment. Despite international efforts to guide the management of suspected infection, accurate diagnosis continues to be a challenge, partly due to the unspecific clinical presentation, presence of confounding factors, variable bacterial virulence, and lack of a 'gold standard' test. The most recent criteria for PJI classify infection as unlikely, likely, or confirmed, based on the information gleaned from clinical examination, levels of various blood and synovial biomarkers, microbiological and histological assessment of synovial fluid and periprosthetic tissue, and nuclear imaging. The system was designed to be more sensitive and accessible than its predecessors; however, problems related to the subjective nature of certain tests, variable performance of biomarkers in patients with confounding comorbidities, and culture-negative infection remain. It is therefore important that any criteria used to diagnose PJI are applied in the context of the individual patient's health. For example, recipients of immunosuppressant therapy have a high risk of infection, but their levels of inflammatory markers may be normal, reducing the utility of serum biomarkers in diagnosing PJI (McNally et al., 2021). Another factor to be wary of is obesity, which may elevate serum inflammatory biomarkers and lead to false-positive results (Saleh et al., 2018). Further studies are needed to determine the appropriate tests and optimal diagnostic thresholds when diagnosing PJI in patients with metallosis or inflammatory arthritis.

The standard first-line tests for PJI—serum CRP and ESR—are not sufficiently specific and often fail to detect low-grade infection. Newer blood biomarkers such as D-dimer and procalcitonin have not demonstrated superior sensitivity, but research is ongoing into other blood molecules and parameters, such as plasma fibrinogen, ratio of platelet count to mean platelet volume, presepsin, osteopontin, CCL2, and soluble urokinase-type plasminogen activator receptor (Marazzi et al., 2018). Using mathematical models to combine serum biomarkers and increase their individual strengths in sensitivity and specificity is attractive in principle but difficult in practice (Klim et al., 2020). There is increasing pressure to define highly accurate synovial biomarkers to allow PJI to be confirmed or excluded preoperatively.

Antimicrobial proteins released locally in response to pathogens seem to be the most promising group of synovial biomarkers for PJI. Alpha-defensin has an established place in the evaluation for PJI, and a positive result confirms infection according to the 2020 EBJIS criteria (McNally et al., 2021).

However, the test is cost-prohibitive for many laboratories and not routinely performed in most countries. Synovial calprotectin, which has the advantages of high specificity, low cost, easy application, and availability in most hospitals is being evaluated as an alternative to alpha-defensin (Wouthuyzen-Bakker et al., 2018). Bactericidal leukocyte enzymes—neutrophil elastase 2, bactericidal/permeability-increasing protein, neutrophil gelatinase-associated lipocalin, and lactoferrin—have previously shown a specificity and sensitivity of 100% for PJI and warrant investigation in larger studies (Deirmengian et al., 2014a; Li et al., 2021). Among synovial cytokines, IL-1β, IL-5, IL-6, IL-8, IL-10, IL-17a, TNF-α, INF-γ, GM-CSF, resistin, and thrombospondin are garnering increasing attention (Deirmengian et al., 2014a). Synovial fluid monocyte percentage and lactate levels, which were surprisingly discriminatory for PJI and aseptic failure in a recent study (AUC, 0.818 and 0.902, respectively), should also be further explored (Sharma et al., 2020). In addition to researching new potential biomarkers for PJI, diagnostic thresholds for some of the existing tests need to be optimised.

The future of biomarkers is likely to shift towards genetic, proteomic, and molecular analyses of blood, synovial fluid, and periprosthetic tissue to identify the cellular and transcriptional changes that occur in response to infection. Flow cytometry, which can be used to analyse physical and chemical properties of individual cells in a sample, shows early promise in the preoperative screening for PJI. Preliminary data suggest that immune cell distribution in the synovial fluid differs between patients with septic and aseptic implant loosening, and that the proportion of natural killer cells, CD3 + cells, monocytes, and neutrophils could reliably predict infection status when used in combination (Korn et al., 2020). Assessing the expression of CD64 on blood monocytes or neutrophils could also prove useful in the setting of suspected PJI (Perry et al., 2013; Qu et al., 2020).

Ultimately, when choosing potential new biomarkers to replace or add to the existing PJI criteria, it will be necessary to not only evaluate their diagnostic accuracy and cost, but also their reliability, applicability to routine clinical practice, and acceptance by the orthopaedic community.

## References

Aggarwal, V.K., Tischler, E., Ghanem, E., Parvizi, J., 2013. Leukocyte esterase from synovial fluid aspirate: a technical note. J. Arthroplasty 28, 193–195.

Alende-Castro, V., Alonso-Sampedro, M., Vazquez-Temprano, N., Tuñez, C., Rey, D., García-Iglesias, C., et al., 2019. Factors influencing erythrocyte sedimentation rate in adults: new evidence for an old test. Medicine 98, e16816.

Archibeck, M.J., Rosenberg, A.G., Sheinkop, M.B., Berger, R.A., Jacobs, J.J., 2001. Gout-induced arthropathy after total knee arthroplasty: a report of two cases. Clin. Orthop. Relat. Res. 392, 377–382.

Arvieux, C., Common, H., 2019. New diagnostic tools for prosthetic joint infection. Orthop. Traumatol. Surg. Res. 105, S23–S30.

Austin, M.S., Ghanem, E., Joshi, A., Lindsay, A., Parvizi, J., 2008. A simple, cost-effective screening protocol to rule out periprosthetic infection. J. Arthroplasty 23, 65–68.

Barrack, R., Bhimani, S., Blevins, J.L., Blevins, K., Demetres, M., Figgie, M., et al., 2019. General assembly, diagnosis, laboratory test: proceedings of international consensus on orthopedic infections. J. Arthroplasty 34, S187–S195.

Barrett, L., Atkins, B., 2014. The clinical presentation of prosthetic joint infection. J. Antimicrob. Chemother. 69, i25–i27.

Beam, E., Osmon, D., 2018. Prosthetic joint infection update. Infect. Dis. Clin. North. Am. 32, 843–859.

Bedenčič, K., Kavčič, M., Faganeli, N., Mihalič, R., Mavčič, B., Dolenc, J., et al., 2016. Does preoperative antimicrobial prophylaxis influence the diagnostic potential of periprosthetic tissues in hip or knee infections? Clin. Orthop. 474, 258–264.

Bingham, J.S., Hassebrock, J.D., Christensen, A.L., Beauchamp, C.P., Clarke, H.D., Spangehl, M.J., 2020. Screening for periprosthetic joint infections with ESR and CRP: the ideal cutoffs. J. Arthroplasty 35, 1351–1354.

Bin, G., Xinxin, Y., Fan, L., Shenghong, W., Yayi, X., 2020. Serum fibrinogen test performs well for the diagnosis of periprosthetic joint infection. J. Arthroplasty 35, 2607–2612.

Bonanzinga, T., Zahar, A., Dütsch, M., Lausmann, C., Kendoff, D., Gehrke, T., 2017. How reliable is the alpha-defensin immunoassay test for diagnosing periprosthetic joint infection? A prospective study. Clin. Orthop. Relat. Res. 475, 408–415.

Bottner, F., Wegner, A., Winkelmann, W., Becker, K., Erren, M., Götze, C., 2007. Interleukin-6, procalcitonin and TNF-alpha: markers of peri-prosthetic infection following total joint replacement. J. Bone Joint Surg. Br. 89, 94–99.

Bozhkova, S., Suardi, V., Sharma, H.K., Tsuchiya, H., Del Sel, H., Hafez, M.A., et al., 2020. The W.A.I.O.T. definition of peri-prosthetic joint infection: a multi-center, retrospective validation study. J. Clin. Med. 9 (6), 1965.

Bozic, K.J., Kurtz, S.M., Lau, E., Ong, K., Vail, T.P., Berry, D.J., 2009. The epidemiology of revision total hip arthroplasty in the United States. J. Bone Joint Surg. Am. 91, 128–133.

Bémer, P., Plouzeau, C., Tande, D., Léger, J., Giraudeau, B., Valentin, A.S., et al., 2014. Evaluation of 16S rRNA gene PCR sensitivity and specificity for diagnosis of prosthetic joint infection: a prospective multicenter cross-sectional study. J. Clin. Microbiol. 52, 3583–3589.

Carli, A.V., Abdelbary, H., Ahmadzai, N., Cheng, W., Shea, B., Hutton, B., et al., 2019. Diagnostic accuracy of serum, synovial, and tissue testing for chronic periprosthetic joint infection after hip and knee replacements: a systematic review. J. Bone Joint Surg. Am. 101, 635–649.

Carroll, F.E., Gooberman-Hill, R., Strange, S., Blom, A.W., Moore, A.J., 2020. What are patients' preferences for revision surgery after periprosthetic joint infection? A discrete choice experiment. BMJ Open 10, e031645.

Chen, Y., Kang, X., Tao, J., Zhang, Y., Ying, C., Lin, W., 2019. Reliability of synovial fluid alpha-defensin and leukocyte esterase in diagnosing periprosthetic joint infection (PJI): a systematic review and meta-analysis. J. Orthop. Surg. Res. 14, 453.

Chernoff, D.J., Barker, J.P., Wingerter, S.A., Shriwise, T.L., 2020. Gout after total knee arthroplasty. Arthroplast. Today 6, 278–282.

Cipriano, C.A., Brown, N.M., Michael, A.M., Moric, M., Sporer, S.M., Della Valle, C.J., 2012. Serum and synovial fluid analysis for diagnosing chronic periprosthetic infection in patients with inflammatory arthritis. J. Bone Joint Surg. Am. 94, 594–600.

Cipriano, C., Maiti, A., Hale, G., Jiranek, W., 2014. The host response: Toll-like receptor expression in periprosthetic tissues as a biomarker for deep joint infection. J. Bone Joint Surg. Am. 96, 1692–1698.

Deirmengian, C., Hallab, N., Tarabishy, A., Della Valle, C., Jacobs, J.J., Lonner, J., et al., 2010. Synovial fluid biomarkers for periprosthetic infection. Clin. Orthop. 468, 2017–2023.

Deirmengian, C., Kardos, K., Kilmartin, P., Cameron, A., Schiller, K., Parvizi, J., 2014a. Diagnosing periprosthetic joint infection: has the era of the biomarker arrived? Clin. Orthop. 472, 3254–3262.

Deirmengian, C., Kardos, K., Kilmartin, P., Cameron, A., Schiller, K., Parvizi, J., 2014b. Combined measurement of synovial fluid α-defensin and C-reactive protein levels: highly accurate for diagnosing periprosthetic joint infection. J. Bone Joint Surg. Am. 96, 1439–1445.

Deirmengian, C., Kardos, K., Kilmartin, P., Gulati, S., Citrano, P., Booth, R.E., 2015. The alpha-defensin test for periprosthetic joint infection responds to a wide spectrum of organisms. Clin. Orthop. 473, 2229–2235.

Di Cesare, P.E., Chang, E., Preston, C.F., Liu, C., 2005. Serum interleukin-6 as a marker of periprosthetic infection following total hip and knee arthroplasty. J. Bone Joint Surg. Am. 87, 1921–1927.

Earll, M.D., Earll, P.G., Rougeux, R.S., 2011. Wound drainage after metal-on-metal hip arthroplasty secondary to presumed delayed hypersensitivity reaction. J. Arthroplasty 338, e5–e7.

Elgeidi, A., Elganainy, A.E., Abou Elkhier, N., Rakha, S., 2014. Interleukin-6 and other inflammatory markers in diagnosis of periprosthetic joint infection. Int. Orthop. 38, 2591–2595.

Erdemli, B., Özbek, E.A., Başarir, K., Karahan, Z.C., Öcal, D., Biriken, D., 2018. Proinflammatory biomarkers' level and functional genetic polymorphisms in periprosthetic joint infection. Acta Orthop. Traumatol. Turc. 52, 143–147.

Fink, B., Gebhard, A., Fuerst, M., Berger, I., Schäfer, P., 2013. High diagnostic value of synovial biopsy in periprosthetic joint infection of the hip. Clin. Orthop. 471, 956–964.

Garfield, K., Noble, S., Lenguerrand, E., Whitehouse, M.R., Sayers, A., Reed, M.R., et al., 2020. What are the inpatient and day case costs following primary total hip replacement of patients treated for prosthetic joint infection: a matched cohort study using linked data from the National Joint Registry and Hospital Episode Statistics. BMC Med. 18, 335.

Ghanem, E., Ketonis, C., Restrepo, C., Joshi, A., Barrack, R., Parvizi, J., 2009. Periprosthetic infection: where do we stand with regard to gram stain? Acta Orthop. 80, 37–40.

Glehr, M., Friesenbichler, J., Hofmann, G., Bernhardt, G.A., Zacherl, M., Avian, A., et al., 2013. Novel biomarkers to detect infection in revision hip and knee arthroplasties. Clin. Orthop. Relat. Res 471, 2621–2628.

Goswami, K., Parvizi, J., 2020. Culture-negative periprosthetic joint infection: is there a diagnostic role for next-generation sequencing? Expert. Rev. Mol. Diagn. 20, 269–272.

Han, X., Xie, K., Jiang, X., Wang, L., Wu, H., Qu, X., et al., 2019. Synovial fluid α-defensin in the diagnosis of periprosthetic joint infection: the lateral flow test is an effective intraoperative detection method. J. Orthop. Surg. Res. 14, 274.

Hughes, J.G., Vetter, E.A., Patel, R., Schleck, C.D., Harmsen, S., Turgeant, L.T., et al., 2001. Culture with BACTEC Peds Plus/F bottle compared with conventional methods for detection of bacteria in synovial fluid. J. Clin. Microbiol. 39, 4468–4471.

Izakovicova, P., Borens, O., Trampuz, A., 2019. Periprosthetic joint infection: current concepts and outlook. EFORT Open. Rev. 4, 482–494.

Jacovides, C.L., Parvizi, J., Adeli, B., Jung, K.A., 2011. Molecular markers for diagnosis of periprosthetic joint infection. J. Arthroplasty 26, 99–103.

Kapadia, B.H., Berg, R.A., Daley, J.A., Fritz, J., Bhave, A., Mont, M.A., 2016. Periprosthetic joint infection. Lancet 387, 386–394.

Kim, T.W., Kim, D.H., Oh, W.S., Sim, J.A., Lee, Y.S., Lee, B.K., 2016. Analysis of the causes of elevated C-reactive protein level in the early postoperative period after primary total knee arthroplasty. J. Arthroplasty 31, 1990–1996.

Klim, S.M., Amerstorfer, F., Glehr, G., Hauer, G., Smolle, M.A., Leitner, L., et al., 2020. Combined serum biomarker analysis shows no benefit in the diagnosis of periprosthetic joint infection. Int. Orthop. 44, 2515–2520.

Klim, S.M., Amerstorfer, F., Gruber, G., Bernhardt, G.A., Radl, R., Leitner, L., et al., 2018. Fibrinogen - a practical and cost efficient biomarker for detecting periprosthetic joint infection. Sci. Rep. 8, 8802.

Ko, P.S., Ip, D., Chow, K.P., Cheung, F., Lee, O.B., Lam, J.J., 2005. The role of intraoperative frozen section in decision making in revision hip and knee arthroplasties in a local community hospital. J. Arthroplasty 20, 189–195.

Korn, M.F., Stein, R.R., Dolf, A., Shakeri, F., Buness, A., Hilgers, C., et al., 2020. High-dimensional analysis of immune cell composition predicts periprosthetic joint infections and dissects its pathophysiology. Biomedicines 8, 358.

Kuiper, J.W.P., Verberne, S.J., Vos, S.J., van Egmond, P.W., 2020. Does the alpha defensin ELISA test perform better than the alpha defensin lateral flow test for PJI diagnosis? A systematic review and meta-analysis of prospective studies. Clin. Orthop. Relat. Res. 478, 1333–1344.

Lee, Y.S., Koo, K.-H., Kim, H.J., Tian, S., Kim, T.-Y., Maltenfort, M.G., et al., 2017. Synovial fluid biomarkers for the diagnosis of periprosthetic joint infection: a systematic review and meta-analysis. J. Bone Joint Surg. Am. 99, 2077–2084.

Lee, Y.S., Lee, Y.-K., Han, S.B., Nam, C.H., Parvizi, J., Koo, K.-H., 2018. Natural progress of D-dimer following total joint arthroplasty: a baseline for the diagnosis of the early postoperative infection. J. Orthop. Surg. 13, 36.

Lenguerrand, E., Whitehouse, M.R., Beswick, A.D., Kunutsor, S.K., Burston, B., Porter, M., et al., 2018. Risk factors associated with revision for prosthetic joint infection after hip replacement: a prospective observational cohort study. Lancet Infect. Dis. 18, 1004–1014.

Liu, K., Fu, J., Yu, B., Sun, W., Chen, J., Hao, L., 2018. Meta-analysis of sonication prosthetic fluid PCR for diagnosing periprosthetic joint infection. PLoS One 13, e0196418.

Li, R., Li, X., Ni, M., Fu, J., Xu, C., Chai, W., et al., 2021. What is the performance of novel synovial biomarkers for detecting periprosthetic joint infection in the presence of inflammatory joint disease? Bone Joint J. 103-B, 32–38.

Li, R., Lu, Q., Zhou, Y.-G., Chai, W., Lu, S.-B., Chen, J.-Y., 2018a. Centrifugation may change the results of leukocyte esterase strip testing in the diagnosis of periprosthetic joint infection. J. Arthroplasty 33, 2981–2985.

Li, C., Renz, N., Trampuz, A., 2018b. Management of periprosthetic joint infection. Hip Pelvis 30, 138–146.

Li, R., Shao, H.-Y., Hao, L.-B., Yu, B.-Z., Qu, P.-F., Zhou, Y.-X., et al., 2019. Plasma fibrinogen exhibits better performance than plasma D-dimer in the diagnosis of periprosthetic joint infection: a multicenter retrospective study. J. Bone Joint Surg. Am. 101, 613–619.

Lu, G., Li, T., Ye, H., Liu, S., Zhang, P., Wang, W., 2020. D-dimer in the diagnosis of periprosthetic joint infection: a systematic review and meta-analysis. J. Orthop. Surg. Res. 15, 265.

Marazzi, M.G., Randelli, F., Brioschi, M., Drago, L., Romanò, C.L., Banfi, G., et al., 2018. Presepsin: a potential biomarker of PJI? A comparative analysis with known and new infection biomarkers. Int. J. Immunopathol. Pharmacol. 31, 394632017749356.

Marín, M., Garcia-Lechuz, J.M., Alonso, P., Villanueva, M., Alcalá, L., Gimeno, M., et al., 2012. Role of universal 16S rRNA gene PCR and sequencing in diagnosis of prosthetic joint infection. J. Clin. Microbiol. 50, 583–589.

McArthur, B.A., Abdel, M.P., Taunton, M.J., Osmon, D.R., Hanssen, A.D., 2015. Seronegative infections in hip and knee arthroplasty: periprosthetic infections with normal erythrocyte sedimentation rate and C-reactive protein level. Bone Joint J. 97-B, 939–944.

McNally, M., Sousa, R., Wouthuyzen-Bakker, M., Chen, A.F., Soriano, A., Vogeley, H.C., et al., 2021. The EBJIS definition of periprosthetic joint infection: a practical guide for clinicians. Bone Joint J. 103-B, 18–25.

Meermans, G., Haddad, F.S., 2010. Is there a role for tissue biopsy in the diagnosis of periprosthetic infection? Clin. Orthop. 468, 1410–1417.

Mikhael, M.M., Hanssen, A.D., Sierra, R.J., 2009. Failure of metal-on-metal total hip arthroplasty mimicking hip infection. A report of two cases. J. Bone Joint Surg. Am. 91, 443–446.

Mirza, S.Z., Richardson, S.S., Kahlenberg, C.A., Blevins, J.L., Lautenbach, C., Demetres, M., et al., 2019. Diagnosing prosthetic joint infections in patients with inflammatory arthritis: a systematic literature review. J. Arthroplasty 34, 1032–1036.

Moser, K.A., Pearson, L.N., Pelt, C.E., Olson, J.D., Goodwin, A.J., Isom, J.A., et al., 2020. Letter to the editor on "The 2018 definition of periprosthetic hip and knee infection: an evidence-based and validated criteria.", J. Arthroplasty 35, 2682–2683.

Oethinger, M., Warner, D.K., Schindler, S.A., Kobayashi, H., Bauer, T.W., 2011. Diagnosing periprosthetic infection: false-positive intraoperative Gram stains. Clin. Orthop. 469, 954–960.

Okroj, K.T., Calkins, T.E., Kayupov, E., Kheir, M.M., Bingham, J.S., Beauchamp, C.P., et al., 2018. The alpha-defensin test for diagnosing periprosthetic joint infection in the setting of an adverse local tissue reaction secondary to a failed metal-on-metal bearing or corrosion at the head-neck junction. J. Arthroplasty 33, 1896–1898.

Omar, M., Ettinger, M., Reichling, M., Petri, M., Guenther, D., Gehrke, T., et al., 2015. Synovial C-reactive protein as a marker for chronic periprosthetic infection in total hip arthroplasty. Bone Joint J. 97-B, 173–176.

Osmon, D.R., Berbari, E.F., Berendt, A.R., Lew, D., Zimmerli, W., Steckelberg, J.M., et al., 2013. Diagnosis and management of prosthetic joint infection: clinical practice guidelines by the Infectious Diseases Society of America. Clin. Infect. Dis. 56, e1–e25.

Ottink, K.D., Strahm, C., Muller-Kobold, A., Sendi, P., Marjan Wouthuyzen-Bakker, M., 2019. Factors to consider when assessing the diagnostic accuracy of synovial leukocyte count in periprosthetic joint infection. J. Bone Joint Infect. 4, 167–173.

Palan, J., Nolan, C., Sarantos, K., Westerman, R., King, R., Foguet, P., 2019. Culture-negative periprosthetic joint infections. EFORT Open. Rev. 4, 585–594.

Partridge, D.G., Gordon, A., Townsend, R., 2017. False-positive synovial fluid alpha-defensin test in a patient with acute gout affecting a prosthetic knee. Eur. J. Orthop. Surg. Traumatol. 27, 549–551.

Parvizi, J., Della, Valle, C.J., 2010. AAOS clinical practice guideline: diagnosis and treatment of periprosthetic joint infections of the hip and knee. J. Am. Acad. Orthop. Surg. 18, 771–772.

Parvizi, J., Gehrke, T., Chen, A.F., 2013. Proceedings of the International Consensus on Periprosthetic Joint Infection. Bone Joint J. 95-B, 1450–1452.

Parvizi, J., Gehrke, T., 2018. Proceedings of the Second International Consensus Meeting on Musculoskeletal Infection, Hip and Knee Section, Data Trace Publishing Company, Brooklandville, MD, USA.

Parvizi, J., Tan, T.L., Goswami, K., Higuera, C., Della Valle, C., Chen, A.F., et al., 2018. The 2018 definition of periprosthetic hip and knee infection: an evidence-based and validated criteria. J. Arthroplasty 33, 1309–1314.

Parvizi, J., Zmistowski, B., Berbari, E.F., Bauer, T.W., Springer, B.D., Della Valle, C.J., et al., 2011. New definition for periprosthetic joint infection: from the workgroup of the musculoskeletal infection society. Clin. Orthop. 469, 2992–2994.

Pérez-Prieto, D., Portillo, M.E., Puig-Verdié, L., Alier, A., Martínez, S., Sorlí, L., et al., 2017. C-reactive protein may misdiagnose prosthetic joint infections, particularly chronic and low-grade infections. Int. Orthop. 41, 1315–1319.

Perry, J., Reed, M.R., Refaie, R., Sprowson, A.P., Rankin, K.S., Refaie, R., 2013. The assessment of neutrophil CD64 count as an early warning marker of joint replacement infection. Arch. Orthop. Trauma. Surg. 133, 1351–1358.

Pisanu, F., Doria, C., Andreozzi, M., Bartoli, M., Saderi, L., Sotgiu, G., et al., 2019. Pleomorphic clinical spectrum of metallosis in total hip arthroplasty. Int. Orthop. 43, 85–96.

Plate, A., Stadler, L., Sutter, R., Anagnostopoulos, A., Frustaci, D., Zbinden, R., et al., 2018. Inflammatory disorders mimicking periprosthetic joint infections may result in false-positive α-defensin. Clin. Microbiol. Infect. 24, 1212.e1–1212.e6.

Prince, N., Penatzer, J.A., Dietz, M.J., Boyd, J.W., 2020. Localized cytokine responses to total knee arthroplasty and total knee revision complications. J. Transl. Med. 18, 330.

Qu, P.F., Li, R., Xu, C., Chai, W., Li, H., Fu, J., et al., 2020. A clinical pilot study to evaluate CD64 expression on blood monocytes as an indicator of periprosthetic joint infection. J. Bone Joint Surg. Am. 102, e99.

Randau, T.M., Friedrich, M.J., Wimmer, M.D., Reichert, B., Kuberra, D., Stoffel-Wagner, B., et al., 2014. Interleukin-6 in serum and in synovial fluid enhances the differentiation between periprosthetic joint infection and aseptic loosening. PLoS One 9, e89045.

Renz, N., Yermak, K., Perka, C., Trampuz, A., 2018. Alpha defensin lateral flow test for diagnosis of periprosthetic joint infection: not a screening but a confirmatory test. J. Bone Joint Surg. Am. 100, 742–750.

Romanò, C.L., Khawashki, H.A., Benzakour, T., Bozhkova, S., Del Sel, H., Hafez, M., et al., 2019. The W.A.I.O.T. definition of high-grade and low-grade peri-prosthetic joint infection. J. Clin. Med. 8, 650.

Romanò, C.L., Petrosillo, N., Argento, G., Sconfienza, L.M., Treglia, G., Alavi, A., et al., 2020. The role of imaging techniques to define a peri-prosthetic hip and knee joint infection: multidisciplinary consensus statements. J. Clin. Med. 9, 2548.

Ryu, S.Y., Greenwood-Quaintance, K.E., Hanssen, A.D., Mandrekar, J.N., Patel, R., 2014. Low sensitivity of periprosthetic tissue PCR for prosthetic knee infection diagnosis. Diagn. Microbiol. Infect. Dis. 79, 448–453.

Sadiq, S., Wootton, J.R., Morris, C.A., Northmore-Ball, M.D., 2005. Application of core biopsy in revision arthroplasty for deep infection. J. Arthroplasty 20, 196–201.

Salari, P., Grassi, M., Cinti, B., Onori, N., Gigante, A., 2020. Synovial fluid calprotectin for the preoperative diagnosis of chronic periprosthetic joint infection. J. Arthroplasty 35, 534–537.

Saleh, A., George, J., Faour, M., Klika, A.K., Higuera, C.A., 2018. Serum biomarkers in periprosthetic joint infections. Bone Joint Res. 7, 85–93.

Samsudin, I., Vasikaran, S.D., 2017. Clinical utility and measurement of procalcitonin. Clin. Biochem. Rev. 38, 59–68.

Shahi, A., Alvand, A., Ghanem, E., Restrepo, C., Parvizi, J., 2019. The leukocyte esterase test for periprosthetic joint infection is not affected by prior antibiotic administration. J. Bone Joint Surg. Am. 101, 739–744.

Shahi, A., Kheir, M.M., Tarabichi, M., Hosseinzadeh, H.R.S., Tan, T.L., Parvizi, J., 2017a. Serum D-dimer test is promising for the diagnosis of periprosthetic joint infection and timing of reimplantation. J. Bone Joint Surg. Am. 99, 1419–1427.

Shahi, A., Parvizi, J., Kazarian, G.S., Higuera, C., Frangiamore, S., Bingham, J., et al., 2016. The alpha-defensin test for periprosthetic joint infections is not affected by prior antibiotic administration. Clin. Orthop. 474, 1610–1615.

Shahi, A., Tan, T.L., Kheir, M.M., Tan, D.D., Parvizi, J., 2017b. Diagnosing periprosthetic joint infection: and the winner is? J. Arthroplasty 32, S232–S235.

Sharma, K., Ivy, M., Block, D.R., Abdel, M.P., Hanssen, A.D., Beauchamp, C., et al., 2020. Comparative analysis of 23 synovial fluid biomarkers for hip and knee periprosthetic joint infection detection. J. Orthop. Res. 38, 2664–2674.

Shohat, N., Goswami, K., Fillingham, Y., Tan, T.L., Calkins, T., Della Valle, C.J., et al., 2018. Diagnosing periprosthetic joint infection in inflammatory arthritis: assumption is the enemy of true understanding. J. Arthroplasty 33, 3561–3566.

Shohat, N., Parvizi, J., 2020. Reply to letter to the editor regarding "The 2018 definition of periprosthetic hip and knee infection: an evidence-based and validated criteria". J. Arthroplasty 35, 2683–2684.

Signore, A., Sconfienza, L.M., Borens, O., Glaudemans, A.W.J.M., Cassar-Pullicino, V., Trampuz, A., et al., 2019. Consensus document for the diagnosis of prosthetic joint infections: a joint paper by the EANM, EBJIS, and ESR (with ESCMID endorsement). Eur. J. Nucl. Med. Mol. Imaging 46, 971–988.

Tarabichi, M., Shohat, N., Goswami, K., Alvand, A., Silibovsky, R., Belden, K., et al., 2018. Diagnosis of periprosthetic joint infection: the potential of next-generation sequencing. J. Bone Joint Surg. Am. 100, 147–154.

Tetreault, M.W., Wetters, N.G., Aggarwal, V., Mont, M., Parvizi, J., Della Valle, C.J., 2014. The Chitranjan Ranawat Award: should prophylactic antibiotics be withheld before revision surgery to obtain appropriate cultures? Clin. Orthop. 472, 52–56.

Toossi, N., Adeli, B., Rasouli, M.R., Huang, R., Parvizi, J., 2012. Serum white blood cell count and differential do not have a role in the diagnosis of periprosthetic joint infection. J. Arthroplasty 27, 51–54.

Trampuz, A., Hanssen, A.D., Osmon, D.R., Mandrekar, J., Steckelberg, J.M., Patel, R., 2004. Synovial fluid leukocyte count and differential for the diagnosis of prosthetic knee infection. Am. J. Med. 117, 556–562.

Trampuz, A., Piper, K.E., Jacobson, M.J., Hanssen, A.D., Unni, K.K., Osmon, D.R., et al., 2007. Sonication of removed hip and knee prostheses for diagnosis of infection. N. Engl. J. Med. 357, 654–663.

Tsaras, G., Maduka-Ezeh, A., Inwards, C.Y., Mabry, T., Erwin, P.J., Murad, M.H., et al., 2012. Utility of intraoperative frozen section histopathology in the diagnosis of periprosthetic joint infection: a systematic review and meta-analysis. J. Bone Joint Surg. Am. 94, 1700–1711.

Villa, F., Toscano, M., De Vecchi, E., Bortolin, M., Drago, L., 2017. Reliability of a multiplex PCR system for diagnosis of early and late prosthetic joint infections before and after broth enrichment. Int. J. Med. Microbiol. 307, 363–370.

Wang, C., Wang, Q., Li, R., Duan, J.-Y., Wang, C.-B., 2016. Synovial fluid C-reactive protein as a diagnostic marker for periprosthetic joint infection: a systematic review and meta-analysis. Chin. Med. J. 129, 1987–1993.

Wasterlain, A.S., Goswami, K., Ghasemi, S.A., Parvizi, J., 2020. Diagnosis of periprosthetic infection: recent developments. J. Bone Joint Surg. Am. 102, 1366–1375.

Wetters, N.G., Berend, K.R., Lombardi, A.V., Morris, M.J., Tucker, T.L., Della Valle, C.J., 2012. Leukocyte esterase reagent strips for the rapid diagnosis of periprosthetic joint infection. J. Arthroplasty 27, 8–11.

Wouthuyzen-Bakker, M., Ploegmakers, J.J.W., Kampinga, G.A., Wagenmakers-Huizenga, L., Jutte, P.C., Muller Kobold, A.C., 2017. Synovial calprotectin: a potential biomarker to exclude a prosthetic joint infection. Bone Joint J. 99-B, 660–665.

Wouthuyzen-Bakker, M., Ploegmakers, J.J.W., Ottink, K., Kampinga, G.A., Wagenmakers-Huizenga, L., Jutte, P.C., et al., 2018. Synovial calprotectin: an inexpensive biomarker to exclude a chronic prosthetic joint infection. J. Arthroplasty 33, 1149–1153.

Wyatt, M.C., Beswick, A.D., Kunutsor, S.K., Wilson, M.J., Whitehouse, M.R., Blom, A.W., 2016. The alpha-defensin immunoassay and leukocyte esterase colorimetric strip test for the diagnosis of periprosthetic infection: a systematic review and meta-analysis. J. Bone Joint Surg. Am. 98, 992–1000.

Xie, K., Dai, K., Qu, X., Yan, M., 2017. Serum and synovial fluid interleukin-6 for the diagnosis of periprosthetic joint infection. Sci. Rep. 7, 1496.

Yi, P.H., Cross, M.B., Moric, M., Levine, B.R., Sporer, S.M., Paprosky, W.G., et al., 2015. Do serologic and synovial tests help diagnose infection in revision hip arthroplasty with metal-on-metal bearings or corrosion? Clin. Orthop. 473, 498–505.

Yoon, H.-K., Cho, S.-H., Lee, D.-Y., Kang, B.-H., Lee, S.-H., Moon, D.-G., et al., 2017. A review of the literature on culture-negative periprosthetic joint infection: epidemiology, diagnosis and treatment. Knee Surg. Relat. Res. 29, 155–164.

Yoon, J.R., Yang, S.H., Shin, Y.S., 2018. Diagnostic accuracy of interleukin-6 and procalcitonin in patients with periprosthetic joint infection: a systematic review and meta-analysis. Int. Orthop. 42, 1213–1226.

Zhang, Q., Dong, J., Zhou, D., Liu, F., 2021. Circulating D-dimer versus fibrinogen in the diagnosis of peri-prosthetic joint infection: a meta-analysis. Surg. Infect. 22, 200–210.

Zheng, Q.-Y., Li, R., Ni, M., Ren, P., Ji, Q.-B., Sun, J.-Y., et al., 2021. What is the optimal timing for reading the leukocyte esterase strip for the diagnosis of periprosthetic joint infection? Clin. Orthop. Relat. Res. 479, 1323–1330.

Chapter 7

# Hip implants and systemic cobalt toxicity: a comprehensive review with case studies

**Ilona Świątkowska[1], Obakanyin J. Akinfosile[2], Ravindra V. Badhe[2], Mark Barba[3], Mathew T. Mathew[2] and Divya Bijukumar[4]**
*[1]Institute of Orthopaedics and Musculoskeletal Science, University College London, Stanmore, United Kingdom, [2]Regenerative Medicine and Disability Laboratory, Department of Biomedical Sciences, University of Illinois College of Medicine Rockford, Rockford, IL, United States, [3]Orthoillinois, Department of Surgery, Rockford, IL, United States, [4]Blazer Nanomedicine Laboratory, Department of Biomedical Sciences, University of Illinois College of Medicine, Rockford, IL, United States*

## 7.1 Introduction

In medical terms, the toxicity of a substance is the degree to which it can damage humans, animals, plants, animal/bacterial cells, or viruses. It can be categorised based on intensity (acute or chronic) and site (local or systemic): in acute toxicity cases, the damage is observed a short time (seconds to hours) after a single exposure to a toxicant, whereas chronic toxicity occurs in response to continuous exposure to a toxicant over a longer period of time (often $>3$ months). Local toxicity involves damage to a particular part of the body and is limited to a small area. In case of systemic toxicity, the damage spreads from the site of toxic exposure to distant parts of the body, often through the bloodstream. Since systemic toxicity can affect multiple organs and systems, it is more serious than local toxicity and its treatment is more complicated. The major types of toxicants that can precipitate systemic adverse reactions include drug molecules, industrial chemicals, poisons, radioactive materials, and heavy metals.

Patients with cobalt-chromium (CoCr) alloy-based hip implants can show both local and systemic symptoms, although systemic complications are much rarer (Fig. 7.1). Repeated cycles of motion lead to the deterioration of implant surfaces and generation of metallic debris, including CoCr wear particles, corrosion products, and metal ions, which interact with the

Biomarkers of Hip Implant Function. DOI: https://doi.org/10.1016/B978-0-12-821596-8.00007-0

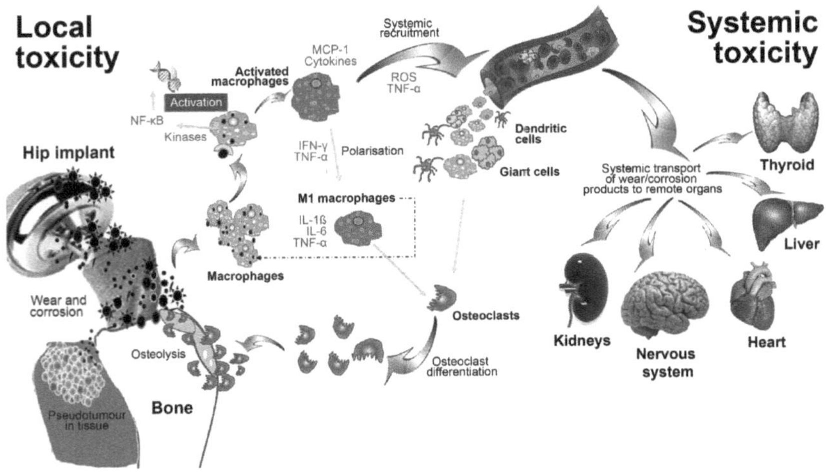


**FIGURE 7.1** A schematic of the adverse local tissue reaction cascade and systemic toxicity associated with metal hip implants. The wear debris activates tissue-resident macrophages, leading to the recruitment and polarisation of systemic macrophages, giant cells, and dendritic cells, and provoking inflammatory reactions in the periprosthetic area. Metal particles and ions disseminated in the blood and lymph can also cause toxicity in distal organs and systems. *IFN*, interferon; *IL*, interleukin; *MCP-1*, monocyte chemoattractant protein-1; *ROS*, reactive oxygen species; *TNF*, tumour necrosis factor (Bijukumar et al., 2018). *Adapted with permission from Bijukumar, D.R., Segu, A., Souza, J.C., Li, X., Barba, M., Mercuri, L.G., 2018. Systemic and local toxicity of metal debris released from hip prostheses: A review of experimental approaches. Nanomedicine: Nanotechnology, Biology and Medicine 14 (3), 951–963.*

physiological environment surrounding the implants. Local toxicity resulting from a gradual build-up of metal debris in the periprosthetic tissue can cause inflammatory masses, tissue necrosis, osteolysis, and, ultimately, implant failure. The adverse local tissue reactions (ALTR) are mediated by monocytes, macrophages, and lymphocytes. Systemic toxicity, on the other hand, is a direct result of metal ions entering the circulation and interacting with the vital organs and systems of the body (Gessner et al., 2015).

The wear pattern of metal-on-metal (MoM) implants is characterised by an initial phase of high material loss and elevated systemic cobalt and/or chromium levels as small irregularities on the bearing surfaces are worn down (1–2 years postimplantation) (Savarino et al., 2014). In patients with well-functioning implants, metal concentrations eventually decrease and plateau, whereas in those with failing prostheses, they continue to increase beyond the initial 'running-in' period. Although both cobalt and chromium levels are often elevated in patients with high-wearing implants, cobalt is the more important contributor to metal ion toxicity. Systemic sequalae resulting from prosthetic cobalt exposure have been described before in relation to occupational exposure, accidental ingestion, and medicinal use of cobalt (Bhardwaj et al., 2011; Kennedy et al., 1981;

Manifold et al., 1978), whereas no analogy exists in cases of chromium intoxication. Further, in vitro studies demonstrated that CoCr wear particles release disproportionally more cobalt ions than chromium ions (Koronfel et al., 2018), and that the former species are more cytotoxic (Apostoli et al., 2013; Catelas et al., 2001).

Systemic cobalt levels are most commonly measured in samples of serum or whole blood using inductively coupled plasma–mass spectrometry (ICP-MS) or graphite-furnace atomic absorption spectroscopy (AAS). The majority of patients with MoM and metal-on-polyethylene (MoP) hip implants display blood cobalt concentrations between 0.1 and 10 μg/L (Paustenbach et al., 2014), while background blood cobalt levels in the general population are usually between 0.04 and 0.09 μg/L (Hahn et al., 2012; Sampson and Hart, 2012). According to the Medicines and Healthcare products Regulatory Agency, blood cobalt or chromium concentrations exceeding 7 μg/L indicate an increased risk of ALTR and implant failure (Hart et al., 2011), but there is no universally accepted threshold level to help guide physicians when evaluating a patient's risk of developing systemic sequelae.

This chapter pertains to systemic cobalt toxicity in the setting of hip arthroplasty, with a focus on the differential diagnosis, treatment options, and predisposing factors. A comprehensive review of the literature is supplemented by case studies that aim to raise awareness of the variable presentation of the condition and expedite its management.

## 7.2 Arthroprosthetic cobaltism

The first report of hip prosthesis-related systemic cobaltism was published in 2001 in an Italian journal (Megaterio et al., 2001). The patient in question experienced accelerated wear of a revision CoCr-alloy femoral head, resulting in a set of generalised complaints, such as motor and sensory neuropathy, pericardial tamponade, and hypothyroidism. Since that time, over 50 cases have been described in which patients with CoCr-based hip prostheses presented with a myriad of systemic symptoms, including haematological, endocrine, dermatological, neurological, cardiac, and constitutional abnormalities, in association with elevated levels of cobalt—a syndrome collectively known as 'arthroprosthetic cobaltism'. Table 7.1 provides an overview of these case reports.

The two affected subpopulations were patients with MoM hips and those with fractured ceramic implants that were subsequently replaced by CoCr-alloy components. Patients in the latter group experienced higher systemic elevations in cobalt levels and more severe adverse effects. The onset of the toxic symptoms was also quicker in those with a history of ceramic fracture, with the majority of the patients presenting with systemic complications within 2 years of revision surgery. It is thought that excessive grinding of residual ceramic particles on the new metal femoral head was responsible for the catastrophically high cobalt release in this subpopulation. Marked

**TABLE 7.1** Summary of the reported cases of systemic toxicity in patients with hip implants and elevated cobalt levels (Swiatkowska, 2019).

| Reference | Age sex | Co level[a] (μg/L) | Toxicity symptoms | | | | | | | Time to first symptoms | Improved after revision? | Comorbidities |
|---|---|---|---|---|---|---|---|---|---|---|---|---|
| | | | C | N | V | H | T | P | Other | | | |
| **Cobalt release via third-body wear of metallic components** | | | | | | | | | | | | |
| Kim et al. (2016) | 53M | 397,800[b] | x | x | | | | | Fatigue, nonspecific weakness | <2 years | Yes, except leg neuropathy | N/R |
| Gilbert et al. (2013); Zywiel et al. (2013) | 46M | 6521[B], 1085[S] | x | x | | | x | x | Severe fatigue, anorexia, weight loss, tinnitus, renal and liver failure | 6 months | Fatal | N/R |
| Grant et al. (2016) | 61F | 2148[P], 1997[B] | x | x | x | x | x | x | Dysgeusia, nausea, vomiting, weight loss | 2 months | Yes | Hyperlipidaemia |
| Griffiths et al. (2015) | 69F | 2006[B] | x | x | x | x | x | | Profound weakness, fatigue, nausea, vomiting, unexplained weight loss | 5 years | Yes, except for lower limb weakness and vision | Hypertension, IHD, pulmonary embolism, myocardial infarct |
| Lecoanet et al. (2019) | 69F | 1464[B] | x | | x | x | | | Fatigue, cognitive decline, memory loss, weight loss, metallic dysgeusia | <1 year | Yes | N/R |

| | | | | | | | | | | | | |
|---|---|---|---|---|---|---|---|---|---|---|---|---|
| Grillo et al. (2016) | 66M | $1078^{S}$ | | | x | x | | | Increasing abdominal distention, ankle swelling, subjective loss of mental clarity, incoordination, rash | 9 months | Yes | Childhood amblyopia of right eye |
| Esteban Sanchez and Pasto-Cardona (2016) | 50M | $1036^{S}$ | x | x | x | x | x | | Fatigue, anorexia, weight loss, loss of smell, anxiety, depression | 1 year | Yes | Hypertension, DM |
| Garcia et al. (2020) | 59F | $>1000^{S}$ | x | x | x | x | x | | Fatigue, hair loss, severe headaches | 9 months | Yes | N/R |
| Stepien et al. (2018) | 58M | $903^{P}$ | x | x | x | x | x | x | Progressive fatigue, vertigo, hepatosplenomegaly | 9 months | Yes, except hearing | Perforated eardrum |
| Dahms et al. (2014) | 55M | $885^{P}$ | x | | x | x | x | | Enlarged lymph nodes | 2 years | Yes, except hearing and vision | N/R |
| Harris et al. (2015) | 57F | 788 | x | x | x | | x | | Fatigue, memory loss | 8 years | Yes | Heavy smoker |
| Ho et al. (2018); Yu (2017) | 63M | $779^{S}$ | x | | x | x | x | | Anorexia, nausea, weight loss | 2 years | Fatal | Hypertension, skin cancer, aortic aneurysm |

*(Continued)*

**TABLE 7.1** (Continued)

| Reference | Age sex | Co level[a] (μg/L) | Toxicity symptoms | | | | | | | Time to first symptoms | Improved after revision? | Comorbidities |
|---|---|---|---|---|---|---|---|---|---|---|---|---|
| | | | C | N | V | H | T | P | Other | | | |
| Sanz Pérez et al. (2019) | 31M | 652[S] | x | | | | | | | 2 years | No, heart transplant required | N/R |
| Oldenburg et al. (2009) | 55M | 625[B] | x | x | | x | x | | Fatigue, headaches, poor concentration, weight loss, nail discoloration, eczema, dysgeusia | 3 months | Yes | N/R |
| Peters et al. (2017) | 71M | 596.5[S] | x | | x | x | x | | Weight loss, vomiting, vertigo | 6 months | Fatal | DM, multiple myeloma, ischaemic cerebrovascular accident |
| Rizzetti et al. (2009) | 58F | 549[B], 90[P] | | x | x | x | x | | Fatigue | 2 months | Yes, except vision | DM |
| Pelclova et al. (2012) | 56M | 506[S] | x | x | | x | x | | Weight loss | 14 months | Yes, except hearing | DM, hypertension |
| Choi et al. (2018) | 52M | 489.5[S] | x | x | | x | | | Fatigue, muscle weakness | 1 year | Yes | Hypertension, alcoholism |

| | | | | | | | | | | | | |
|---|---|---|---|---|---|---|---|---|---|---|---|---|
| Gautam et al. (2019) | 37M | $373^{S}$ | x | | | | | | Shortness of breath, easy fatigability | 3 years | Fatal | N/R |
| Apel et al. (2013) | 65M | $355^{P}$ | x | x | x | | x | | General malaise | 4 years | Yes | Hypertension, hypercholesterolaemia, DM, obesity |
| Fox et al. (2016) | 60F | $424^{B}$ | x | | | x | | | Metallic dysgeusia, weight loss, fatigue | 10 months | Fatal | N/R |
| Weber et al. (2015) | 66F | $412^{S}$ | x | x | x | x | x | x | Recurrent depression, anorexia, weight loss, fatigue | 15 months | Yes, except vision | Coronary artery disease, secondary hypothyroidism, a heavy smoker |
| Ikeda et al. (2010) | 56F | $>400^{B}$ | | x | | x | x | | Fatigue | 2 years | Yes | N/R |
| Steens et al. (2006) | 53M | $398^{S}$ | | x | x | x | | | Generalised dermatitis | 2 years | Yes | N/R |
| Giampreti et al. (2014) | 75M | $352.6^{S}$ | x | | | | | | | 5 years | Yes | N/R |
| Choi et al. (2018) | 46M | $112^{S}$ | x | | | | | | Kidney failure | 6 years | Yes | Hypertension, CKD |
| Olmedo-Garcia and Zagra (2018) | 32M | $>59^{B}$ | x | | | x | | | Pancreatitis | 6 years | No | N/R |

(*Continued*)

**TABLE 7.1** (Continued)

| Reference | Age sex | Co level[a] (μg/L) | Toxicity symptoms | | | | | | | Time to first symptoms | Improved after revision? | Comorbidities |
|---|---|---|---|---|---|---|---|---|---|---|---|---|
| | | | C | N | V | H | T | P | Other | | | |
| Vasukutty and Minhas (2016) | 40M | 45[S] | x | | | | | | Cognitive impairment, poor memory, mood swings, encephalomyelitis | 7 years | Yes | Hemochromatosis |
| **Cobalt release via wear of metal-on-metal surfaces** | | | | | | | | | | | | |
| Ho et al. (2017) | 60M | 1096[S] | | x | x | x | x | x | Tinnitus | 16 years | N/R | N/R |
| Allen et al. (2014) | 59F | 287.6[S] | x | | | | x | | Fatigue, foot swelling | 3–4 years | Yes | Polyarthritis |
| Leikin et al. (2013) | 61F | 254[B] | | x | | | | | | 3 months | Yes | N/R |
| Tilney et al. (2017) | 40M | 246[S] | x | | | | | | | 4 years | Yes | Occasional cocaine use |
| Khan et al. (2015) | 69F | 200–300[S] | x | | | | | | | N/R | Fatal | Hypertension, mild renal impairment |
| Day et al. (2017) | 69F | 199[S] | x | | | | | | | 5 years | Fatal | N/R |
| Martin et al. (2015) | 64F | 192[S] | x | | | | | | Impaired renal function | 2 years | Fatal | N/R |

| | | | | | | | | | | | | |
|---|---|---|---|---|---|---|---|---|---|---|---|---|
| Moniz et al. (2017) | 58F | $169^{S}$ | x | | | | | | Acute renal failure | 10 years | Yes | N/R |
| Ho et al. (2017) | 56M | $165^{S}$ | | x | | | | | Extreme lethargy | 7 years | Yes | N/R |
| Charette et al. (2017) | 46M | $156^{S}$ | x | | | | | | | 2 years | Yes | N/R |
| Connell et al. (2013) | 42M | $156^{S}$ | x | | | | | x | | Several years | Not revised | N/R |
| Singh et al. (2020) | 67F | 121 | x | | | | | | | N/R | No, heart transplant required | N/R |
| Mosier et al. (2016); Samar et al. (2015) | 54M | $120^{S}$ | x | x | | | | | Fatigue | 4–5 years | No, LVAD required | Obesity, mild renal impairment |
| Bonilla and Bhimaraj (2018) | 39M | $>100^{S}$ | x | | | | | | | N/A | Yes | Thyroid cancer, thyroidectomy, low albumin |
| Tower (2012) | N/A | $64^{S}$ | | x | | x | | | New-onset anxiety and depression, tinnitus, cognitive decline | 15 months | Not revised | N/R |
| Tower (2010a, b) | 49M | $50^{S}$ | x | x | x | x | | | Fatigue, headaches, anxiety, tinnitus, cognitive decline, poor memory, lassitude, depression, axillary rashes | 3–11 months | Yes, except for tinnitus and vision | N/R |

*(Continued)*

**TABLE 7.1** (Continued)

| Reference | Age sex | Co level[a] (μg/L) | Toxicity symptoms | | | | | | | Time to first symptoms | Improved after revision? | Comorbidities |
|---|---|---|---|---|---|---|---|---|---|---|---|---|
| | | | C | N | V | H | T | P | Other | | | |
| Ng et al. (2013) | 39F | $45^{S}$ | | | x | | | | Occasional metallic gustation, morning nausea | 5 years | Not revised | N/R |
| Mao et al. (2011) | 73F | $24^{S}$ | | | | | | | Fatigue, cognitive decline, poor memory, depression, weight loss, severe headaches, metallic gustation | 5 years | Yes | Stroke |
| Tower (2010a, b) | 49M | $23^{S}$ | x | | | x | | | Cognitive decline, mental fog, memory loss, vertigo, rash, dyspnoea | 1 year | Yes | N/R |
| Machado et al. (2012) | 75M | $14^{P}$ | x | | | | | | | 6 years | Yes | DM, renal impairment, hypercholesterolaemia, prostate cancer, obesity |

| | | | | | | | | | | | | |
|---|---|---|---|---|---|---|---|---|---|---|---|---|
| Mao et al. (2011) | 60M | $11^{S}$ | | | | | | | Fatigue, cognitive decline, poor memory, poor exercise tolerance, dyspnoea, painful muscle fatigue | 4 years | Yes | Hypertension |
| Reich et al. (2019) | 45F | $10^{S}$ | | | | | | | Fatigue, headaches, brain fog, leg paraesthesia, intermittent, disseminated, and pruritic facial, hand, and arm rash | 10 years | Yes, except leg paraesthesia | Polymyalgia, polyarthralgia, lupus erythematosus, POTS |
| Zeynalov et al. (2018) | 56M | $1.6^{S}$ | | | | | | | Insomnia, daytime sleepiness, severe obstructive sleep apnoea, headaches, crying spells, depression, decreased concentration, loss of appetite | 3 years | Yes | |

*C*, cardiomyopathy; *CKD*, chronic kidney disease; *DM*, diabetes mellitus; *F*, female *H*, hearing loss; *IHD*, ischaemic heart disease; *LVAD*, left ventricular assist device; *M*, male; *N*, peripheral neuropathy; *N/R*, not recorded; *P*, polycythaemia; *POTS*, postural orthostatic tachycardia syndrome; *T*, hypothyroidism; *V*, visual loss.

[a] *First measurement in the* $^{S}$*serum,* $^{P}$*plasma, or* $^{B}$*blood after the onset of systemic symptoms.*

[b] *Likely a reporting error.*

*Source*: Adapted from Swiatkowska, I., 2019. Toxicity of Metal Debris From Hip Implants (Ph.D. thesis). University College London, London, UK/CC-BY-4.0.

blackening of implant-adjacent tissue was a common intraoperative finding, particularly in patients with prior ceramic failure. Removal of the offending prosthesis was usually accompanied by gradual resolution of systemic symptoms, though some patients are thought to have developed irreversible neurological damage, several required a heart transplant or placement of a left ventricular assist device (LVAD), and at least eight suffered fatal complications. While the majority of the patients underwent formal testing to confirm the adverse effects, some reports were based on subjective self-reporting and were not detailed enough to judge the contribution of cobalt to the reported symptoms. Selected cases, in which the diagnosis of arthroprosthetic cobaltism was confirmed or considered highly likely, are described in detail in Section 7.2.1.

### 7.2.1 Case studies

Arthroprosthetic cobaltism presents a diagnostic challenge because it has no pathognomonic features, and there has been considerable variation in its manifestation and the associated cobalt levels in patients with metal hip implants. The following case studies are included to illustrate the range of presenting symptoms and emphasise the importance of timely diagnosis and treatment.

#### *7.2.1.1 Case study 1*

A 73-year-old woman treated for osteoarthritis with the MoM Articular Surface Replacement (ASR) system presented with symptoms of cognitive decline, memory deficits, and depression 5 years after the primary procedure (Mao et al., 2011). The neurological symptoms had been present since a stroke 7 months previously. She also complained of a continuous metallic taste in her mouth, severe headaches, anorexia, and weight loss. Save for mild groin pain, the patient experienced no local hip symptoms, and pelvic X-rays showed the implant to be well aligned and well fixed. Her serum cobalt and chromium levels were 24.1 and 12.5 μg/L, respectively. Because of the elevated metal levels and systemic manifestations, the patient underwent revision surgery to replace the metal bearing with an all-polyethylene cemented cup and a ceramic head. Intraoperatively, there was marked tissue metallosis; cobalt and chromium concentration in the aspirated joint fluid was 249 and 11,283 μg/L, respectively. Both cobalt and chromium could also be detected in the cerebrospinal fluid (0.5 and 0.7 μg/L, respectively). Eight weeks after the revision surgery, serum cobalt level had dropped to 3.5 μg/L, and the patient reported a subjective improvement: she gained weight, had more energy, and no longer experienced dysgeusia or groin pain.

### *7.2.1.2 Case study 2*

A 60-year-old male complained of systemic adverse effects 4 years after surgical implantation of the ASR XL Acetabular System (Mao et al., 2011). Within 3 years of the primary surgery, the patient developed several symptoms whose severity steadily increased: painful muscle fatigue associated with cramps in the arms and legs, dyspnoea and feeling faint while performing daily tasks, inability to climb a flight of stairs without needing to rest, and a subjective decline in cognitive function. He had difficulty remembering names and concentrating, and his previously stable hypertension had become uncontrolled, necessitating additional medication. At that time, the patient's serum cobalt level was 10.9 μg/L and rose to 15.2 μg/L over the following 5 months. His serum chromium concentration was normal at 5.2 μg/L. Although no local symptoms were present, the patient underwent revision surgery to replace the metal head and cup with a ceramic-on-polyethylene (CoP) bearing. Intraoperatively, there was no metallic debris around the implant or evidence of ALTR. At the 8-week follow-up, the patient reported substantial subjective improvement to energy levels and exercise tolerance, as well as decreased muscle fatigue. His serum cobalt level had decreased to 2.5 μg/L.

### *7.2.1.3 Case study 3*

A 49-year-old male developed auxiliary dermatitis, hip pain, and dyspnoea within 3–11 months of receiving the ASR XL device (Tower, 2010a,b,c). His serum cobalt level at the time was 50 μg/L. Over the following 2 years, he experienced increasing pain at the prosthesis and noted diffuse behavioural, neurocognitive, and mood disturbances, including anxiety, headaches, irritability, fatigue, tinnitus, audiometry-confirmed high-frequency hearing loss, hand tremor, incoordination, cognitive decline, depression, and sleep apnoea requiring continuous positive airway pressure ventilation. Three years after the initial presentation, optic nerve atrophy was found, and his serum cobalt concentration was 122 μg/L. In the next few months, a 'subtle right temporal superior quadrantanopia' and 'a flashing light in the same quadrant' had also developed. At 42 months, echocardiography revealed diastolic dysfunction (not present on the echocardiogram taken before initial surgery). The patient had the MoM implant revised a few weeks later. Marked joint metallosis, tissue necrosis, and lymphocytic infiltrate were noted during surgery. In the year following the procedure, the serum cobalt fell to 1.2 μg/L, and the patient's hip pain, affect, cognition, hearing, exercise tolerance, tremor, and professional productivity improved, while the tinnitus and visual symptoms were stable. Diastolic dysfunction had resolved at 63 months and the mood, cognition, and exercise tolerance further improved, though the latter did not fully recover and 'subtle visual field deficits' persisted.

### *7.2.1.4 Case study 4*

A 39-year-old female presented with a 3-week history of a paracentral scotoma in her left eye coupled with bilateral ocular discomfort, occasional metallic dysgeusia, and early-morning nausea, without any signs of cardiac or thyroid involvement (Ng et al., 2013). The patient had received bilateral MoM ASR implants 2 and 5 years previously. Ocular examination showed good central visual acuity (20/16) in both eyes, without pupillary defects or altered intraocular pressure. High-resolution optical coherence tomography (OCT) revealed degenerative alterations of the retinal pigment epithelium and photoreceptors, and the findings of indocyanine green angiography suggested choroidal infarction. The patient denied medication use and was not aware of any occupational exposure to toxic substances or a family history of ocular defects. As the serum cobalt level at presentation was elevated (44.7 μg/L), the chorioretinal degeneration was attributed to toxic cobalt release from the hip implants. The patient's vision was regularly followed up for 6 months and, given the lack of clinical progression, stable serum cobalt concentration, and good functional performance of both implants, surgical intervention was deemed unnecessary.

### *7.2.1.5 Case study 5*

A 65-year-old man with a history of ceramic-on-ceramic (CoC) total hip replacement (THR) complicated by femoral head fracture and subsequent conversion to a metal-on-ceramic (MoC) bearing presented to his ophthalmologist 5 years postrevision with worsening vision over 1 year (Apel et al., 2013). The vision decline was associated with a multisystem illness, including motor axonopathy, pericardiomyopathy, diabetes, antibody-negative hypothyroidism, and bulbar palsy. The patient's medical history was notable for obesity and hypertension, and he had an extensive medication list, including thyroxine, bisoprolol, and rosuvastatin. Optical examination revealed markedly reduced unaided visual acuity (6/60 in both eyes) along with a very poor colour vision. Although the patient's cornea, fundi, and optic discs were morphologically normal, bilateral moderate nuclear sclerosis was detected; brain magnetic resonance imaging (MRI) was largely unremarkable. The results of blood tests, including vitamin B and folate levels, and mitochondrial studies were all normal. The patient was provisionally diagnosed with optic neuritis secondary to medication toxicity and commenced on high-dose steroids with limited success. Subsequent cataract surgery brought slight improvements to his visual acuity and colour vision but all phases of the electroretinogram (ERG) remained abnormal, indicating inner retinal pathology. Laboratory testing showed plasma cobalt and blood chromium of 446 μg/L and 46 μg/L, respectively, prompting a revision of the MoC implant to a MoP bearing. Following the procedure, the patient's cobalt and chromium levels decreased and his vision and general health improved significantly.

### 7.2.1.6 Case study 6

A 66-year-old man presented with visual and hearing disturbances 9 months after revision of a shattered ceramic bearing using a MoM prosthesis (Grillo et al., 2016). He complained of deficits in colour vision and visual acuity (both at a distance and near) and required hearing aids for normal conversation. He also reported progressive abdominal distention and ankle swelling, pale diarrhoea, dermatitis, incoordination, pain in the left hip, and subjective cognitive decline. Examination revealed decreased visual acuity (20/80 and 20/30 in the right and left eye, respectively) and colour vision, and visual testing in the left eye showed cecocentral depression (results from the right eye were inconclusive owing to childhood amblyopia). Although these findings were consistent with toxic optic neuropathy, full-field and multifocal ERG detected no abnormalities and spectral-domain OCT showed a normal thickness of the retinal nerve fibre layer and retinal ganglion cell-inner plexiform layer. The serum cobalt level measured at the time was extremely elevated (1078 μg/L), prompting the decision to revise the metal hip. At surgery, the CoCr femoral head was significantly worn and ceramic shards were found in the joint. After the procedure, the patient experienced a sharp drop in serum cobalt levels and his visual acuity, colour vision, and visual field returned to near-normal values by the 18-month follow-up.

### 7.2.1.7 Case study 7

A 61-year-old woman with hyperlipidaemia and anxiety complained of insomnia, excessive thirst, and unintentional 10-pound weight loss since her broken ceramic implant was revised to a MoM bearing 2 months prior (Grant et al., 2016). On examination, she was tachycardic (141 beats/minute) and had mildly limited left leg abduction. Laboratory studies were notable for elevated haemoglobin (16.4 g/dL), haematocrit (50.5%), and glucose (156 mg/dL) levels, but haemoglobin A1C, thyroid, and renal tests all returned normal results. The patient was prescribed a sleep aid and an antidepressant. Within weeks, she developed crying spells, decreased appetite with dysgeusia, nausea, vomiting, paraesthesia, hearing impairment, and unspecified vision deficits that prevented her from driving. She was diagnosed with profound audiometry-confirmed bilateral hearing loss, sinus tachycardia, persistent hyperglycaemia requiring metformin, and severe gastritis on esophagogastroduodenoscopy; brain MRI was unremarkable. Within 3 months of the initial presentation, she developed new hypothyroidism and could only hear loud voices and see shadows. Transthoracic echocardiogram suggested mild diastolic dysfunction with preserved systolic function (ejection fraction [EF], >60%) and left ventricle concentric remodelling. The plasma cobalt level measured at the time was 2184 μg/L. Immediate implant removal could not be performed, so haemodialysis and therapeutic plasma exchange (TPE) were initiated in an attempt to rapidly lower the extremely elevated systemic cobalt levels and ease worsening symptoms. While haemodialysis had no discernible effect on the blood metal levels, a single TPE procedure markedly lowered the plasma cobalt and chromium content, albeit only temporarily. It was only after removal of the MoM prosthesis that the cobalt levels had plateaued below 300 μg/L. The patient was discharged 15 days after the surgery with mild symptomatic improvement in vision and hearing (no further details provided).

### *7.2.1.8 Case study 8*

A 60-year-old woman with a history of right CoC THR complicated by acetabular cup fracture and subsequent conversion to a MoP prosthesis presented to the emergency department with right hip pain and progressively worsening dyspnoea over 3 weeks (Fox et al., 2016 ). On examination, she was tachycardic, tachypnoeic, and hypertensive, with bilateral pulmonary emboli evident on computed tomography (CT) pulmonary angiography. The patient also reported progressively worsening hearing and 50-pound weight loss over the past 6–8 months, which she attributed to dyspepsia and metallic dysgeusia. She did not have recurrent hip pain until shortly prior to presentation but noticed changes to taste and hearing within months after revision surgery. The prosthesis was intact, with no evidence of loosening or infection, but extensive heterotopic ossification was noted on plain radiographs. Transthoracic echocardiogram revealed new cardiomyopathy with global left ventricular hypokinesis and an EF of 35%–40%, inconsistent with heart strain from a pulmonary embolism. Following medical treatment, the patient reported subjective improvement and was discharged. Ten days later, she returned with a right hip dislocation that was successfully treated with a joint reduction. At this time, her blood cobalt level was 817 μg/L (up from 424.3 μg/L on initial admission) while random and 24-hour total urine cobalt was 5677.1 μg/L and 4830.5 μg/L, respectively. The patient was discharged but re-admitted the next day with increased fatigue, generalised weakness, and difficulty ambulating. Two days later, she developed worsening signs of congestive heart failure and was diagnosed with post-hip arthroplasty cobalt toxicity. Complete excision of the hip was recommended, and N-acetylcysteine protocol was started. Prior to prosthesis removal, the patient decompensated, with worsening hypotension and tachycardia and eventually developed metabolic acidosis and respiratory failure. Despite intensive treatment, the patient's clinical condition deteriorated and she expired 4 days after admission. The blood cobalt level measured 2 days before death was 641.6 μg/L. An autopsy revealed extensive cobalt accumulation around the right hip prosthesis (41,000 μg/L) and in the heart muscle (2.5 μg/g). The heart was globular in appearance, with dilated atrial and ventricular chambers.

### 7.2.1.9 Case study 9

A 37-year-old male complained of dull hip pain, shortness of breath, and easy fatigability 3 years after his fractured ceramic hip implant was revised to a MoP bearing (Gautam et al., 2019). He was diagnosed with dilated cardiomyopathy with severe left ventricular dysfunction (EF, 20%), which was attributed to a highly elevated serum cobalt concentration (373 μg/L). Imaging studies revealed a range of abnormalities in the periprosthetic area, including severe osteolysis, collections, and oedema of the iliopsoas muscle. The implant was scheduled to be exchanged for a CoC articulation. Aspiration of the hip before capsulotomy yielded blackish, non-purulent fluid with granular material, and deposits of blackish metal debris could be seen in the hypertrophied synovium. The retrieved femoral head and polyethylene liner were grossly deformed, with scratches all over the articulating surface. Following the procedure, the patient's serum cobalt level declined to 173 μg/L but his cardiac function continued to worsen. He expired 3 months after the revision surgery while waiting for a heart transplant.

### 7.2.1.10 Case study 10

A previously healthy 55-year-old male with a history of revision MoP hip arthroplasty for ceramic head breakage developed diffuse systemic symptoms 3–5 months after the revision procedure, including poor concentration, fatigue, headaches, peripheral paraesthesia, convulsions, drastic unintended weight loss, nail discolouration, eczema, dysgeusia, muscle mass reduction, progressive hearing loss, and persistent sinus tachycardia (Oldenburg et al., 2009). He was admitted to an intensive care unit and underwent examinations that revealed hypothyroidism, increased distal sensory latency, reduced conduction velocity of the peroneal nerve, and moderately reduced systolic left ventricular function with concentric left ventricular hypertrophy. Myocardial biopsy demonstrated interstitial fibrosis. Increasing hip joint pain and abnormal cobalt levels (625 μg/L in the blood and 16,500 μg/L in the urine) necessitated a second revision surgery. Intraoperatively, a grey-black liquid, massive tissue metallosis and necrosis, and catastrophic wear of the CoCr femoral head were revealed. A new metal head and polyethylene liner were inserted, prompting a drastic drop in the blood/urine cobalt content. The patient's medical condition improved significantly within 6 months of the procedure; however, the neurological symptoms have persisted in the long-term follow-up and the cobalt levels had begun to rise again, indicating abrasion of the newly implanted femoral head.

### 7.2.1.11 Case study 11

A 53-year old male with a history of revision left MoP hip arthroplasty for ceramic liner fracture presented to the emergency department with progressive shortness of breath (Kim et al., 2016). He also complained of non-specific fatigue and general weakness but had no other systemic symptoms. Chest X-ray and enhanced heart MRI showed bilateral pericardial and pleural effusion. There was evidence of heart failure with rapidly worsening left ventricular EF, and a heart transplant was recommended. During work-up, the patient reported progressive left hip pain, as well as weakness and numbness in the left leg. Simple radiographs and CT revealed heterotrophic ossification of the joint and a huge cystic mass in the left groin. The CoCr femoral head was severely deformed, producing massive tissue metallosis. A heavy metal screen performed at the time indicated cobalt levels of 397,800 μg/L and chromium levels of 236,000 μg/L (the authors did not specify the sample type). Ten cycles of chelation with ethylenediaminetetraacetic acid (EDTA) led to a reduction in metal ion levels and symptomatic improvement, allowing the patient to undergo revision surgery. Meticulous removal of all visible ceramic fragments and aggressive soft-tissue debridement were performed, and a new ceramic head was implanted. Three months after the surgery, the cobalt levels decreased drastically and the patient's general condition improved; however, his left leg neuropathy persisted.

### 7.2.1.12 Case study 12

A 56-year-old male with an indwelling MoM arthroplasty developed insomnia, headaches, crying spells, depression, decreased concentration, and loss of appetite 3 years after his hip replacement surgery (Zeynalov et al., 2018). He had been prescribed fluoxetine for depression and propranolol for headaches, without benefit. The patient also complained of snoring and daytime sleepiness, and the results of a home sleep apnoea test indicated severe obstructive sleep apnoea. Automatic positive airway pressure therapy did not produce symptomatic improvement. A heavy metal screen was ordered and the patient's serum cobalt level was measured at 1.6 μg/L. Based on the systemic symptoms and laboratory results, cobalt intoxication was diagnosed and 4 years later, the patient underwent revision surgery with a ceramic implant. Following the procedure, his cobalt levels decreased and all symptoms, including the sleep-wake complaints, resolved.

### 7.2.1.13 Case study 13

A female with developmental dysplasia of the right hip who had undergone multiple reconstructive surgeries as a child received a right THA at the age of 22 (Reich et al., 2019). Despite doing well initially, she went on to develop acetabular osteolysis and had the acetabular component exchanged for a CoCr femoral head. Two years later, the 35-year-old patient sought care for rheumatoid and cardiovascular health issues, including polymyalgia, polyarthralgia, oral ulcers, headaches, fevers, facial rash, lightheadedness, fatigue, and syncope (fall in blood pressure leading to temporary loss of consciousness). The symptoms were ascribed to systemic lupus erythematosus, mixed connective tissue disease, fibromyalgia, and postural orthostatic tachycardia syndrome (POTS). Treatments including oral and topical steroids, anti-inflammatory agents, doxycycline, escitalopram, tacrolimus ointment, hydroxychloroquine, and midodrine provided little symptomatic relief. Ten years after the revision surgery, she developed 'brain fog' and aching pain in the lower and upper extremities. At that time, her dermatological complaints included an intermittent, disseminated, and pruritic facial, hand, and arm rash that had features of spongiotic dermatitis and worsened with sun exposure. Extensive laboratory and endocrinological studies were unremarkable and the symptoms were attributed to elevated serum cobalt content (10 μg/L). The patient underwent second revision surgery with a ceramic head and titanium-alloy stem; intraoperative findings included clear evidence of metallosis at the trunnion. Six months after the surgery, her cobalt levels had normalised and she reported complete resolution of her skin rash and an improved ability to think. At the 15-month follow-up, she noted subtle improvement regarding her POTS symptoms; however, the lower extremity paraesthesia did not resolve.

### *7.2.1.14 Case study 14*

A previously healthy 59-year-old female presented with a 2-week history of cough, exertional dyspnoea, and foot swelling that had developed while she was on vacation (Allen et al., 2014). She denied the use of tobacco, illicit drugs, herbal supplements, and alcohol, and had no known occupational exposures. Her medical history was unremarkable except for bilateral MoM ASR hip replacements implanted 3 and 4 years previously. The symptoms were attributed to pneumonia and 'travel-related oedema', and the patient was treated with a course of antibiotics. The following week, she returned to her physician with progressive dyspnoea, fatigue, and oedema, and was found to have a biventricular cardiac failure with a left ventricular EF of 25%, global hypokinesis, and moderate circumferential pericardial effusion on echocardiography. Despite intensive treatment, her cardiac function continued to deteriorate, necessitating implantation of an LVAD and, subsequently, orthotopic heart transplant 11 and 14 months after the initial presentation, respectively. The serum cobalt level measured 8 months after the presentation was 287.6 μg/L. Two months after the heart transplant, the patient was diagnosed with thyroglobulin antibody-positive hypothyroidism (initially attributed to amiodarone) and started on thyroxine supplementation. Five months later, her serum cobalt remained elevated at 374.3 μg/L and pelvic MRI revealed fluid collections at both hip joints. Electron microscopic analysis of the patient's native heart tissue showed abnormal mitochondrial forms and electron-dense deposits consistent with cobalt cardiomyopathy, and the decision to remove both MoM implants was made. Before the first hip surgery, the patient developed progressive fatigue and the left ventricular EF in the cardiac graft had decreased from 60% to 46% despite normal coronary angiography and no evidence of graft rejection. Intraoperatively, a large pseudotumour and gross metallosis of both hip joints were noted. The sharp fall in serum cobalt that followed the revision operations was accompanied by steady symptomatic improvement, with a left ventricular EF of 58% at the 5-month follow-up.

### *7.2.1.15 Case study 15*

A healthy 54-year-old male with bilateral, large-diameter MoM hip replacements inserted 4 and 5 years prior complained of exertional chest pressure, fatigue, and decreased sensation in both hands and feet that had developed over the previous year (Mosier et al., 2016). He denied any tinnitus or hearing/vision deficits and had normal haematocrit and thyroid hormone levels. An echocardiogram showed mitral regurgitation and stage II diastolic dysfunction with an EF of 55%; the coronary angiogram was normal. Although both hips were painless and well fixed, with no cystic changes evident on radiographs, the patient elected to proceed with the revision of both MoM implants for fears of systemic toxicity from elevated serum cobalt levels (120 μg/L). Preoperative contrast-enhanced cardiac MRI revealed severe biatrial enlargement with diffuse oedema, left ventricular EF of 36%, and right ventricular EF of 39%. Additionally, diffuse myocardial hyperenhancement of the anterior, lateral, and apical walls, with sparing of the base and mid septum, was noted (Fig. 7.2). Cardiac catheterisation showed that the patient's left ventricular EF had decreased to 30%. He was also noted to have mild renal insufficiency that was likely caused by prerenal azotaemia secondary to decreased cardiac output. Histopathological analysis of a cardiac tissue sample harvested during catheterisation revealed myocyte hypertrophy, interstitial fibrosis, myofiber disarray, and increased cytoplasmic vacuolation. Electron microscopy showed reduced contractile elements, vacuolar spaces, and increased lipofuscin deposits – findings consistent with cobalt cardiomyopathy (Fig. 7.2).

Intraoperative findings included pseudotumours in the periprosthetic soft tissues and abundant brown, creamy fluid indicative of metallic debris in the joint capsule. Both implants were well fixed, without significant visible wear of the metal heads; however, the trunnion area on both femoral parts was heavily corroded. After the procedure, the cobalt levels had dropped to 16 μg/L, but a repeat cardiac MRI performed at 5 months showed no signs of improvement. The patient continued to decline clinically (left and right ventricular EF of 23% and 25%, respectively) and required the implantation of an LVAD. He has since received a heart transplant and is doing well.

### *7.2.1.16 Case study 16*

A previously healthy 64-year-old female presented with progressive shortness of breath, impaired renal function (creatinine, 4.1 mg/dL), increasing pain in both hips, and markedly elevated cobalt levels (192 μg/L) 2 years

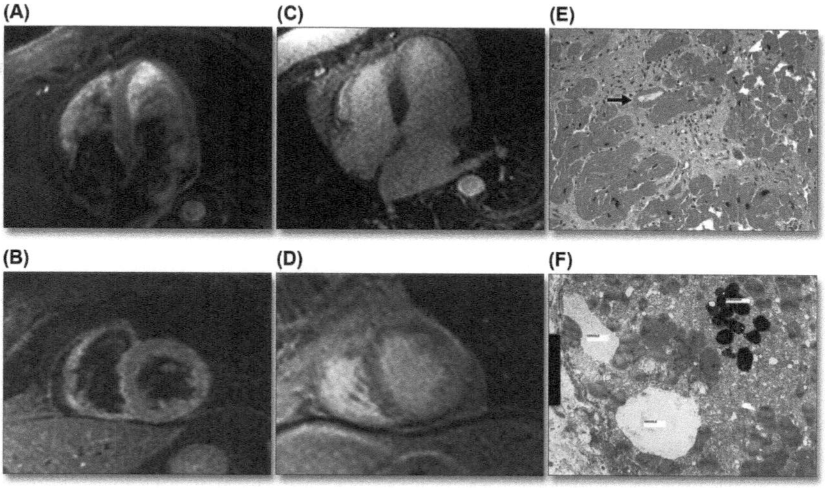


**FIGURE 7.2** (A and B) T2-weighted spin echocardiogram images in four-chamber and short-axis views, respectively, showing diffusely increased signal intensity in both ventricles. The left ventricular myocardial to skeletal muscle T2 ratio was 3 (normal $<1.9$). (C and D) Late gadolinium enhancement images in four-chamber and short-axis views, respectively, revealing marked diffuse biventricular hyperenhancement in a transmural pattern with sparing of the basal and mid septum. (E) Endomyocardial biopsy showing changes consistent with interstitial fibrosis (groups of myocytes separated by collagen) and myocardial hypertrophy (enlarged and hypertrophic nuclei). The sarcoplasm contains focal vacuoles (arrow). The resected apical core showed identical findings. Haematoxylin and eosin 200×. (F) Electron microscopic analysis of the cardiac biopsy showing vacuolar spaces and increased lipofuscin, consistent with cobalt cardiomyopathy (Samar et al., 2015). *Reproduced with permission from Samar, H.Y., Doyle, M., Williams, R.B., Yamrozik, J.A., Bunker, M., Biederman, R.W.W., et al., 2015. Novel use of cardiac magnetic resonance imaging for the diagnosis of cobalt cardiomyopathy. JACC Cardiovasc. Imaging 8, 1231–1232.*

after bilateral MoM THR (Martin et al., 2015). Echocardiography and cardiac MRI revealed severely depressed right and left ventricular function (EF, 10%–15%) and 'an unusual delayed enhancement pattern and hyperenhancement of the atria'. The patient was hospitalised and underwent implantation of a biventricular assist device before bilateral hip revision with CoP bearings. Intraoperative findings included metallosis with grey-stained fluid and tissue and a cystic mass that infiltrated each psoas bilaterally. The postoperative course was complicated by increased drain output, declining renal function, and embolic cerebrovascular event, and she expired 8 days after the revision surgery. An autopsy revealed diffuse cobalt deposition throughout the body, including in the heart (4.75 μg/g tissue), liver, and kidneys.

#### *7.2.1.17 Case study 17*

A 55-year-old man was admitted to the hospital with a 1-year history of deteriorating cardiac function (EF, 25%), progressive hearing and vision loss, and hypothyroidism (Dahms et al., 2014). His medical history was uneventful except for bilateral THRs; his left MoP prosthesis was inserted 18 months prior, following ceramic failure. Cobalt levels in the blood and urine (24-hour collection) measured at admission were 885 and 362 μg/L, respectively. The patient was started on chelation therapy and referred to an orthopaedic clinic where his left hip implant was revised to a new ceramic bearing. On account of severe heart failure, he was also implanted with a cardioverter-defibrillator. Fourteen months after revision surgery, the patient's blood cobalt concentration had reduced to 86.1 μg/L and his EF had improved to 40%, but the vision and hearing remained severely impaired.

#### *7.2.1.18 Case study 18*

A healthy 52-year-old man with bilateral THRs complained of progressive left hip pain, fatigue, dyspnoea, anorexia, and 10-kg weight loss 6 months after receiving a MoP implant for a previously fractured left CoC prosthesis (Gilbert et al., 2013; Zywiel et al., 2013). On examination, he had a painful mass at the left hip that, when aspirated, drained a tarry fluid abundant in cobalt (33,713 μg/L). Investigations revealed polycythaemia (haemoglobin, 190 g/L) and profound hypothyroidism (TSH, 92 mIU/L; thyroxine, undetectable) requiring thyroxine supplementation. Some 8 months after the initial presentation, he was admitted to the hospital with dilated cardiomyopathy. He had normal valvular function; however, the left ventricular EF was severely reduced (30%) and a large pericardial effusion with impending tamponade was noted. Following pericardiocentesis, the patient went into cardiogenic shock and developed liver failure. Blood cobalt and chromium levels were extremely elevated at 6521 and 23.6 μg/L, respectively, prompting a 3-day course of chelation therapy (dimercaprol) coupled with resection of the offending prosthesis. The patient's cobalt levels decreased substantially following the procedure, but his left ventricular function remained profoundly depressed and he died of multiorgan failure. An autopsy revealed massive cobalt saturation of the vital organs, including the liver, heart, and thyroid gland.

### 7.2.2 Mechanisms of cobalt toxicity

Divalent cobalt ions ($Co^{2+}$) are the most toxicologically relevant products of hip implant degradation. In the blood and tissue, they are readily sequestered by large proteins, such as albumin, and certain small molecular complexes, which prevents them from entering cells and triggering the events that lead to adverse effects. However, a fraction of the total cobalt load remains in an unbound state that is free to interact with target biomolecules. The molecular mechanisms thought to underlie systemic cobalt toxicity are characterised below.

### *7.2.2.1 Induction of oxidative stress*

Cobalt ions catalyse the generation of highly toxic hydroxyl radicals ($\cdot OH$) from hydrogen peroxide ($H_2O_2$) in a Fenton-like reaction (Eq. 7.1) (Beyersmann and Hartwig, 2008).

$$Co^{2+} + H_2O_2 \rightarrow Co^{3+} + \cdot OH + OH^- \quad (7.1)$$

Reactive oxygen species (ROS) such as $\cdot OH$ promote the oxidation of proteins, lipids, carbohydrates, and DNA, which can lead to cell death. Though Fenton chemistry occurs naturally in vivo, its impact is generally negligible owing to the work of antioxidants and specialised enzymes that are designed to 'mop up' excess ROS. When metal ions are present at an increased concentration, as is often the case in patients with metal hips, the rate of $\cdot OH$ formation can overwhelm the protective mechanisms—a condition known as oxidative stress (Catalani et al., 2012). The brain is a prime target for oxidative injury because it contains high concentrations of readily oxidisable polyunsaturated fatty acids and relatively modest antioxidant levels (Matés et al., 2010). In the nervous system, excess ROS can induce demyelination, axonal loss, and toxic damage to oligodendrocytes, which are particularly vulnerable to oxidative stress (Cheung et al., 2016).

### *7.2.2.2 Disruption of mitochondrial function*

Cellular respiration is a tightly regulated metabolic pathway that turns glucose into usable cellular energy. The process consists of three main stages—glycolysis, citric acid cycle, and oxidative phosphorylation – and relies on a number of enzymes and co-factors to run smoothly. Cobalt ions bind to one of the co-factors (lipoic acid), hindering its function and inhibiting the reactions that rely on it, namely oxidative decarboxylation of pyruvate to acetyl coenzyme A and of alpha-ketoglutarate to succinate. This effectively interrupts the citric acid cycle and depletes the cell's energy source (Paustenbach et al., 2013). In parallel, cobalt ions can induce the phenomenon of mitochondrial permeability transition through the opening of the transition pore, leading to mitochondrial swelling and electrical membrane potential collapse (Battaglia et al., 2009). The build-up of metabolic intermediates in the mitochondria increases the osmotic pressure, resulting in oedema and structural disruption (Rona, 1971). Disturbed cellular respiration promotes the production of ROS, release of various apoptogenic factors, caspase activation, and, ultimately, cell death.

In the heart, interference with cellular energy production leads to oxygen deprivation, reduced contractility, and myocardial dysfunction. Because of its inherent reliance on oxidative metabolism, the nervous system is another target for this toxicity mechanism. The retina, being one of the highest oxygen-consuming tissues of the body, is particularly vulnerable (Apostoli et al., 2013).

### 7.2.2.3 *Simulation of cellular hypoxia*

Cobalt ions induce a hypoxia-like state in the cell, even in the presence of normal molecular oxygen pressure (Nyga et al., 2015). The underlying mechanism involves the stabilisation of hypoxia-inducible factor 1 (HIF-1) – a transcription factor that is usually degraded when sufficient oxygen is present. HIF-1 directs the transcription of an array of genes that promote cell survival during times of low oxygen, including those encoding angiogenic growth factors, glucose transporters, glycolytic enzymes, and proteins involved in the regulation of cell proliferation and apoptosis (Fig. 7.3) (Semenza, 2000). Polycythaemia associated with elevated systemic cobalt load is likely caused by HIF-1-mediated stimulation of the erythropoietin gene and consequent overproduction of red blood cells. HIF-1 activation may also be partly responsible for the ocular defects noted in some cases of arthroprosthetic cobaltism. Animal studies have shown that intraperitoneal or intravitreal administration of cobalt chloride produces choroidal infractions that lead to hypoxic injury of the outer retina and photoreceptor degeneration (Hara et al., 2006; Monies and Prost, 1994).

While HIF-1-mediated responses serve to increase cell survival in times of low oxygen perfusion, they can have tumourigenic consequences in tissues with normal oxygen supply. HIF-1 signalling inhibits p53 (an important tumour suppressor) (Lee et al., 2001), increases the expression of vascular endothelial growth factor (a key player in angiogenesis and bone/tissue remodelling), as well as stimulating cell proliferation and neovascularisation. These adaptations

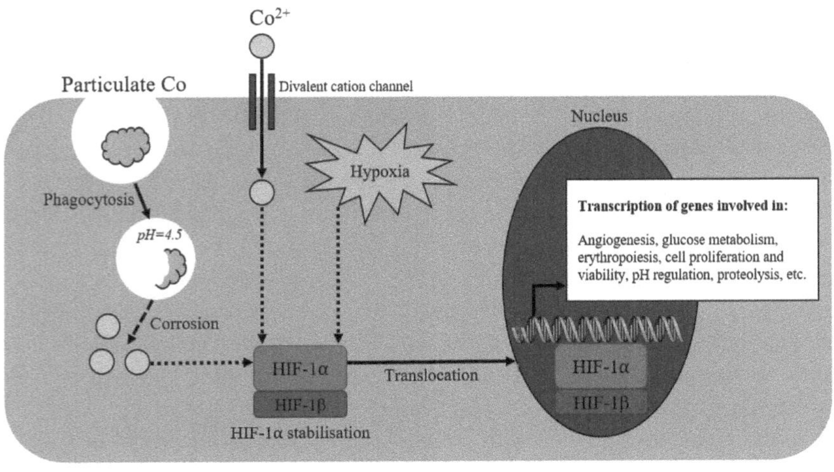


**FIGURE 7.3** Pathways of HIF-1 induction. While HIF-1 signalling is normally induced by low tissue oxygen (hypoxia), cobalt ions can activate this pathway even in times of normal oxygen pressure, leading to the transcription of genes that promote cell survival. *HIF-1(α/β), hypoxia-inducible factor 1 (α/β subunit). Adapted from Swiatkowska, I., 2019. Toxicity of Metal Debris From Hip Implants (Ph.D. thesis). University College London, London, UK/ CC-BY-4.0.*

are important for neoplastic growth and tumour progression. It follows that cobalt may generate a series of potentially carcinogenic responses, though their clinical relevance in human cancers is unclear. A recent review of the topic concluded that 'in vitro cobalt-stimulated HIF-1 overexpression does not correlate with cancer risk from cobalt exposure in humans' (Smith and Perfetti, 2019). Due to a lack of convincing in vivo evidence, the International Agency for Research on Cancer currently classifies cobalt and its compounds as Group 2B substances (possibly carcinogenic to humans) (WHO, 2006).

#### *7.2.2.4 Interference with calcium signalling*

Cobalt ions compete with calcium ions for binding sites on enzymes and can 'hijack' calcium channels to gain entry into cells. Displacement of calcium binding in the peripheral nervous system blocks synaptic transmission, and suppressed communication between neurons can result in weakness, paraesthesia, hearing loss, and vision loss (Leikin et al., 2013). A similar effect is seen in individuals who ingest large amounts of magnesium and suffer weakness due to the displacement of calcium.

#### *7.2.2.5 Displacement of divalent metal cations from metalloproteins*

Cobalt ions can displace other divalent ions from their enzymatic active sites, leading to the inhibition or modification of enzyme function. Most notably, the substitution of $Co^{2+}$ for $Zn^{2+}$ in the zinc finger domain of transcription factors, and for $Mg^{2+}$ in magnesium-dependent DNA repair proteins, has been postulated to contribute to cobalt-induced DNA damage (Kopera et al., 2004).

#### *7.2.2.6 Inhibition of iodine uptake*

The main function of the thyroid gland is the synthesis of thyroid hormones – thyroxine and triiodothyronine. The process relies on an efficient uptake of iodine and the amino acid tyrosine, and their subsequent reaction. Cobalt ions hinder this process by directly inhibiting iodine uptake by the thyroid or by binding to the enzymes or co-factors necessary for tyrosine iodination, such as tyrosine iodinase (Cheung et al., 2016). When insufficient quantities of triiodothyronine and thyroxine are generated, the growth of the thyroid gland is stimulated as a means of increasing its hormone production. When iodine binding is inhibited, the thyroid is unable to produce more triiodothyronine and thyroxine, but the gland continues to grow, forming a goitre (Schirrmacher, 1967).

### 7.2.3 Differential diagnosis

#### *7.2.3.1 Cardiomyopathy*

Although there is no evidence to advise routine cardiac screening of patients with MoM implants (Manisty et al., 2018), individuals presenting with new-onset shortness of breath, exertional chest tightness, and/or heart palpitations several months or years after hip replacement may benefit from a cardiological review, including electrocardiogram (ECG), echocardiography, and contrast-enhanced cardiac MRI (Khan et al., 2015). The most frequently discussed metric for cobalt cardiomyopathy in previous case reports was the left ventricular EF (<40% in all affected patients). Pericardial effusion, left ventricular dilation, and/or cardiomegaly without ischaemia were also commonly noted. Packer (2016) emphasised that the diagnosis of cobalt cardiomyopathy requires (1) demonstration of biventricular dilatation and systolic dysfunction at a time when blood/tissue concentrations of cobalt are increased, and (2) normalisation of cardiac structure and function when exposure ceases and blood/tissue cobalt levels return to normal (in the absence of other interventions for the treatment of cardiomyopathy). Although some have suggested that cobalt can impair diastolic filling without affecting systolic function, echocardiographic evidence of diastolic dysfunction is commonly found in middle-aged and elderly people and, as such, its presence cannot reliably distinguish those who have been exposed to cardiotoxic substances from unexposed individuals (Packer, 2016).

Differential diagnosis may also be difficult in the presence of other preexisting risk factors for cardiomyopathy, such as protein/thiamine deficiency, malnutrition, heavy alcohol consumption, metabolic disorders (obesity, diabetes), or renal impairment (Umar et al., 2020). The following features have been proposed to distinguish cobalt-related cardiomyopathy from cobalt-unrelated non-ischaemic dilated cardiomyopathy: a recent history of anorexia and weight loss, abrupt onset and rapid progression of heart failure, cyanosis, low voltage across all ECG leads, sinus tachycardia, absence of cardiac arrhythmias, and co-existence of polycythaemia, hypothyroidism, and/or lactic acidosis (Packer, 2016). Cardiac MRI findings can help exclude alternative aetiologies, such as post-infarct, infiltrative, acquired, viral, or alcoholic/drug-related (Samar et al., 2015). In previous reports, cobalt cardiomyopathy was associated with oedema on T2 images, prominent enhancement of the subepicardial lateral wall, and biventricular hyperenhancement in transmural pattern, with sparing of the basal and mid-septum in contrast-enhanced images (Khan et al., 2015; Martin et al., 2015; Samar et al., 2015).

Histopathological analysis of cardiac tissue samples and measurement of their cobalt content may shed additional light on the root cause of cardiomyopathy. Myocardial hypertrophy and interstitial fibrosis were reported in all cases of cobalt cardiomyopathy where either cardiac biopsy or autopsy was performed. In addition, some authors noted cobalt-specific features such as increased vacuolation and lipofuscin (Fig. 7.4) (Martin et al., 2015; Mosier et al., 2016; Tilney et al., 2017), myofiber disarray (Moniz et al., 2017; Mosier et al., 2016), loss of myofibril filaments (Barborik and Dusek, 1972; Kennedy et al., 1981; Manifold et al., 1978), and structurally altered mitochondria with electron-dense deposits (Allen et al., 2014; Choi et al., 2018;

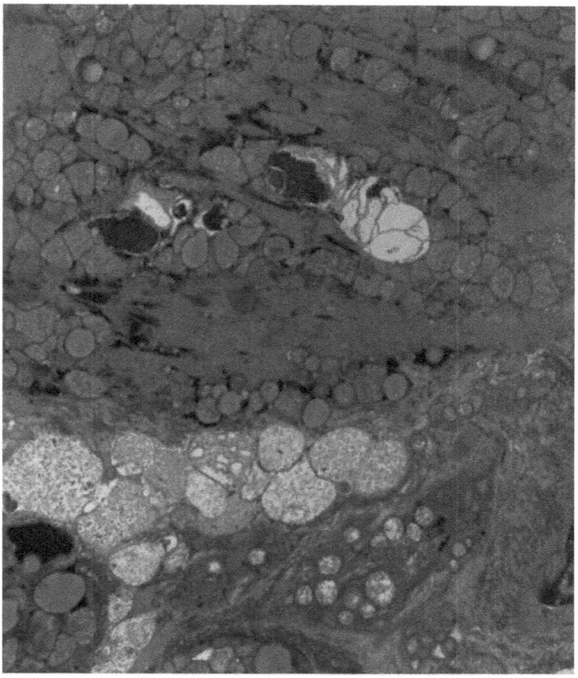

**FIGURE 7.4** Electron micrograph of a cardiac biopsy from a patient with fatal cobalt cardiomyopathy. Increased vacuolation and lipofuscin are consistent with cobalt-related cardiotoxicity (Martin et al., 2015). *Reproduced with permission from Martin, J., Spencer-Gardner, L., Camp, C., Stulak, J., Sierra, R., 2015. Cardiac cobaltism: a rare complication after bilateral metal-on-metal total hip arthroplasty. Arthroplast. Today 1, 99–102.*

Fox et al., 2016; Zywiel et al., 2013). Dense, osmophilic intramitochondrial particles 0.3–0.4 μm in diameter are regarded as a pathognomonic feature of systemic cobalt toxicity (Choi et al., 2018; Rona, 1971). The absence of calcified fibrils or other deposits within myofibrils can further differentiate cobalt-induced cardiomyopathy from other toxic aetiologies (Cheung et al., 2016).

In previous reports, the mean blood cobalt level associated with histopathologically confirmed cobalt cardiomyopathy was 425 μg/L (range, 112–1556 μg/L) in patients with pre-existing risk factors for cardiomyopathy, and 1217 μg/L (range, 122–6521 μg/L) in those without such risk factors (Umar et al., 2020).

### *7.2.3.2 Neurotoxic and psychiatric symptoms*

Often, patients with cobalt cardiomyopathy also exhibit peripheral neuropathy (diminished reflexes, tremor, numbness in the hands and feet), varying degrees of hearing and/or visual impairment, and psychiatric symptoms such as emotional lability, depression, cognitive decline, forgetfulness, and reduced ability to concentrate. Cobalt-related auditory dysfunction manifests as tinnitus and/or progressive sensorineural hearing loss evident upon audiometric examination. Ophthalmological findings of cobalt intoxication include decreased visual acuity, deficient colour vision, central visual field defects, abnormal ERG, optic atrophy, decreased choroidal perfusion, macular dysfunction, irregular visual evoked potentials (VEPs), and hyperintensity of optic nerves on MRI (Garcia et al., 2020). A normal ERG in a patient with a history of MoM arthroplasty does not rule out cobalt toxicity, particularly when accompanied by other common symptoms. Individuals losing sight due to suspected cobalt intoxication should be offered a complete ophthalmological work-up, including standard full-field ERG, multifocal ERG, multifocal VEPs, and OCT (Grillo et al., 2016). Of note, cobalt-related optic neuropathy and vitamin B12 deficiency-related optic neuropathy can have similar clinical manifestations (Jefferis et al., 2019); thus, clinicians would be prudent to request vitamin B12 measurement in addition to blood cobalt levels in patients with suspected arthroprosthetic cobaltism – particularly in those following a vegan diet and/or disclosing heavy alcohol use. The diagnosis should also take into consideration the presence of comorbidities, such as advanced age, diabetes (Pinazo-Durán et al., 2016), occupational exposures, and use of oculotoxic/ototoxic medications or dietary supplements (Leyssens et al., 2020).

Exposure to low circulating concentrations of cobalt typical of well-functioning MoM prostheses is not thought to cause clinically detectable defects in visual or auditory function, but such effects may become clinically present with longer-term exposure (Prentice et al., 2014). The majority of patients experiencing hip prosthesis-related neurotoxicity in previous case reports had blood/serum cobalt levels of 400 μg/L and higher, although mild/moderate symptoms were sometimes observed at much lower levels.

Nominal cobalt exposure from the degradation of MoM bearings or modular junctions may cause a subtler presentation of neurologic, psychiatric, and constitutional sequelae that is easily misattributed to aging. Bridges et al. (2020) noted a pathognomonic pattern of quantitative brain hypometabolism in neurologically symptomatic patients with cobalt levels of $\geq$ 0.4 μg/L in whole blood and $\geq$ 1 μg/L in urine. Of the 57 evaluated patients, 38 received MoP hips, which are not presently considered problematic. The distribution of hypometabolism, as assessed with quantitative $^{18}$F-FDG-PET/CT brain imaging, was similar to that in toxic encephalopathy found in solvent-exposed workers (Callender et al., 1993) and differed from that seen in classical dementias. These results suggest that systemic cobalt toxicity may be more common and associated with lower blood cobalt levels than presently expected. Tower et al. (2020) reported that corrosion at the stem–cement interface of polished CoCr femoral stems in MoP implants can provoke reversible neurological and psychiatric symptoms consistent with arthroprosthetic cobalt encephalopathy (ACE). Quantitative $^{18}F$-FDG-PET/CT brain imaging, which can help confirm the diagnosis of ACE, may provide an early adjunct to the clinical assessment of patients with elevated cobalt levels and systemic toxicity symptoms and guide treatment decisions. The use of self-assessment questionnaires at follow-up, such as the Neurotoxic Symptom Checklist-60 (Hooisma and Emmen, 1992), can also be useful when evaluating for clinically significant cobalt neurotoxicity (Leyssens et al., 2020; Swiatkowska et al., 2020; Van Lingen et al., 2013).

#### *7.2.3.3 Thyroid abnormalities*

Patients with elevated systemic cobalt load after THR sometimes report symptoms of progressive fatigue, poor concentration, depression, and/or muscle weakness. Determining the specific aetiology of these complaints can be difficult due to the large number of genetic, environmental, nutritional, and immune-related root causes and contributing factors for hypothyroidism. In cases where thyroid insufficiency is accompanied by unexplained vision/hearing loss and/or cardiomyopathy, hip implant-related cobalt toxicity should certainly be suspected.

Thyroid function is evaluated using radioimmune assays of thyroid-stimulating hormone (TSH), thyroxine, and, less commonly, triiodothyronine. Thyroid antibody tests may be additionally requested to exclude autoimmune thyroiditis, such as Hashimoto's disease. In patients with an underactive thyroid, the serum levels of TSH are usually elevated, while those of triiodothyronine and thyroxine are either within the normal range (subclinical thyroid insufficiency) or lower (clinical hypothyroidism). If many signs of hypothyroidism are present but biochemical tests return normal values, CT or ultrasonography of the thyroid gland can aid in the diagnosis. Cobalt ions are toxic to follicular cells and inhibit iodine uptake by the thyroid, producing features of destructive thyroiditis on CT (Yu, 2017) (Fig. 7.5).

Thyroid dysfunction and increased red blood cell production are believed to be the most sensitive endpoints of cobalt toxicity, and some authors advise monitoring patients with metal hip implants for signs of hypothyroidism and polycythaemia starting at blood/serum cobalt concentrations of 100 μg/L (Paustenbach et al., 2013).

### 7.2.4 Treatment

In all cases of cobalt toxicity, the physician's main objective is to reduce or eliminate the source of cobalt exposure, as well as treating the systemic symptoms.

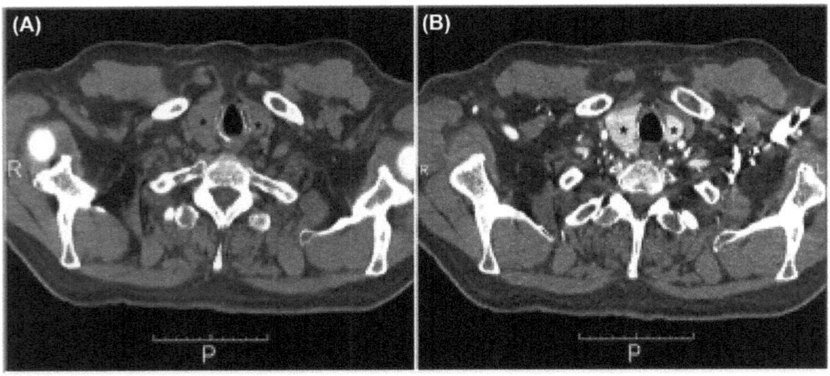


**FIGURE 7.5** Computed tomography scan of the thyroid gland in a patient with arthroprosthetic cobaltism. (A) Low thyroid density without intravenous contrast, suggesting low iodine content (Hounsfield units, 52). (B) Strong thyroid enhancement with intravenous contrast, implying high blood flow (Hounsfield units, 170–200) (Yu, 2017)/CC-BY-4.0.

### *7.2.4.1 Revision surgery*

Treatment of hip implant-related cobalt intoxication largely relies on a revision of the causative prosthesis as soon as the problem is recognised. This is often combined with a debridement of the affected tissues to lower the systemic cobalt load. The use of MoP constructs to replace failed ceramic hips is favoured by some surgeons because it is thought to decrease the risk of ceramic re-fracture. However, it is now apparent that even a meticulous synovectomy and extensive joint lavage are unable to completely remove all of the ceramic debris, and revision of a fractured ceramic bearing with a CoCr head can lead to catastrophic failure and potentially life-threatening cobalt toxicity (Griffiths et al., 2015). Thus, it is strongly recommended that fractured ceramic arthroplasties are only revised using new ceramic components (Peters et al., 2017; Umar et al., 2020; Zywiel et al., 2013). When performed correctly and in a timely manner, revision operation will be followed by a sharp drop in systemic cobalt levels and marked improvement, or even complete reversal, of the associated systemic complaints. Conversely, delayed diagnosis can lead to symptomatic deterioration and neurological/cardiac damage that persists even after the problematic hardware had been removed (Dahms et al., 2014; Mosier et al., 2016). To minimise the risk of permanent injury, patients with a metallic bearing inserted following a ceramic fracture may benefit from long-term surveillance for clinical, biochemical, and/or echocardiographic features of cobalt toxicity (Fox et al., 2016; Gautam et al., 2019; Umar et al., 2020). Presenting symptoms of malaise, lethargy, nausea, vomiting, metallic dysgeusia, and/or significant unintentional weight loss should raise suspicion of arthroprosthetic cobaltism, particularly if accompanied by cardiac/neuro-ocular complaints.

### 7.2.4.2 Chelation therapy

In cases where blood cobalt concentration is elevated but immediate implant removal cannot be performed or is contraindicated, surgeons may wish to opt for chelation therapy. Metal chelators, such as EDTA and N-acetylcysteine (NAC), are compounds that bind free metal ions and render them chemically inert. The metal–chelator complexes are excreted from the body, primarily in the urine, without affecting the vital organs.

Although chelation therapy is an established treatment for heavy metal poisoning stemming from environmental or occupational exposure, its effectiveness in the setting of arthroprosthetic cobaltism and the permanency of its results are debated. Oral administration of high-dose NAC was shown to markedly reduce the systemic cobalt load in patients with MoM implants and mildly elevated blood cobalt levels (Ambrosi and Ursino, 2020; Giampreti et al., 2014); however, the low cobalt levels did not always persist, with some of the patients experiencing a progressive increase in blood cobalt content after cessation of therapy. Dietary supplements containing metal chelators are marketed to promote the process of heavy metal detoxification (Coetzee and Mari, 2017), but solid evidence for their efficacy and long-term benefits in patients with prosthetic cobalt exposure is so far lacking.

Metal chelation may be more useful as an adjunct to revision surgery in cases of extreme cobalt intoxication accompanied by systemic toxicity symptoms. Pazzaglia et al. (2011) used multiple cycles of EDTA treatment to bring down highly elevated cobalt levels in a patient with peripheral neuropathy and severe vision and hearing loss associated with third-body wear of a MoP prosthesis. Each chelation cycle was followed by a drop in cobalt levels that was more pronounced in the plasma than in whole blood. A few days after the initial reduction, plasma cobalt levels increased again, albeit to levels lower than before, and a permanent reduction was only achieved after the MoP prosthesis was revised to a ceramic bearing.

An important consideration when deciding whether chelation should be employed is that metal chelators depend on efficient renal excretion, which makes them unsuitable for individuals with impaired kidney function.

### 7.2.4.3 Therapeutic plasma exchange

TPE is an established treatment for a number of neurological and autoimmune diseases that can be used in the setting of renal insufficiency. During TPE, the patient's blood plasma is separated and removed before red blood cells and replacement fluid, such as plasma and/or human serum albumin, are re-infused (Nieto-Aristizabal et al., 2020). Since soluble cobalt ions are albumin bound, the procedure provides a means to rapidly lower the systemic cobalt load. In Grant et al. (2016), TPE was applied to a patient with extremely elevated plasma cobalt concentration (2148 μg/L) and a constellation of systemic toxicity symptoms. The treatment reduced the cobalt levels by two-thirds, but only temporarily, as the cobalt concentration rebounded to pre-TPE values within 8 hours. It was only after revision operation that a stable reduction was achieved and the clinical symptoms improved. It is evident that without the elimination of the source of cobalt exposure, chelation and TPE treatments will be futile as metal ions that are removed from the plasma are replaced by those accumulated in red blood cells or tissue stores as a new equilibrium is reached (Grant et al., 2016).

### 7.2.5 Individual susceptibility to systemic cobalt toxicity

Factors known to increase the risk of elevated cobalt levels include loose or malpositioned implants, large-diameter MoM arthroplasty, hip resurfacing, and revision of a fractured ceramic implant using metallic components (Zywiel et al., 2016). Importantly, not all patients with high prosthetic cobalt release go on to develop generalised toxicity symptoms, and it is thought that the co-existence of patient-specific predisposing factors is required to induce systemic adverse effects. These factors are discussed in the following paragraphs.

#### *7.2.5.1 Decreased albumin–cobalt binding capacity*

Human serum albumin has over 20 divalent metal-binding sites, of which 2–3 have a relatively high affinity for $Co^{2+}$ ions (Bar-Or et al., 2001; Nandedkar et al., 1973). It is estimated that in healthy adults with serum cobalt concentrations up to 146 μg/L, 94%–96% of serum cobalt is sequestered by albumin and other large proteins, while the remaining 4%–6% remains labile or bound to small proteins such as glutathione and lipoic acid (Kerger et al., 2013a, b; Paustenbach et al., 2013). Compared to cobalt–protein complexes, free cobalt ions have higher tissue bioavailability and toxic potential. Thus, reduced albumin binding and the resultant increase in the proportion of unbound $Co^{2+}$ could predispose an individual to adverse responses. This situation may arise if the synthesis of serum albumin is impaired, if its structure/function is altered, or if cobalt ions are displaced from their albumin binding sites by competing species.

Normal albumin level in the serum is between 3.4 and 5.4 g/dL and remains fairly stable, though chronic malnutrition, alcoholism, and certain disease states can decrease albumin synthesis or increase its loss. Besides analbuminemia – a rare congenital disease – the main pathologies known to lower albumin concentration are the nephrotic syndrome (Harris and Ismail, 1994), liver cirrhosis, carcinoma, protein-losing enteropathy, major burns, Cushing's syndrome, thyroid disorders, and stress secondary to trauma, infection, or radiation (Paustenbach et al., 2013). The serum albumin concentration is also thought to fall with age and cigarette smoking (Phillips et al., 1989).

The magnitude of albumin decrease required to cause a clinically important increase in free $Co^{2+}$ requires further investigation.

Pathologies, such as trauma, acute infection, end-stage renal disease, diabetes, peripheral vascular disease, brain/myocardial ischaemia, acidosis, obesity, and malignancy can induce post-translational mutations at the N-terminus of albumin and generate ischaemia-modified albumin (IMA) – a deformed variant of the protein with decreased metal-binding capacity (Aslan and Apple, 2005; Chen et al., 2011; Kiyici et al., 2010; Paustenbach et al., 2013). The mechanism of IMA generation has not been fully elucidated, but it likely involves ROS-mediated damage during ischaemic events (Bar-Or et al., 2000). Diabetic patients, who experience chronic hypoxia as a result of hyperglycaemia and oxidative stress, exhibit markedly higher IMA levels and lower albumin–cobalt binding capacity than healthy subjects (Piwowar et al., 2008), which may predispose them to arthroprosthetic cobaltism. Elevated IMA levels have also been reported in patients with beta-thalassaemia, systemic sclerosis, cirrhosis, and hypercholesterolaemia (Awadallah et al., 2012; Borderie et al., 2004; Chen et al., 2011; Duarte et al., 2009). While underlying disease alone is probably insufficient information to judge a patient's susceptibility to cobalt toxicity, high accumulation of IMA (as can be the case in acute and severe inflammatory states) in conjunction with hypoalbuminemia might be expected to increase the risk of adverse responses (Paustenbach et al., 2013). The minimum concentration of IMA required to cause a clinically significant shift in albumin–cobalt binding capacity in exposed individuals is, however, unknown.

Three of the major cobalt binding sites on albumin also bind cadmium, zinc, copper, and nickel ions. An appreciable increase in the concentration of any of these ions could prevent cobalt from binding to albumin, or displace it from its binding sites, shifting the equilibrium towards free $Co^{2+}$. However, given the relatively high affinity of $Co^{2+}$ for these binding sites, significant displacement is unlikely to occur unless blood concentrations of competing metals are in excess of 1 mg/L (Paustenbach et al., 2013).

#### *7.2.5.2 Kidney disease*

In addition to elevated IMA production, kidney disease can impede the clearance of metal ions generated from hip implants, increasing their systemic levels (Hur et al., 2008). This is especially important for cobalt, which is more soluble and reliant on urinary excretion than chromium and titanium. Accumulation of cobalt in the blood, combined with increased levels of oxidative stress and impaired antioxidant defence which often accompany renal failure, can translate into a higher incidence of systemic complications, particularly in patients on dialysis, who are susceptible to protein malnutrition and hypoalbuminemia (Curtis et al., 1976; Paustenbach et al., 2013). Thus, not only are anephric patients expected to exhibit higher blood cobalt levels than healthy individuals, they are also likely to have a higher proportion of unbound $Co^{2+}$ ions. For these reasons, MoM bearings are contraindicated in individuals with chronic renal disease (Chandran and Giori, 2011).

#### *7.2.5.3 Nutritional and hormonal deficiencies*

Animal studies have revealed that cobalt toxicity can become manifest or more pronounced in the presence of malnutrition, hypothyroidism, and/or heavy alcohol consumption (Grice et al., 1969; Morvai et al., 1993; Rona, 1971; Sandusky et al., 1981). In particular, the ultrastructural changes in the cardiac tissue elicited by cobalt can be reproduced experimentally by protein restriction, thiamine deficiency, and blocking thyroxine production. Amino acids and proteins, especially if they are rich in sulphydryl and amino groups, sequester free cobalt ions, curbing their reactivity and preventing their interaction with important biomolecules and receptors – a protective mechanism that is absent in the setting of insufficient protein intake (Rona, 1971). Deficiency of thiamine and thyroxine is thought to act at the same enzymatic sites as cobalt to interfere with the normal functioning of the citric acid cycle (Macho and Palkovic, 1963; Shulman et al., 1985). Synergistic interactions between the above risk factors are illustrated by the Quebec beer-drinkers' cardiomyopathy epidemic of the 1960s, during which habitual consumers of large amounts of cobalt-fortified beer were admitted to various hospitals with an unusual type of cardiomyopathy (Morin et al., 1967). While it is likely that several overlapping factors, including chronic ischaemia, oxidative stress, acidosis, and alcohol-induced cumulative organ damage, may best explain the increased susceptibility of this group to trace doses of cobalt, it is clear that a poor diet was a large contributor to the severe presentation (Paustenbach et al., 2013). Individuals with a grossly inadequate diet, particularly lacking in protein, suffered more profound cardiac dysfunction and higher mortality than did well-nourished beer drinkers who ingested similar doses of cobalt (approximately 0.09 mg/kg/day) (Kesteloot et al., 1968).

### 7.2.6 Systemic toxicity and free cobalt ion levels

Blood/serum cobalt levels quoted in the case reports of systemic cobalt toxicity seldom correlate with the observed toxicity symptoms. The issue was highlighted by Olmedo-Garcia and Zagra (2018), who reported on two cases of severe complications following revision of a fractured ceramic implant with a MoP prosthesis. Even though both patients had markedly elevated blood cobalt content 10 years after the procedure (60–100 μg/L), one suffered severe symptoms of cobalt poisoning (cardiomyopathy and hearing loss) for years, without suspecting an association with the implant, while the other patient did not have any systemic complaints. As discussed in Chapter 4, several different analytical approaches can be employed to measure systemic cobalt levels in patients with metallic hip implants. The two most commonly applied techniques, namely ICP-MS and graphite furnace AAS, measure the total cobalt load, including large cobalt–protein complexes ( $>$ 50 kDa), small cobalt–protein complexes ($<$1 kDa), and free $Co^{2+}$; however, only the latter species are associated with clinically important toxicity. Since the toxicologically relevant cobalt ions are not quantified, references to 'cobalt metal ion concentrations' often seen in the literature are technically misleading. Considering that perturbed cobalt kinetics (i.e., reduced albumin–cobalt binding capacity and/or equilibrium shift towards sustained high $Co^{2+}$ levels in the blood and tissues) are thought to predispose patients with hip implants to adverse reactions, free $Co^{2+}$ levels might be a more accurate dose metric for predicting systemic adverse effects and understanding their variability than total blood/serum cobalt concentrations, which have been relied upon thus far (Paustenbach et al., 2013). Methods that enable separate quantification of various protein-bound fractions and free ions in undiluted human serum have been developed by combining size exclusion chromatography with ICP-MS detection (Kerger et al., 2018). Studies utilising these new methods are expected to provide novel insights into dose-response anomalies and individual susceptibility to systemic cobalt toxicity in the setting of hip arthroplasty.

## 7.3 Summary and future directions

High-wearing hip implants release massive amounts of cobalt into the circulation, which, in rare cases, can cause widespread toxicity involving several organ systems. Currently, diagnosis of arthroprosthetic cobaltism is based on the presence of symptoms consistent with the known cardiotoxic, neurotoxic, and endocrine effects of cobalt and increased blood/serum cobalt concentrations several months/years after hip replacement, albeit no universally accepted threshold for blood/serum cobalt has been defined. Timely revision of the causative implant usually leads to a drop in systemic cobalt levels and marked improvement, or complete reversal, of the systemic symptoms.

There is no clear dose-response relationship between blood/serum cobalt levels and adverse effects. The risk of developing systemic symptoms appears to be influenced by underlying disease states and events that impair albumin–cobalt binding and increase the proportion of free cobalt ions in the blood. Functional alterations of albumin as a result of genetic mutations, single-nucleotide polymorphism, or post-translational modifications warrant further study as potential markers of individual susceptibility to cobalt (Facchin et al., 2017).

Recently developed methods that enable separate quantification of unbound cobalt ions in undiluted human serum may provide further insights into the effects of, and individual susceptibility to, sustained elevations in free blood $Co^{2+}$ levels (Kerger et al., 2018). In the meantime, individuals with a surgical history that predisposes them to high prosthetic cobalt release (MoM arthroplasty or cobalt-containing component implanted for ceramic failure) ought to be questioned about possible systemic complaints at follow-up and referred for cardiac, ophthalmological, auditory, and/or neurocognitive evaluation as appropriate, particularly if serial blood/serum cobalt measurements are elevated (typically $Co > 20\ \mu g/L$) and there is no other plausible explanation for the observed symptoms. Laboratory investigations should include a full blood count with differential, erythrocyte sedimentation rate, C-reactive protein, creatinine, urea and electrolytes, liver enzymes, liver function tests (total bilirubin, albumin, international normalised ratio), and TSH (Zywiel et al., 2016). Closer surveillance of patients who exhibit chronic disease states leading to clinically important hypoalbuminemia and/or severe IMA elevations should be considered (Paustenbach et al., 2013).

Although the use of large-diameter CoCr femoral heads and hip resurfacing implants has now waned, approximately a million patients worldwide have an indwelling MoM prosthesis and over five million received hip implants with cobalt-containing components. Reversible cobaltism was recently reported in association with MoP implants and blood cobalt levels as low as 0.4 μg/L (Bridges et al., 2020; Tower et al., 2020), prompting the notion that systemic cobalt toxicity is underappreciated in the literature. Thus, it is important to raise awareness of the long-term as well as short-term risks of exposure to implant degradation products and develop sensitive methods to facilitate timely diagnosis and treatment. While there are no unique neurocognitive markers for heavy metal toxicity, the pattern of hypometabolism on quantitative $^{18}F$-FDG-PET/CT brain imaging shows promise in the differential diagnosis of ACE in neurologically symptomatic patients (Bridges et al., 2020; Tower et al., 2020).

## References

Allen, L.A., Ambardekar, A.V., Devaraj, K.M., Maleszewski, J.J., Wolfel, E.E., 2014. Missing elements of the history. N. Engl. J. Med. 370, 559–566.

Ambrosi, R.D., Ursino, N., 2020. N-acetyl-cysteine reduces blood chromium and cobalt levels in metal-on-metal hip arthroplasty. Arthroplast. Today 6, 149–152.

Apel, W., Stark, D., Stark, A., O'Hagan, S., Ling, J., 2013. Cobalt-chromium toxic retinopathy case study. Doc. Ophthalmol. 126, 69–78.

Apostoli, P., Catalani, S., Zaghini, A., Mariotti, A., Poliani, P.L., Vielmi, V., et al., 2013. High doses of cobalt induce optic and auditory neuropathy. Exp. Toxicol. Pathol. 65, 719–727.

Aslan, D., Apple, F., 2005. Clinical and analytical review of ischemia-modified albumin measured by the albumin cobalt binding test. Adv. Clin. Chem. 39, 1–10.

Awadallah, S., Atoum, M., Nimer, N., Saleh, S., 2012. Ischemia modified albumin: an oxidative stress marker in beta-thalassemia major. Clin. Chim. Acta 413, 907–910.

Bar-Or, D., Lau, E., Winkler, J.V., 2000. A novel assay for cobalt-albumin binding and its potential as a marker for myocardial ischemia—a preliminary report. J. Emerg. Med. 19, 311–315.

Bar-Or, D., Winkler, J., Vanbenthuysen, K., Harris, L., Lau, E., Hetzel, F., 2001. Reduced albumin–cobalt binding with transient myocardial ischemia after elective percutaneous transluminal coronary angioplasty: a preliminary comparison to creatinine kinase-mb, myoglobin, and troponin. Am. Heart J. 14, 985–991.

Barborik, M., Dusek, J., 1972. Cardiomyopathy accompaning industrial cobalt exposure. Br. Heart J. 34, 113–116.

Battaglia, V., Compagnone, A., Bandino, A., Bragadin, M., Alberto, C., Zanetti, F., et al., 2009. Cobalt induces oxidative stress in isolated liver mitochondria responsible forpermeability transition and intrinsic apoptosis in hepatocyte primary cultures. Int. J. Biochem. Cell Biol. 41, 586–594.

Beyersmann, D., Hartwig, A., 2008. Carcinogenic metal compounds: recent insight into molecular and cellular mechanisms. Arch. Toxicol. 82, 493–512.

Bhardwaj, N., Perez, J., Peden, M., 2011. Optic neuropathy from cobalt toxicity in a patient who ingested cattle magnets. Neuroophthalmology 35, 24–26.

Bijukumar, D.R., Segu, A., Souza, J.C., Li, X., Barba, M., Mercuri, L.G., 2018. Systemic and local toxicity of metal debris released from hip prostheses: A review of experimental approaches. Nanomedicine: Nanotechnology, Biology and Medicine 14 (3), 951–963.

Bonilla, H.M.G., Bhimaraj, A., 2018. A case of cobalt cardiomyopathy. J. Am. Coll. Cardiol. 71, A2386.

Borderie, D., Allanore, Y., Meune, C., Devaux, J., Ekindjian, O., Kahan, A., 2004. High ischemia-modified albumin concentration reflects oxidative stress but not myocardial involvement in systemic sclerosis. Clin. Chem. 50, 2190–2193.

Bridges, R., Cho, C., Beck, M., Gessner, B., Tower, S., 2020. F-18 FDG PET brain imaging in symptomatic arthroprosthetic cobaltism. Eur. J. Nucl. Med. Mol. Imaging 47, 1961–1970.

Callender, T., Morrow, L., Subramanian, K., Duhon, D., Ristovv, M., 1993. Three-dimensional brain metabolic imaging in patients with toxic encephalopathy. Environ. Res. 60, 295–319.

Catalani, S., Rizzetti, M., Padovani, A., Apostoli, P., 2012. Neurotoxicity of cobalt. Hum. Exp. Toxicol. 31, 421–437.

Catelas, I., Petit, A., Zukor, D.J., Huk, O.L., 2001. Cytotoxic and apoptotic effects of cobalt and chromium ions on J774 macrophages- implication of caspase-3 in the apoptotic pathway. J. Mater. Sci. Mater. Med. 12, 949–953.

Chandran, S.E., Giori, N.J., 2011. Nine-year incidence of kidney disease in patients who have had total hip arthroplasty. J. Arthroplasty 26, 24–27.

Charette, R.S., Neuwirth, A.L., Nelson, C.L., 2017. Arthroprosthetic cobaltism associated with cardiomyopathy. Arthroplast. Today 3, 225–228.

Chen, C., Tsai, W., Lin, P., Shiesh, S., 2011. The value of serum ischemia-modified albumin for assessing liver function in patients with chronic liver disease. Clin. Chem. Lab. Med. 49, 1817–1821.

Cheung, A.C., Banerjee, S., Cherian, J.J., Wong, F., Butany, J., Gilbert, C., et al., 2016. Systemic cobalt toxicity from total hip arthroplasties: review of a rare condition. Part 1 - history, mechanism, measurements and pathophysiology. Bone Joint J. 98-B, 6–13.

Choi, H.-I., Hong, J.A., Kim, M.-S., Lee, S.E., Jung, S.-H., Yoon, P.W., et al., 2018. Severe cardiomyopathy due to arthroprosthetic cobaltism: report of two cases with different outcomes. Cardiovasc. Toxicol. 19, 82–89.

Coetzee, O., Mari, A., 2017. Prosthetic hip-associated cobalt toxicity (Phact) reversal through nutrition supplementation and dietary intervention. J. Hum. Nutr. 1, 14–18.

Connell, E.O., Fesniak, H., Mead, N., 2013. Idiopathic cardiomyopathy following metal-on-metal hip arthroplasty: the new face of "beer drinker's cardiomyopathy. Eur. J. Cardiovasc. Med. 11, 210–211.

Curtis, J., Goode, G., Herrington, J., Urdaneta, L., 1976. Possible cobalt toxicity in maintenance hemodialysis patients after treatment with cobaltous chloride: a study of blood and tissue cobalt concentrations in normal subjects and patients with terminal and renal failure. Clin. Nephrol. 5, 61–65.

Dahms, K., Sharkova, Y., Heitland, P., Pankuweit, S., Schaefer, J.R., 2014. Cobalt intoxication diagnosed with the help of Dr House. Lancet 383, 574.

Day, P.L., Eckdahl, S.J., Maleszewski, J.J., Wright, T.C., Murray, D.L., 2017. Establishing human heart chromium, cobalt and vanadium concentrations by inductively coupled plasma mass spectrometry. J. Trace Elem. Med. Biol. 41, 60–65.

Duarte, M., Rocha, J., Moresco, R., Duarte, T., Da Cruz, I., Loro, V., et al., 2009. Association between ischemia-modified albumin, lipids and inflammation biomarkers in patients with hypercholesterolemia. Clin. Biochem. 42, 666–671.

Esteban Sanchez, C., Pasto-Cardona, L., 2016. Cobalt intoxication in a patient with hip prosthesis. Eur. J. Clin. Pharm. 18, 189–190.

Facchin, F., Catalani, S., Bianconi, E., Pasquale, D., De, Stea, S., et al., 2017. Albumin as marker for susceptibility to metal ions in metal-on-metal hip prosthesis patients. Hum. Exp. Toxicol. 36, 319–327.

Fox, K.A., Phillips, T.M., Yanta, J.H., Abesamis, M.G., 2016. Fatal cobalt toxicity after total hip arthroplasty revision for fractured ceramic components. Clin. Toxicol. 54, 874–877.

Garcia, M.D., Hur, M., Chen, J.J., Bhatti, M.T., 2020. Cobalt toxic optic neuropathy and retinopathy: case report and review of the literature. Am. J. Ophthalmol. Case Rep. 17, 100606.

Gautam, D., Pande, A., Malhotra, R., 2019. Fatal cobalt cardiomyopathy following revision total hip arthroplasty – a brief report with review of literature. Arch. Bone Joint Surg. 7, 379–383.

Gessner, B.D., Steck, T., Woelber, E., Tower, S.S., 2015. A systematic review of systemic cobaltism after wear or corrosion of chrome-cobalt hip implants. J. Patient Saf. 15, 97–104.

Giampreti, A., Lonati, D., Locatelli, C.A., 2014. Chelation in suspected prosthetic hip-associated cobalt toxicity. Can. J. Cardiol. 30, 465.

Gilbert, C.J., Cheung, A., Butany, J., Zywiel, M.G., Syed, K., McDonald, M., et al., 2013. Hip pain and heart failure: the missing link. Can. J. Cardiol. 29, 639.

Grant, M.L., Karp, J.K., Palladino, M., Le, N., Hall, N., Herman, J.H., 2016. Does therapeutic plasma exchange have a role in the treatment of prosthetic hip–associated cobalt toxicity? A case report and literature review. Hemapheresis 56, 2368–2373.

Grice, H., Munro, I., Wiberg, G., Hegg-Tveit, H., 1969. The pathology of experimentally induced cobalt cardiomyopathies. A comparison with beer drinkers' cardiomyopathy. Clin. Toxicol. 2, 273.

Griffiths, J., Colvin, A., Yates, P., Meyerkort, D., Kop, A., Prosser, G., 2015. Extreme cobalt toxicity: bearing the brunt of a failed ceramic liner. JBJS Case Connect. 5, e92.

Grillo, L., Nguyen, H., Tsang, S., Hood, D., Odel, J., 2016. Cobalt-chromium metallosis with normal ERG: a case report and review. J. Neuroopthalmol. 36, 383–388.

Hahn, A., Fuhlrott, J., Loos, A., Barcikowski, S., 2012. Cytotoxicity and ion release of alloy nanoparticles. J. Nanopart. Res. 14, 1–10.

Hara, A., Niwa, M., Aoki, H., Kumada, M., Kunisada, T., Oyama, T., et al., 2006. A new model of retinal photoreceptor cell degeneration induced by a chemical hypoxia-mimicking agent, cobalt chloride. Brain Res. 1109, 192–200.

Harris, R.C., Ismail, N., 1994. Extrarenal complications of the nephrotic syndrome. Am. J. Kidney Dis. 23, 477–497.

Harris, A., Johnson, J., Mansuripur, P.K., Limbird, R., 2015. Cobalt toxicity after revision to a metal-on-polyethylene total hip arthroplasty for fracture of ceramic acetabular component. Arthroplast. Today 1, 89–91.

Hart, A.J., Sabah, S.A., Bandi, A.S., Maggiore, P., Tarassoli, P., Sampson, B., et al., 2011. Sensitivity and specificity of blood cobalt and chromium metal ions for predicting failure of metal-on-metal hip replacement. J Bone Joint Surg. Br. 93-B, 1308–1313.

Hooisma, J., Emmen, H., 1992. Onderzoek naar de grenswaarden voor de Neurotoxic Symptom Checklist (NSC-60). Research on limit values for the Neurotoxic Symptom Checklist 60 (NSC-60). Report ARB 0094.9203. Amsterdam: Stichting Arbouw.

Ho, V., Arac, A., Shieh, P., 2018. Hearing and vision loss in an older man. JAMA Neurol. 75, 1439–1440.

Ho, J.H., Leikin, J.B., Dargan, P.I., Archer, J.R.H., Wood, D.M., Brent, J., 2017. Metal-on-metal hip joint prostheses: a retrospective case series investigating the association of systemic toxicity with serum cobalt and chromium concentrations. J. Med. Toxicol. 13, 321–328.

Hur, C.I., Yoon, T.R., Cho, S.G., Song, E.K., Seon, J.K., 2008. Serum ion level after metal-on-metal THA in patients with renal failure. Clin. Orthop. Relat. Res. 466, 696–699.

Ikeda, T., Takahashi, K., Kabata, T., Sakagoshi, D., Tomita, K., Yamada, M., 2010. Polyneuropathy caused by cobalt-chromium metallosis after total hip replacement. Muscle Nerve 42, 140–143.

Jefferis, J.M., Hickman, S.J., 2019. Treatment and outcomes in nutritional optic neuropathy. Curr. Treat. Options. Neurol. 21, 5.

Kennedy, A., Dornan, J., King, R., 1981. Fatal myocardial disease associated with industrial exposure to cobalt. Lancet 317, 412–414.

Kerger, B., Gerads, R., Gurleyuk, H., Thuett, K., Finley, B., Paustenbach, D., 2013a. Cobalt speciation assay for human serum, Part I. Method for measuring large and small molecular cobalt and protein-binding capacity using size exclusion chromatography with inductively coupled plasma-mass spectroscopy detection. Toxicol. Environ. Chem. 95, 687–708.

Kerger, B., Gerads, R., Gurleyuk, H., Tsuji, J., 2018. Metals measurements in body tissues and fluids: toxicological and clinical importance of standardizing quality analytical methods for differentiating cobalt partitioning on a molecular level. In: Mihalko, W., Lemons, J., Greenwald, A., Kurtz, S. (Eds.), Beyond the Implant: Retrieval Analysis Methods for Implant Surveillance. ASTM International, West Conshohocken, PA, pp. 229–244.

Kerger, B., Tvermoes, B., Unice, K., Finley, B., Paustenbach, D., Galbraith, D., 2013b. Cobalt speciation assay for human serum, Part II. Method validation in a study of human volunteers ingesting cobalt(II) chloride dietary supplement for 90 days. Toxicol. Environ. Chem. 95, 709–718.

Kesteloot, H., Roelandt, J., Willems, J., 1968. An enquiry into the role of cobalt in the heart disease of chronic beer drinkers. Circulation 37, 854–864.

Khan, A.H., Verma, R., Bajpai, A., Mackey-Bojack, S., 2015. Unusual case of congestive heart failure: cardiac magnetic resonance imaging and histopathologic findings in cobalt cardiomyopathy. Circ. Cardiovasc. Imaging 8, e003352.

Kim, C., Choi, Y.H., Jeong, M.Y., 2016. Cobalt intoxication heart failure after revision total hip replacement for ceramic head fracture: a case report. Hip Pelvis 28, 259–263.

Kiyici, A., Mehmetoglu, I., Karaoglan, H., 2010. Ischemia-modified albumin levels in patients with end-stage renal disease patients on hemodialysis: does albumin analysis method affect albumin-adjusted ischemia-modified albumin levels? J. Clin. Lab. Anal. 24, 273–277.

Kopera, E., Schwerdtle, T., Hartwig, A., Bal, W., 2004. Co(II) and Cd(II) substitute for Zn(II) in the zinc finger derived from the DNA repair protein XPA, demonstrating a variety of potential mechanisms of toxicity. Chem. Res. Toxicol. 17, 1452–1458.

Koronfel, M.A., Goode, A.E., Weker, J.N., Tay, S.E.R., Stitt, C.A., Simoes, T.A., et al., 2018. Understanding the reactivity of CoCrMo-implant wear particles. npj Mater. Degrad. 2, 1–5.

Lecoanet, P., Blangis, M., Garcia, M., Legallois, Y., Fabre, T., 2019. Chromium-cobalt intoxication with intense systemic complications following total hip revision after per-operative ceramic fracture. Case Rep. Orthop. 2019, 4209796.

Lee, S.G., Lee, H., Rho, H.M., 2001. Transcriptional repression of the human p53 gene by cobalt chloride mimicking hypoxia. FEBS Lett. 507, 259–263.

Leikin, J.B., Karydes, H.C., Whiteley, P.M., Wills, B.K., Cumpston, K.L., Jacobs, J.J., 2013. Outpatient toxicology clinic experience of patients with hip implants. Clin. Toxicol. 51, 230–236.

Leyssens, L., Vinck, B., Straeten, C., Van Der, Dhooge, I., Wuyts, F.L., Maes, L.K., 2020. The ototoxic potential of cobalt from metal- on-metal hip implants: a pilot study on the patient-reported auditory, vestibular, and general neurological outcome. Int. J. Audiol. 60, 44–53.

Machado, C., Appelbe, A., Wood, R., 2012. Arthroprosthetic cobaltism and cardiomyopathy. Hear. Lung Circ. 21, 759–760.

Macho, L., Palkovic, M., 1963. The influence of endocrine glands on carbohydrate metabolism: VII.–The effect of thyroxine on citric and pyruvic acid levels in blood during the pyruvate tolerance test. Arch. Int. Physiol. Biochim. 71, 758–761.

Manifold, I.H., Platts, M.M., Kennedy, A., 1978. Cobalt cardiomyopathy in a patient on maintenance haemodialysis. Br. Med. J. 2, 1609.

Manisty, C., Skinner, J., Moon, J.C., 2018. Metal-on-metal hips and heart failure – Can we relax? Int. J. Cardiol. 284, 65–66.

Mao, X., Wong, A.a, Crawford, R.W., 2011. Cobalt toxicity-an emerging clinical problem in patients with metal-on-metal hip prostheses? Med. J. Aust. 194, 649–651.

Martin, J., Spencer-Gardner, L., Camp, C., Stulak, J., Sierra, R., 2015. Cardiac cobaltism: a rare complication after bilateral metal-on-metal total hip arthroplasty. Arthroplast. Today 1, 99–102.

Matés, J.M., Segura, J.A., Alonso, F.J., Márquez, J., 2010. Roles of dioxins and heavy metals in cancer and neurological diseases using ROS-mediated mechanisms. Free. Radic. Biol. Med. 49, 1328–1341.

Megaterio, S., Galetto, F., Alossa, E., Capretto, S., 2001. Effetti a distanza del rilascio di ioni metallo in usura della testa protesica: presentazione di un caso. Sci. Total. Environ. 27, 173–175.

Monies, A., Prost, M., 1994. Experimental studies on lesions of eye tissues in cobalt intoxication. Klin. Ocz. 96, 135–139.

Moniz, S., Hodgkinson, S., Yates, P., 2017. Cardiac transplant due to metal toxicity associated with hip arthroplasty. Arthroplast. Today 3, 151–153.

Morin, Y., Foley, A., Martineau, G., Roussel, J., 1967. Quebec beer-drinkers' cardiomyopathy: forty-eight cases. Can. Med. Assoc. J. 97, 881–883.

Morvai, V., Szakmary, E., Tatrai, E., 1993. The effects of simultaneous alcohol and cobalt chloride administration on the cardiovascular system of rats. Acta Physiol. Hung. 81, 253–261.

Mosier, B.A., Maynard, L., Sotereanos, N.G., Sewecke, J.J., 2016. Progressive cardiomyopathy in a patient with elevated cobalt ion levels and bilateral metal-on-metal hip arthroplasties. Am. J. Orthop. 45, 132–135.

Nandedkar, A., Basu, P., Friedberg, F., 1973. Co(II) binding by plasma proteins. Bioinorg. Chem. 2, 149–157.

Ng, S.K., Ebneter, A., Gilhotra, J.S., 2013. Hip-implant related chorio-retinal cobalt toxicity. Indian. J. Ophthalmol. 61, 35–37.

Nieto-Aristizabal, I., Vivas, A., Ruiz-Montaño, P., Aragon, C., Posso-Osorio, I., Quiñones, J., et al., 2020. Therapeutic plasma exchange as a treatment for autoimmune neurological disease. Autoimmune Dis. 2020, 3484659.

Nyga, A., Hart, A., Tetley, T.D., 2015. Importance of the HIF pathway in cobalt nanoparticle-induced cytotoxicity and inflammation in human macrophages. Nanotoxicology 5390, 905–917.

Oldenburg, M., Wegner, R., Baur, X., 2009. Severe cobalt intoxication due to prosthesis wear in repeated total hip arthroplasty. J. Arthroplasty 24, 825.

Olmedo-Garcia, N.I., Zagra, L., 2018. High risk of complications using metal heads after ceramic fracture in total hip arthroplasty. Hip Int. 29, 373–378.

Packer, M., 2016. Cobalt cardiomyopathy: a critical reappraisal in light of a recent resurgence. Circ. Heart Fail. 9, e003604.

Paustenbach, D.J., Galbraith, D.A., Finley, B.L., 2014. Interpreting cobalt blood concentrations in hip implant patients. Clin. Toxicol. 52, 98–112.

Paustenbach, D.J., Tvermoes, B.E., Unice, K.M., Finley, B.L., Kerger, B.D., 2013. A review of the health hazards posed by cobalt. Crit. Rev. Toxicol. 43, 316–362.

Pazzaglia, U.E., Apostoli, P., Congiu, T., Catalani, S., Marchese, M., Zarattini, G., 2011. Cobalt, chromium and molybdenum ions kinetics in the human body: data gained from a total hip replacement with massive third body wear of the head and neuropathy by cobalt intoxication. Arch. Orthop. Trauma. Surg. 131, 1299–1308.

Pelclova, D., Sklensky, M., Janicek, P., Lach, K., 2012. Severe cobalt intoxication following hip replacement revision: clinical features and outcome. Clin. Toxicol. 50, 262–265.

Peters, R., Willemse, P., Rijk, P., Hoogendoorn, M., 2017. Fatal cobalt toxicity after a non-metal-on-metal total hip arthroplasty. Case Rep. Orthop. 2017, 9123684.

Phillips, A., Gerald Shaper, A., Whincup, P., 1989. Association between serum albumin and mortality from cardiovascular disease, cancer, and other causes. Lancet 334, 1434–1436.

Pinazo-Durán, M.D., Arévalo, J.F., García-Medina, J.J., Zanón-Moreno, V., Gallego-Pinazo, R., Nucci, C., 2016. Ocular comorbidities and the relationship between eye diseases and systemic disorders. Biomed. Res. Int. 2016, 9519350.

Piwowar, A., Knapik-Kordecka, M., Warwas, M., 2008. Ischemia-modified albumin level in type 2 diabetes mellitus- preliminary report. Dis. Markers 24, 311–317.

Prentice, J.R., Blackwell, C.S., Raoof, N., Bacon, P., Ray, J., Hickman, S.J., et al., 2014. Auditory and visual health after ten years of exposure to metal-on-metal hip prostheses: a cross-sectional study follow up. PLoS One 9, e90838.

Reich, M.S., Javidan, P., Garg, V.K., Copp, S.N., 2019. Chronic systemic metal ion toxicity from wear on a revised cobalt- chromium trunnion. J. Orthop. Case Rep. 9, 48–51.

Rizzetti, M.C., Liberini, P., Zarattini, G., Catalani, S., Pazzaglia, U., Apostoli, P., et al., 2009. Loss of sight and sound. Could it be the hip? Lancet 373, 1052.

Rona, G., 1971. Experimental aspects of cobalt cardiomyopathy. Br. Heart J. 33, 171–174.

Samar, Huma Y., Doyle, M., Williams, R.B., Yamrozik, J.A., Bunker, M., Biederman, R.W.W., et al., 2015. Novel use of cardiac magnetic resonance imaging for the diagnosis of cobalt cardiomyopathy. JACC Cardiovasc. Imaging 8, 1231–1232.

Sampson, B., Hart, A., 2012. Clinical usefulness of blood metal measurements to assess the failure of metal-on-metal hip implants. Ann. Clin. Biochem. 49, 118–131.

Sandusky, G., Crawford, M., Roberts, E., 1981. Experimental cobalt cardiomyopathy in the dog: a model for cardiomyopathy in dogs and man. Toxicol. Appl. Pharmacol. 60, 263–278.

Sanz Pérez, M.I., Rico Villoras, A.M., Moreno Velasco, A., Bartolomé García, S., Campo Loarte, J., 2019. Heart transplant secondary to cobalt toxicity after hip arthroplasty revision. Hip Int. 29, 1–5.

Savarino, L., Cadossi, M., Chiarello, E., Fotia, C., Greco, M., Baldini, N., et al., 2014. How do metal ion levels change over time in hip resurfacing patients? A cohort study. 2014, 291925.

Schirrmacher, U.O.E., 1967. Case of cobalt poisoning. Br Med J. 1, 544–545.

Semenza, G., 2000. HIF-1: mediator of physiological and pathophysiological responses to hypoxia. J. Appl. Physiol. 88, 1474–1480.

Shulman, G., Ladenson, P., Wolfe, M., Ridgway, E., Wolfe, R., 1985. Substrate cycling between gluconeogenesis and glycolysis in euthyroid, hypothyroid, and hyperthyroid man. J. Clin. Invest. 76, 757–764.

Singh, M., Ghazzal, A., Krishnan, M., Kadakkal, A., Hofmeyer, M., Lam, P., et al., 2020. From hip to heart: cobalt-induced cardiomyopathy requiring cardiac transplantation. J. Card. Fail. 26, S161.

Smith, C.J., Perfetti, T.A., 2019. In vitro cobalt-stimulated hypoxia-inducible factor-1 overexpression does not correlate with cancer risk from cobalt exposure in humans. Toxicol. Res. Appl. 3.

Steens, W., von Foerster, G., Katzer, A., 2006. Severe cobalt poisoning with loss of sight after ceramic-metal pairing in a hip- a case report. Acta Orthop. 77, 830–832.

Stepien, K.M., Abidin, Z., Lee, G., Cullen, R., Logan, P., Pastores, G.M., 2018. Metallosis mimicking a metabolic disorder: a case report. Mol. Genet. Metab. Rep. 17, 38–41.

Swiatkowska, I., 2019. Toxicity of Metal Debris From Hip Implants (Ph.D. thesis). University College London, London, UK.

Swiatkowska, I., Henckel, J., Sabah, S.A., Hart, A.J., 2020. Self-reported neurotoxic symptoms in hip arthroplasty patients with highly elevated blood cobalt: a case-control study. J. Patient Saf. 18, 1–8.

Tilney, R., Burg, M.R., Sammut, M.A., 2017. Cobalt cardiomyopathy secondary to hip arthroplasty: an increasingly prevalent problem. Case Rep. Cardiol. 2017, 1–4.

Tower, S., 2010a. Arthroprosthetic cobaltism: neurological and cardiac manifestations in two patients with metal-on-metal arthroplasty: a case report. J. Bone Joint. Surg. 92, 2847–2851.

Tower, S., 2010b. Cobalt toxicity in two hip replacement patients. State Alsk. Epidemiol. Bull. 14, 1.

Tower, S., 2010c. Arthroprosthetic cobaltism: identification of the at-risk patient. Alsk. Med. 52, 28–32.

Tower, S., Medlin, D., Bridges, R., Cho, C., 2020. Corrosion of polished cobalt-chrome stems presenting as cobalt encephalopathy. Arthroplast. Today 6, 1022–1027.

Tower, S.S., 2012. Arthroprosthetic cobaltism associated with metal on metal hip implants. J. Bone Joint Surg. Am. 344, e430.

Umar, M., Jahangir, N., Khan, M.F., Saeed, Z., Sultan, F., Sultan, A., 2020. Cobalt-related cardiomyopathy: a real concern! A review of published evidence. J. Orthop. Surg. 28.

Van Lingen, C.P., Ettema, H.B., Timmer, J.R., De Jong, G., Verheyen, C.C.P.M., 2013. Clinical manifestations in ten patients with asymptomatic metal-on-metal hip arthroplasty with very high cobalt levels. Hip Int. 23, 441–444.

Vasukutty, N.L., Minhas, T.H.A., 2016. Systemic effects of cobalt toxicity after revision hip replacement can manifest in intermediate to long term follow-up. Hip Int. 26, e31–e34.

Weber, K.P., Schweier, C., Kana, V., Guggi, T., Byber, K., Landau, K., 2015. Wear and tear vision. J. Neuroopthalmol. 35, 82–85.

WHO, 2006. IARC monographs on the evaluation of carcinogenic risks to humans: cobalt in hard metals and cobalt sulfate, gallium arsenide, indium phosphide and vanadium pentoxide, World Health Organization International Agency For Research On Cancer, Vol. 86. IRAC, Lyon, France, pp. 3–5.

Yu, R., 2017. Cobalt toxicity—an overlooked cause of hypothyroidism. J. Endocrinol. Thyroid. Res. 1, 1–4.

Zeynalov, E., Cutrufello, N., Pierce, A., 2018. Cobalt blues: a case report of cobalt intoxication associated obstrictive sleep apnea (OSA). Case Rep. Clin. Trainees 41, 2–3.

Zywiel, M., Brandt, J.-M., Overgaard, C., Cheung, A., Turgeon, T., Syed, K., 2013. Fatal cardiomyopathy after revision total hip replacement for fracture of a ceramic liner. Bone Joint J. 95-B, 31–37.

Zywiel, M.G., Cherian, J.J., Banerjee, S., Cheung, A.C., Wong, F., Butany, J., et al., 2016. Systemic cobalt toxicity from total hip arthroplasties: review of a rare condition Part 1 - history, mechanism, measurements, and pathophysiology. Bone Joint. J. 98B, 14–20.

Part III

# Translational effect of biomarker research on clinical practice

Chapter 8

# Clinical guidelines on the use of biomarkers for surveillance of hip replacements

**Harry Hothi[1], Reshid Berber[2], Shiraz A. Sabah[3] and Alister J. Hart[1]**
*[1]Royal National Orthopaedic Hospital, Stanmore, United Kingdom, [2]Nottingham University Hospital NHS Trust, Nottingham, United Kingdom, [3]Nuffield Department of Orthopaedics, Rheumatology, and Musculoskeletal Sciences, University of Oxford, Oxford, United Kingdom*

## 8.1 Introduction

Hip replacement surgery is one of the most commonly performed operations worldwide. Annually, approximately 500,000 people receive a hip implant for the first time in the United States and more than 100,000 in the United Kingdom (Maradit Kremers et al., 2015; National Joint Registry, 2019). The procedure is widely regarded as one of the most successful surgical interventions owing to the considerable improvement in quality of life, joint pain, and function reported by patients. The National Joint Registry reports that over 95% of primary hip replacements remain in situ at a 10-year follow-up. However, more than 7000 patients undergo revision total hip replacement (THR) each year. While the majority of cases are performed for aseptic loosening, lysis, or implant wear, approximately 10% are indicated to treat adverse reactions to metal debris (ARMD) (National Joint Registry, 2019).

The complications associated with wear and corrosion in metal-on-metal (MoM) hip implants include local destruction of soft tissue and bone, and the potential for systemic effects from circulating metal ions (Berber et al., 2016). These have resulted in several high-profile recalls of MoM hip designs, including the ASR hip manufactured by DePuy: the device was implanted in 90,000 patients worldwide, of whom 50% have subsequently undergone revision surgery (Galea et al., 2018). Although the use of MoM bearing couples in hip replacements has dramatically declined, CoCrMo alloy remains widely used in orthopaedic implants, with wear and corrosion at other implant interfaces being increasingly well recognised. Importantly, CoCrMo is not the only metal alloy with potential for local or systemic

**Biomarkers of Hip Implant Function. DOI: https://doi.org/10.1016/B978-0-12-821596-8.00004-5**

toxicity. Titanium release from joint replacements and dental implants was also linked to adverse local tissue reactions (ALTR), allergy symptoms, and systemic complications (McPherson et al., 2014; Moran et al., 1991; Tarpada et al., 2020; Thomas et al., 2006).

As life expectancy increases worldwide, the requirement for hip replacement procedures is projected to rise (Kurtz et al., 2007). Additionally, surgery is increasingly being offered to younger patients. A growing number of hip implant recipients combined with higher activity levels and longer life expectancy translates into greater potential exposure to metal debris. The risk of revision due to periprosthetic joint infection also appears to be increasing (Lenguerrand et al., 2017). Improving clinical investigation tools and techniques will help identify and manage at-risk individuals.

The previous chapters have discussed research into the utility of and potential for biomarkers to objectively inform clinical decision-making. This chapter provides an overview of how biomarkers have become incorporated into clinical guidelines for managing patients with hip implants.

## 8.2 Evaluating implant wear and risk of local adverse reactions to metal debris in patients with metal-on-metal hips

Approximately 71,000 patients in the United Kingdom have received MoM THR or hip resurfacing (HR) (National Joint Registry, 2019). Globally, this number exceeds 1.5 million, and for approximately 1 million patients, the MoM device has not been revised and requires routine surveillance. Guidelines for the management of patients with MoM hips have been published by the United Kingdom's Medicines and Healthcare products Regulatory Agency (MHRA, 2017), the European Federation of National Associations of Orthopaedics and Traumatology (EFORT, 2012), Australia's Therapeutic Goods Administration (TGA, 2017), US Food and Drug Administration (FDA, 2019), and Health Canada (Health Canada, 2012). Each of the five regulators/organisations provides their recommendations on how often each patient should be followed up, depending on if they are asymptomatic or symptomatic and if they are in a patient group with a higher risk of developing ARMD. Blood metal ion measurement is a prominent feature of each set of guidelines; however, nearly always as an adjunct to clinical and radiological assessment rather than a binary classifier. The remainder of this section will outline current clinical guidelines, focusing on those produced by the MHRA.

The most recent MHRA guidelines from 2017 identified the MoM implant designs at high risk of subsequent failure. Patients implanted with these devices were recommended for annual surveillance (while the MoM device remains in situ) regardless of whether or not they were symptomatic. The identified devices were:

1. THR with a femoral head diameter $\geq 36$ mm.
2. The recalled DePuy ASR resurfacing.
3. Any design of HR in a female patient.
4. Any design of HR with a femoral head diameter $\leq 48$ mm in a male patient.

Male recipients of HR implants with femoral head diameter >48 mm and all patients with MoM THR with femoral head diameter <36 mm are considered to be at lower risk of ARMD. However, the MHRA guidelines recommend annual review for patients in these categories if they are symptomatic.

Asymptomatic patients outside of these high-risk categories are triaged to less regular follow-up, dependent on the track record of their implant. Patients with implants rated <10 A by the Orthopaedic Data Evaluation Panel (ODEP), or with THRs, should be seen annually for the first 5 years, then at a reduced frequency if they remain asymptomatic. Patients with implants rated $\geq 10$ A by the ODEP are recommended for follow-up at 1, 7, and every 3 years thereafter.

The MHRA guidance is that all patients with a MoM hip undergo blood metal level testing at each follow-up visit, and complete a questionnaire to assess the clinical function of their implant (Oxford Hip Score). Cobalt and chromium concentrations should be measured in whole blood, and if the results are $\geq 7$ μg/L for either metal (119 nmol/L for cobalt and 134.5 nmol/L for chromium), then cross-sectional imaging and closer patient follow-up are warranted. In patients with MoM THRs, blood metal levels <7 μg/L may be associated with wear and/or corrosion at surfaces other than the metal bearing, such as the head/stem junction. Depending on the design of the MoM hip implanted, some patients will also require a plain radiograph, metal artefact reduction sequence (MARS)-MRI, or ultrasound imaging of their hip at every follow-up. In other patient groups (e.g., women with HR), MARS-MRI is recommended if blood metal ion levels have risen since the last follow-up and if there has been deterioration in the Oxford Hip Score. Rising blood metal levels may indicate soft-tissue reactions and should be taken into account when considering the need to revise a MoM implant. Importantly, blood metal levels are only one part of the diagnostic work-up and should be used in conjunction with the clinical and radiological evaluations. Imaging results showing bone or muscle loss are of the most significant clinical concern. Following revision surgery, an improvement in symptoms and a clear fall in blood metal levels are expected. If these do not occur, further investigation is required to identify the underlying cause.

Table 8.1 gives an overview of the guidance provided by the five major health regulators/organisations as to the frequency of follow-up and use of blood metal testing and imaging in asymptomatic and symptomatic patients with MoM hips.

**TABLE 8.1** Recommendations for follow-up in patients with MoM hip implants.

| Organisation | Asymptomatic patients | | Symptomatic patients | | Comments | Metal ion testing | |
|---|---|---|---|---|---|---|---|
| | Follow-up frequency | Investigations | Follow-up frequency | Investigations | | Method | Thresholds defined |
| MHRA | Annually for high-risk groups | Blood metal testing; MARS-MRI or ultrasound if blood metals elevated or poor function | Annually | Blood metal testing; MARS-MRI or ultrasound | High risk = large-head THRs, all ASR HR, women with HR, and men with HR ≤48 mm | Co and Cr concentration measured in whole blood | Levels ≥7 μg/L require further investigation and cross-sectional imaging |
| EFORT | Annually for large-head (≥36 mm) THRs and high-risk HR | Blood metal testing and plain radiographs; further imaging if necessary | Undefined, but presumed at least annually | Blood metal testing; radiographs, MARS-MRI, CT, or ultrasound | High risk = female patients, head size <50 mm, low coverage arc | Co concentration measured in whole blood | Levels <2 μg/L not of concern in asymptomatic patients. Clinical concern when levels between 2 and 7 μg/L. Levels >7 μg/L require additional imaging. Levels exceeding 20 μg/L are an indication for revision |
| TGA | Annually for all large-head THRs and HR ≤45 mm | Plain radiographs or MARS-MRI; blood metal testing | Annually | Blood metal testing; radiographs, MARS-MRI, CT, or ultrasound | Follow-up frequency of all other asymptomatic patients the same as for non-MoM hips | Co and Cr concentrations measured in whole blood or serum | No specific threshold (the trend in values should be monitored instead) |

| | | | | | | | |
|---|---|---|---|---|---|---|---|
| FDA | Every 12 years | Plain radiographs only to evaluate for loosening or osteolysis | Every 6 months | Blood metal testing; radiographs, MARS-MRI, CT, or ultrasound | Advice given on high-risk patients, such as women or those with small-head HR | Co and Cr concentrations measured in whole blood | No specific threshold (measures should be used in conjunction with other investigations) |
| Health Canada | Annually for at least the first 5 years | Not specified | Undefined, but presumed at least annually | Blood metal testing; radiographs, MARS-MRI, CT, or ultrasound | Increased risk = female patients, very high BMI, increased activity, bilateral implants | Co and Cr concentrations measured in whole blood and serum | Levels ≥7 μg/L require further investigation |

*BMI*, body mass index; *CT*, computed tomography; *EFORT*, European Federation of National Associations of Orthopaedics and Traumatology; *FDA*, US Food and Drug Administration; *HR*, hip resurfacing; *MARS-MRI*, metal artefact reduction sequence-magnetic resonance imaging; *MHRA*, Medicines and Healthcare products Regulatory Agency; *MoM*, metal-on-metal; *TGA*, Therapeutic Goods Administration; *THR*, total hip replacement.

### 8.2.1 Circulating cobalt and chromium levels

In non-occupationally exposed individuals without hip implants, blood cobalt and chromium levels are expected to be $<1\ \mu g/L$ (Grassin-Delyle et al., 2019). For most individuals with MoM implants, they range from 0.2 to 10 μg/L, with steady-state median concentrations between 1.5 and 2.3 μg/L, and reported levels exceeding 20 μg/L in some asymptomatic patients (Prentice et al., 2013; Wretenberg, 2008).

There is currently no consensus as to the definition of thresholds of blood metal ion levels that warrant clinical concern. Indeed, some regulators, such as the TGA, question the usefulness of this measure to determine ARMD in patients.

In 2010, the MHRA advised that whole blood cobalt/chromium levels over 7 μg/L indicate the potential for adverse soft-tissue reactions and increased risk of implant failure. This threshold was reported to have a high specificity (89%) but poor sensitivity (52%) for detecting failed MoM HR (Hart et al., 2011, 2014). The updated MHRA guidance issued in 2017 states that 'there is no agreed threshold value for whole blood metal levels that either predicts outcome, or mandates revision'.

The European Consensus stated that in MoM hips, 'blood cobalt levels of under 2 μg/L are probably devoid of clinical concern, whereas the threshold value for clinical concern is expected to be within the range of 2–7 μg/L' (Hannemann et al., 2013).

The Mayo Clinic advised that patients with serum cobalt and chromium exceeding 10 and 15 μg/L, respectively, are likely to display 'significant implant deterioration'. They also noted that while elevated metal ion concentrations may indicate implant wear, they are not considered a health hazard. Similarly, the FDA concluded that there is 'insufficient evidence to correlate the presence of localised lesions, clinical outcomes, and/or the need for revision with specific metal ion levels for individual patients'.

Although blood cobalt and chromium levels are approximate indicators of low, moderate, and high implant wear rates, physiological differences between patients and analytical approaches used by different testing laboratories contribute to uncertainty in this measure.

### 8.2.2 The decision to revise

There is consensus amongst regulatory and advisory bodies that the decision to revise should be made following an assessment of several parameters (e.g., blood metal levels, cross-sectional imaging, and physical examination) rather than any one of them in isolation. The MHRA and EFORT guidelines indicate that revision should be considered if there are abnormal findings on imaging that may be explained by elevated blood metal levels. EFORT further define levels of 20 μg/L and above as being a 'case of excessive elevation', with increased risks to long-term health as well as osteolysis and tissue necrosis; they recommend that these risks and the option of revision are discussed with the patient. Health Canada places a greater emphasis on MRI findings, advising that symptomatic patients with evidence of a soft-tissue mass, or even asymptomatic patients with evidence of a soft-tissue mass that is increasing in size, should be seriously considered for revision.

Evidently, there is considerable variation in the published guidance, which can lead to uncertainty among surgeons when setting revision thresholds and making decisions regarding patient management. To avoid this, there has been a greater push for a multidisciplinary team (MDT) approach to the MoM hip burden. MDT meetings combine the tacit knowledge of an expert panel, regulatory guidance, and up-to-date evidence to improve decision-making among surgeons, including the use of investigations (imaging and biomarkers) and their interpretation. Berber et al. (2015), who proposed this approach, demonstrated a high level of concordance between the recommendation from the meeting and the actual treatment offered to the patient, suggesting that the method effectively reduced uncertainty among surgeons and could improve patient outcomes. This approach has been extended to cover all complex hip and knee revision procedures, and revision networks are now planned for all areas of England and Wales, supported by the Department of Health and national specialist surgical societies.

### 8.2.3 Conclusion

Blood cobalt and chromium levels are important markers of implant wear rate and useful adjuncts to other investigative tools when evaluating the function of MoM implants.

## 8.3 Investigation for systemic toxicity

Chronic exposure to extremely high circulating levels of cobalt can lead to systemic toxicity in rare cases. One case report, published in 2001, described the experience of a patient with a high-wearing CoCrMo femoral head and a constellation of symptoms, including peripheral, motor, and sensory neuropathy, pericardial tamponade, and hypothyroidism—a syndrome collectively referred to as 'arthroprosthetic cobaltism' (Megaterio et al., 2001). Subsequent accounts of patients with metal hips and elevated systemic cobalt load documented haematological, psychiatric, endocrine, neurological, dermatological, and cardiac adverse effects (see Chapter 7 for a comprehensive review of the subject). The two patient groups impacted have been those with MoM hips and those with fractured ceramic bearings that were subsequently revised to metal-on-polyethylene (MoP) implants with CoCrMo femoral heads. Circulating biomarkers of essential organ function are readily available and may inform the diagnosis of arthroprosthetic cobaltism in recipients of metal hip implants. The use of implant-related parameters, most notably blood cobalt level, to predict the risk of arthroprosthetic cobaltism is also an active area of research.

### 8.3.1 Circulating cobalt levels

It is important to emphasise that while many reports do appear to show an association between high implant wear/elevated blood metal levels (typically Co >20 μg/L) and adverse systemic effects, the understanding of toxic levels of cobalt remains incomplete. To generate insights, several groups have studied the potential effects of cobalt on different body organs.

Researchers from the consulting group ChemRisk carried out a series of cobalt supplementation studies to better define blood cobalt kinetics (Finley et al., 2013; Tvermoes et al., 2013, 2014). One of the studies found that blood cobalt levels of up to 117 μg/L (average, 10–70 μg/L) were not associated with clinically significant changes in vision, hearing, cardiac/neurological function, or haematological parameters after daily administration of 1 mg of $CoCl_2$ for 90 days (Tvermoes et al., 2014). Using the collected pharmacokinetics data, Unice et al. (2014, 2012) developed a biokinetic model to estimate tissue cobalt levels in patients with MoM hips exposed to sustained blood cobalt of 10 μg/L for 10 years. The authors reported that the peak cardiac, kidney, and liver cobalt concentrations in these patients were similar to those estimated following 90-day cobalt supplementation, and concluded that most MoM implant recipients are unlikely to suffer systemic health effects related to their blood cobalt load. Further calculations indicated that biologically important systemic adverse effects were very unlikely to occur with blood cobalt levels below 300 μg/L (Paustenbach et al., 2014). This threshold does not seem to be supported by case report data, where systemic sequelae were often associated with much lower systemic cobalt concentrations.

Berber et al. (2017) employed cardiac MRI, echocardiography, and blood biomarker sampling to assess cardiac function in three age- and sex-matched patient groups: (1) non-MoM hips, (2) MoM hips with blood cobalt levels below 7 μg/L, and (3) MoM hips with blood cobalt levels above 7 μg/L. There were no significant between-group differences in B-type natriuretic peptide or troponin levels, or in myocardial cobalt deposition (measured as T2* values). Furthermore, no correlation was demonstrated between circulating cobalt concentration and the ejection fraction. The authors found that blood cobalt of up to 118 μg/L did not exert noticeable cardiotoxic effects. In line with these results, large-scale population studies showed that MoM THA is not associated with cardiac complications (Goodnough et al., 2018; Sabah et al., 2018). The cardiology community has been reassured by these findings and acknowledges that routine cardiac monitoring of patients with MoM hips is impractical and unnecessary; however, clinicians need to be vigilant that cobalt cardiomyopathy is a reversible cause of heart failure and should be considered in patients with highly elevated blood cobalt levels (Manisty et al., 2018).

As discussed in Section 8.2, healthcare regulators such as the MHRA and FDA advise on how to manage asymptomatic and symptomatic patients with MoM hips. Importantly, the focus in these recommendations is the assessment of local symptoms and indicators of poor implant function. Guidance on the specific blood cobalt levels above or below which systemic toxicity can be identified or eliminated as a cause of concern is not available. Evidence derived from case reports of confirmed arthroprosthetic cobaltism shows cobalt levels to be markedly elevated; however, the values are highly variable and have been as low as 14 μg/L in those experiencing mild neurotoxicity. In the absence of official guidance, several authors have suggested their own criteria regarding investigations for systemic toxicity (Table 8.2).

**TABLE 8.2 Proposed blood cobalt thresholds to warrant investigation for systemic toxicity in patients with metal hip implants.**

| References | Proposed threshold | Basis | Recommended action |
|---|---|---|---|
| Paustenbach et al. (2013, 2014) | 100 μg/L in whole blood | Review of 122 studies and case reports involving cobalt-exposed individuals (not including patients with hip implants); oral cobalt dosing studies (Finley et al., 2013; Tvermoes et al., 2013, 2014); cobalt biokinetic model (Unice et al., 2012). The proposed threshold includes an uncertainty factor of 3 to account for the fact that persons in the various studies were exposed to cobalt for less than a year | Monitor for signs of hypothyroidism and polycythaemia |
| Van Der Straeten et al. (2013) | 20 μg/L in blood serum | Existing case reports of neurological, ototoxic, and cardiotoxic symptoms in THA recipients | Consider revision, even with minor clinical symptoms |
| Tower (2010a,b) | 7 μg/L in blood serum | MHRA cut-off; two Alaskan case reports of arthroprosthetic cobaltism | Assess for possible cardiac and neurological symptoms |

*MHRA*, Medicines and Healthcare products Regulatory Agency.

### 8.3.2 The decision to revise

The advice from the FDA to orthopaedic surgeons continues to consider the outcome of different investigations to build up an understanding of the patient's overall health and the function of their implant. The FDA recommends using blood metal testing as one of these investigative tools but does not specify any thresholds. Instead, rises or falls in serial testing should be used to aid the interpretation of patient-reported symptoms. The regulator has also issued broader guidance for non-surgeon clinicians (e.g., family doctors) about the possible systemic symptoms to be aware of when treating patients with MoM hips. These include general hypersensitivity reactions, cardiomyopathy, neurological changes, psychological status changes, renal function impairment, and thyroid dysfunction. When a clinician suspects that these symptoms may be due to the metal implant, they are to advise the patient to follow up with their treating surgeon for further investigations.

Bradberry et al. (2014), who reviewed 28 cases of arthroprosthetic cobaltism, recommended the following criteria for determining if patients presenting with symptoms characteristic of systemic toxicity do indeed have issues related to excessive metal release:

1. Clinical history: MoM hip or revision to a metal-containing implant following ceramic hip fracture. The patient is presenting with localised pain and symptoms that align with what is known about the neurological, cardiac, or thyroid impacts of high metal release.
2. Clinical investigation: Investigations confirm that there are clear indicators of the neurological, cardiac, or thyroid impacts of high metal release, and there is no other apparent explanation for these to have occurred. Imaging or operative findings show evidence of high wear and metallosis.
3. Timing: The symptoms of systemic toxicity present between several months or years after implantation.
4. Cobalt levels: These are found to be significantly higher than in patients with well-functioning hip implants (no specific cut-off value given).
5. Changes post-revision: Following removal or revision of the metal component, there is a reduction in blood metal levels and symptomatic improvement.

### 8.3.3 Conclusion

There is no universally accepted threshold value in the measures of blood cobalt concentration to help guide surgeons when evaluating the risk of systemic toxicity in orthopaedic patients. This mainly stems from a lack of historical concern regarding the issue, and minimal data relating specific blood cobalt levels to different adverse effects. Systemic toxicity secondary to a metal hip implant is extremely rare and nearly always associated with very high circulating cobalt concentrations. Since the majority of reports of organ dysfunction involve patients with severe circulating blood metal levels, these patients are the focus of clinical attention. There is currently little evidence to support regular screening of asymptomatic patients for systemic toxicity.

## 8.4 Monitoring of patients with titanium-based hip implants

Titanium alloys are widely used in medical devices, including cementless THR constructs. The metal is renowned for its biocompatibility and corrosion resistance; however, studies have shown that patients with titanium-based implants can have measurably raised blood titanium levels (Gofton and Beaule, 2015; Swiatkowska et al., 2020; Vendittoli et al., 2010). Several authors have called for the measurement of circulating titanium levels as a biomarker of implant performance (Ho et al., 2018; Jacobs et al., 2004; Leopold et al., 2000; McAlister and Abdel, 2016). This is based on the observation that loose implants, as well as those exhibiting wear or polyethylene wear-through, generally liberate higher concentrations of the metal than do well-functioning implants (Grosse et al., 2014; Jacobs et al., 1991; McAlister and Abdel, 2016; Omlor et al., 2013; Quitmann et al., 2006; Stulberg et al., 1994).

### 8.4.1 Circulating titanium levels

Several groups have attempted to set reference values for circulating titanium levels. The Mayo Clinic Laboratory guidelines recommend that serum titanium of up to 3 μg/L indicates a prosthetic device in good condition, whereas concentrations exceeding 10 μg/L suggest implant wear. These values are based on two small-scale studies in which graphite-furnace atomic absorption spectroscopy was used for trace metal analysis (Bartolozzi and Black, 1985; Liu et al., 1998). Analyses using high-resolution inductively coupled plasma–mass spectrometry (ICP-MS) support that blood titanium levels in patients with well-functioning hip implants are generally below 3 μg/L (Yao et al., 2020). A recent study of 95 patients with well-functioning ceramic-on-ceramic implants, who were matched for implant type and operated on by the same surgeon, concluded that the upper reference limits (95th percentile) for blood and plasma titanium were 2.20 and 2.56 μg/L, respectively (Swiatkowska et al., 2020). These findings should be an important platform for future studies to build on to further understand the utility of blood titanium testing to identify patients with problematic hips. Larger studies will be necessary to account for potential surgeon-, implant-, and patient-specific factors when interpreting the titanium levels measured.

The current guidelines from the MHRA relating to the measurement of cobalt and chromium concentrations recommend that these are determined in whole blood rather than in serum/plasma. It is unclear whether this advice is also applicable to titanium measurements. Swiatkowska et al. (2020) recommend analysis of whole blood when high-resolution ICP-MS instrumentation is used, because the additional sample preparation steps required when analysing serum samples increase the contamination risk, potentially leading to measurement error.

### 8.4.2 The decision to revise

The clinical value of regular monitoring of titanium levels in THA recipients will be realised if it can be used to identify and treat patients with malfunctioning implants before failure takes place. Similar to blood cobalt and chromium, titanium is likely an adjunct to clinical and radiological assessment rather than a biomarker to be interpreted in isolation. Management of symptomatic patients should continue in line with regulator guidance, focusing on imaging findings and Oxford Hip Scores. In symptomatic patients with prostheses containing titanium components and those made of CoCr alloy, it may be prudent to request measurement of all three metals, particularly in those with a known sensitivity to titanium. Although CoCr alloy components have a known risk of leading to adverse tissue reactions, in rare cases, wear and corrosion of titanium components can also produce local toxicity (McPherson et al., 2014; Sakamoto et al., 2016; Tarpada et al., 2020).

### 8.4.3 Conclusion

At the time of writing, there have been very few large-scale studies investigating systemic levels of titanium in patients with optimally functioning and problematic hip implants. Additionally, there is considerable variability amongst laboratories in the analytical approach used, meaning that the results derived from different centres may not be directly comparable. Compounding this issue is that instrumentation capable of reliably quantifying trace levels of titanium in biological samples is not available in most routine testing laboratories. With these factors considered, regulators are currently unable to define official guidelines for titanium level thresholds that may be used to inform clinical decision-making.

## 8.5 Investigation for periprosthetic infection

Periprosthetic joint infection (PJI) poses a serious and growing challenge for orthopaedic surgeons. Multiple investigative tests may be performed to inform the diagnosis of infection, and there is certainly no single strategy for this. As discussed in Chapter 6, blood and synovial biomarkers, including C-reactive protein (CRP), erythrocyte sedimentation rate (ESR), alpha-defensin, white blood cell (WBC) count, and polymorphonuclear neutrophil percentage (PMN%), can all play a role, with known advantages and limitations regarding their diagnostic accuracy and ease of implementation. The following paragraphs summarise a series of consensus statements for the diagnosis of PJI that were published jointly by the European Association of Nuclear Medicine (EANM), European Society of Radiology (ESR), European Bone and Joint Infection Society (EBJIS), and European Society of Clinical Microbiology and Infectious Diseases (ESCMID). The reader is advised that only statements directly relevant to the use of biomarkers have been selected; the full joint consensus document is available in the publication by Signore et al. (2019).

*Statement 1*

All patients with suspected PJI should undergo screening for inflammatory markers, including CRP and ESR. Although these tests have wide variability in their sensitivity and specificity, they are still considered clinically useful, with the advantage of being performed quickly and cheaply. For diagnostic purposes, a threshold of 10 mg/L for CRP and 30 mm/h for ESR is recommended. Clinicians should be mindful that normal values are not necessarily indicative of an absence of PJI as several factors, such as patient age and analytical technique, can impact the results obtained. Measuring both the CRP level and ESR is recommended, followed by further investigations if either of these markers is elevated.

*Statement 2*

Patients with elevated ESR and CRP levels should undergo joint aspiration. Joint fluid should be tested for WBC count and PMN%. WBC count above 3000 cells/μL and PMN% exceeding 70% are strong indicators of PJI.

*Statement 3*

Synovial fluid biomarkers alpha-defensin, leukocyte esterase, interleukin (IL)-6, and CRP are all useful in detecting the presence of PJI. Alpha-defensin in particular has been shown to have a very high sensitivity and specificity; however, the test is expensive to perform and not routinely available clinically.

The American Academy of Orthopaedic Surgeons (AAOS) has also published clinical practice guidelines for diagnosing and preventing PJI (AAOS, 2019). Specifically, they advised that there is strong clinical evidence to support measuring serum ESR, CRP, and IL-6 levels. The guidance cautions that CRP and ESR, which are most routinely used, are not always accurate in their diagnosis of PJI and may produce false-positive results in patients with concurrent inflammatory conditions. Therefore, the results should be used in conjunction with other diagnostic tests. The guidance adds that there is moderate evidence for the effectiveness of aerobic and anaerobic bacterial cultures, nucleic acid amplification testing, and the following synovial biomarkers: WBC count, PMN%, leukocyte esterase, alpha-defensin, and CRP.

The diagnostic criteria for PJI have evolved over the past decade, and at least six different definitions of PJI have been proposed, underlining the diagnostic challenge posed by the condition (see Chapter 6) (McNally et al.,

2021; Osmon et al., 2013; Parvizi et al., 2011, 2013, 2018; Romanò et al., 2019). The definitions provide reference standards for clinical and diagnostic studies, but the issue remains that no single definition of PJI has reached widespread acceptance amongst orthopaedic societies and clinicians as the 'gold standard'. There are many factors at play here, namely the clinical complexity, geographical variation, cost-effectiveness of tests, and disagreements about the sensitivity and specificity of various biomarkers. There is also the ongoing concern that not all cases of suspected or low-grade infection fit within a binary definition of PJI (infected or not infected). To address the latter issue, the most recent guidance from the EBJIS, MSIS, and ESCMID features a three-level traffic-light system that classifies suspected PJI as confirmed, likely, or unlikely (Table 8.3). The criteria consider the significance of each test when making the diagnosis, where in order for a single positive test to diagnose PJI, the specificity of the test must be high (generally >90%). Importantly, patients with a suspected infection may have some positive diagnostic tests (imaging, blood biomarkers, etc.), which are associated with infection but are not specific enough to confirm it. These patients are those with potential low-grade infections, which do not fit into the traditional binary definitions published by the Musculoskeletal Infection Society (MSIS) and International Consensus Meeting on Musculoskeletal Infection (ICM) (Parvizi et al., 2011, 2013, 2018). In the suspected infection group, multiple positive suggestive tests do not confirm infection as in the other definitions. Instead, a positive test from the confirmatory tests is required before the infection is confirmed.

### 8.5.1 Conclusion

The burden of PJI is rising and there is a need for accurate diagnosis and evidence-based management. Over the past decade, considerable progress has been made to standardise and improve the diagnosis of PJI, but there is no 'gold standard' definition or a single test with absolute diagnostic accuracy. Combination tests may be needed to improve the performance of currently available biomarkers (Abdelbary et al., 2020). Due to the complexity of treating PJI, MDT working and delivery of care in specialist units are important (Kalson et al., 2020).

## 8.6 Summary and future directions

Hip replacement surgery is highly successful in improving joint function and quality of life of patients with end-stage hip disease. However, the introduction of MoM hip implants and, in most cases, their subsequent withdrawal, has highlighted the importance of a robust strategy for clinical follow-up of new devices. Over the past decade, the role of orthopaedic registries and

**TABLE 8.3 The 2020 EBJIS definition of periprosthetic joint infection.**

| | |
|---|---|
| RED | 1. Sinus tract with evidence of communication to the joint or visualisation of the prosthesis<br>2. Increased synovial leukocyte count (>3000 cells/μL)[a,b]<br>3. Elevated PMN% (>80%)[a,b]<br>4. Positive synovial alpha-defensin immunoassay or lateral flow test[c]<br>5. >50 CFU/mL of any organism in sonication fluid[d,e]<br>6. ≥2 positive samples with the same microorganism in intraoperative fluid/tissue[e]<br>7. Positive histology (≥5 neutrophils/HPF in ≥5 HPFs)[a,f]<br>8. Presence of visible microorganisms in periprosthetic tissue[a,f] |
| ORANGE | 1. Radiological signs of loosening within the first 5 years after implantation OR previous wound healing problems OR history of recent fever or bacteraemia OR purulence around the prosthesis[g]<br>2. Elevated serum CRP (>10 mg/L)[a]<br>3. Elevated synovial WBC count (>1500 cells/μL)[a,b]<br>4. Elevated PMN% (>65%)[a,b]<br>5. Positive aspiration fluid culture[e]<br>6. Single positive culture[e,h]<br>7. >1 CFU/mL of any organism in sonication fluid[d,e,h]<br>8. Positive histology (≥5 neutrophils in a single HPF)[a,f]<br>9. Positive WBC scintigraphy[i] |
| GREEN | 1. Clear alternative reason for implant dysfunction (e.g., fracture, implant breakage, malposition, tumour)<br>2. Synovial WBC count ≤1500 and PMN% ≤65%[a,b]<br>3. All intraoperative cultures negative[e]<br>4. No growth in sonication fluid culture[d,e]<br>5. Negative histology[a,f]<br>6. Negative three-phase isotope scan[a] |
| PJI confirmed if ≥1 RED test positive | |
| PJI likely if there is a positive clinical feature, or elevated serum CRP, together with another ORANGE test; consider further comprehensive investigation | |
| PJI unlikely if all GREEN criteria met | |

*ALTR*, adverse local tissue reaction; *CFU*, colony-forming unit(s); *CRP*, C-reactive protein; *ESR*, erythrocyte sedimentation rate; *HFP*, high-power field; *LE*, leukocyte esterase; *PMN%*, synovial polymorphonuclear cell percentage; *WBC*, white blood cell.
[a] *Interpret with caution when other possible causes of inflammation are present, such as crystal deposition disease, metallosis, periprosthetic fracture, active inflammatory joint disease (e.g., rheumatoid arthritis), or early postoperative period.*
[b] *Parameters are only valid when clear fluid is obtained and no lavage has been performed. Volume for the analysis should be >250 μL, ideally 1 mL, collected in an EDTA-containing tube and analysed in <1 hour, preferentially using automated techniques. For viscous samples, pre-treatment with hyaluronidase improves the accuracy of optical or automated techniques. In case of bloody samples, the adjusted synovial WBC = synovial WBC observed − [WBC blood/RBC blood × RBC synovial fluid] should be used.*
[c] *Not valid in cases of ALTR, haematomas, acute inflammatory disease, or gout.*
[d] *If centrifugation is applied then the suggested cut-off is 200 CFU/mL to confirm infection—if other variations to the protocol are used then the published cut-offs for each protocol must be applied.*

*(Continued)*

**TABLE 8.3 (Continued)**

[e]*Results of microbiological analysis may be compromised by prior antibiotic treatment (not simple prophylaxis). In these cases, molecular techniques may have a place. Results of culture may be obtained from preoperative synovial aspiration, preoperative synovial biopsies or (preferred) from intraoperative tissue samples.*
[f]*Histological analysis can be from preoperative biopsy or intraoperative tissue samples, with either paraffin or frozen section preparation.*
[g]*Except in ALTR or crystal arthropathy cases.*
[h]*Interpretation of single positive culture (or <50 CFU/mL in sonication fluid) must be cautious and taken together with other evidence. If a preoperative aspiration identified the same microorganism, they should be considered as two positive confirmatory samples. Uncommon contaminants or virulent organisms (e.g., S. aureus or gram-negative rods) are more likely to represent infection than common contaminants (such as coagulase-negative staphylococci, micrococci, or C. acnes).*
[i]*Uptake at the 20-hour scan is increased compared with the earlier scans (especially when combined with complementary bone marrow scan).*
*Source*: Based on McNally et al. (2021): McNally, M., Sousa, R., Wouthuyzen-Bakker, M., Chen, A., Soriano, A., Vogely, H., et al., 2021. The EBJIS definition of periprosthetic joint infection. Bone Joint J. 103-B, 18–25.

other bodies for post-market surveillance (such as ODEP and Beyond Compliance) has expanded.

In most cases, a failing implant is identified based on the development of symptoms from the joint replacement. However, in a very small proportion of patients, a failing hip implant may be minimally symptomatic. For these patients, routine clinical assessment, patient-reported outcome measure instruments (and other symptom questionnaires), plain radiographs and cross-sectional imaging, and laboratory biomarkers may allow earlier detection of a malfunctioning device.

This chapter has summarised the common modes of hip implant failure. MoM hip implants rapidly gained popularity for their perceived benefits, including improved wear characteristics, increased stability afforded through large-diameter femoral bearings, and bone conservation. However, cobalt and chromium wear particles were subsequently deemed toxic to the local tissues and, in very rare cases, to the vital body organs and systems. Raised blood metal levels were quickly established as indicators of poor performance of MoM hip implants and were adopted into regulatory guidance for the monitoring of patients. Debate continues regarding the true sensitivity and specificity of blood cobalt and chromium concentrations to differentiate between optimally functioning and poorly functioning devices and estimate the risk of ALTR. Further research is necessary to refine thresholds for concern and revision surgery, particularly in patients with bilateral implants, and establish whether action levels should differ between implant designs (Matharu et al., 2015).

Greater adoption of titanium-based uncemented hip implants raises the need to assess for potential toxicity related to the metal and evaluate whether circulating levels of titanium have value as a marker of implant wear and function. While titanium levels can be reliably measured in biological samples, the necessary instrumentation is not routinely available, and the variable analytical approaches used in different testing laboratories hinder meaningful inter-study comparisons. As a result of these factors, combined with the widely held view that titanium implants are not associated with significant toxicity, measurement of blood titanium is not commonplace in clinical practice.

Systemic toxicity secondary to metal hip implants appears to be extremely rare and nearly always associated with severely elevated circulating cobalt levels. For patients manifesting these symptoms, early clinical detection and management are crucial. On a population level, there does not appear to be an increased risk of cancer, cardiotoxicity, or neurotoxicity at early follow-up (Sabah et al., 2018; Smith et al., 2012; Swiatkowska, 2020).

Periprosthetic osteolysis and aseptic implant loosening were historically assumed to only occur in response to polyethylene wear debris; however, it is now clear that they can also occur in response to the failure of osseointegration, metal toxicity, unfavourable biomechanics, and infection. As yet, there is no routine use of specific molecular biomarkers to indicate periprosthetic osteolysis, and physicians still rely on imaging techniques to assess the fixation status of the prosthesis. Unfortunately, radiographic evaluation often fails to identify loosening until late in the disease process, after substantial peri-implant bone loss has already occurred. Various urinary, blood, and synovial biomarkers of bone formation/resorption and inflammation have been investigated and some show promise, such as crosslinked N-terminal telopeptide, crosslinked C-terminal telopeptide of type I collagen, deoxypyridinoline, TRAP 5b, and IL-6 (Sumner et al., 2014); however, many are marred by poor specificity since they can be elevated in other conditions that lead to inflammation or high bone turnover. The future of osteolysis biomarkers will likely involve the use of biomarker panels, which may be able to identify at-risk patients with higher accuracy than any individual test (He et al., 2013; Ross et al., 2018). Long-term prospective studies are required to validate the clinical utility of molecular biomarkers to detect impending implant failure, i.e., before clinical symptoms and radiographic evidence of periprosthetic osteolysis emerge.

PJI is a devastating complication of hip replacement and a major topic of discussion amongst orthopaedic communities. Much time and energy has been invested into understanding and diagnosing the condition, with particular attention paid to developing new laboratory biomarkers and repurposing existing ones. Although several parameters of blood and synovial fluid have been included within international definitions of PJI, most come with significant practical limitations such as expense and availability, and none have the specificity required to be truly diagnostic. Work continues to develop a test, or a combination of tests, that couples excellent diagnostic accuracy with reliability, cost-effectiveness, and ease of use.

The understanding of molecular and imaging biomarkers of implant function continues to evolve, and in the future, we can expect better long-term surveillance of orthopaedic implants.

## References

AAOS, 2019. Diagnosis and prevention of periprosthetic joint infections clinical practice guideline, American Academy of Orthopaedic Surgeons Board.

Abdelbary, H., Cheng, W., Ahmadzai, N., Carli, A.V., Shea, B.J., Hutton, B., et al., 2020. Combination tests in the diagnosis of chronic periprosthetic joint infection: systematic review and development of a stepwise clinical decision-making tool. J. Bone Joint Surg. Am. 102, 114–124.

Bartolozzi, A., Black, J., 1985. Chromium concentrations in serum, blood clot and urine from patients following total hip arthroplasty. Biomaterials 6, 2–8.

Berber, R., Abdel-Gadir, A., Rosmini, S., Captur, G., Nordin, S., Culotta, V., et al., 2017. Assessing for cardiotoxicity from metal-on-metal hip implants with advanced multimodality imaging techniques. J. Bone Joint Surg. Am. 99, 1827–1835.

Berber, R., Pappas, Y., Khoo, M., Miles, J., Carrington, R., Skinner, J., et al., 2015. New approach to managing patients with problematic metal hip implants: the use of an internet-enhanced multidisciplinary team meeting. AAOS exhibit selection. J. Bone Joint Surg. Am. 97 (4), e:42.

Berber, R., Skinner, J.A., Hart, A.J., 2016. Management of metal-on-metal hip implant patients: Who, when and how to revise? World J. Orthop. 7, 272–279.

Bradberry, S.M., Wilkinson, J.M., Ferner, R.E., 2014. Systemic toxicity related to metal hip prostheses. Clin. Toxicol. 52, 837–847.

EFORT, 2012. Consensus statement current evidence on the management of metal-on-metal bearings. Hip Int. 23, 2–5.

FDA, 2019. Information for Orthopaedic Surgeons: general recommendations for metal hip resurfacing surgery.

Finley, B.L., Unice, K.M., Kerger, B.D., Otani, J.M., Paustenbach, D.J., Galbraith, D.A., et al., 2013. 31-day study of cobalt(II) chloride ingestion in humans: pharmacokinetics and clinical effects. J. Toxicol. Environ. Heal. A 76, 1210–1224.

Galea, V.P., Laaksonen, I., Connelly, J.W., Matuszak, S.J., Nortje, M., Madanat, R., et al., 2018. What is the clinical presentation of adverse local tissue reaction in metal-on-metal hip arthroplasty? An MRI study. Clin. Orthop. Relat. Res. 477, 353–360.

Gofton, W., Beaule, P.E., 2015. Serum metal ions with a titanium modular neck total hip replacement system. J. Arthroplasty 30, 1781–1786.

Goodnough, L.H., Bala, A., Huddleston III, J.I., Goodman, S.B., Maloney, W.J., Amanatullah, D.F., 2018. Metal-on-metal total hip arthroplasty is not associated with cardiac disease. Bone Joint J. 100-B, 28–32.

Grassin-Delyle, S., Martin, M., Hamzaoui, O., Lamy, E., Jayle, C., Sage, E., et al., 2019. A high-resolution ICP-MS method for the determination of 38 inorganic elements in human whole blood, urine, hair and tissues after microwave digestion. Talanta 199, 228–237.

Grosse, S., Haugland, H.K., Lilleng, P., Ellison, P., Hallan, G., Johan, P., 2014. Wear particles and ions from cemented and uncemented titanium-based hip prostheses—a histological and chemical analysis of retrieval material. J. Biomed. Mater. Res. B Appl. Biomater. 103, 709–717.

Hannemann, F., Hartmann, A., Schmitt, J., Lützner, J., Seidler, A., Campbell, P., et al., 2013. European multidisciplinary consensus statement on the use and monitoring of metal-on-metal bearings for total hip replacement and hip resurfacing. Orthop. Traumatol. Surg. Res. 99, 263–271.

Hart, A.J., Sabah, S.A., Bandi, A.S., Maggiore, P., Tarassoli, P., Sampson, B., et al., 2011. Sensitivity and specificity of blood cobalt and chromium metal ions for predicting failure of metal-on-metal hip replacement. Bone Joint J. 93-B, 1308–1313.

Hart, A.J., Sabah, S.A., Sampson, B., Skinner, J.A., Powell, J.J., Palla, L., et al., 2014. Surveillance of patients with metal-on-metal hip resurfacing and total hip prostheses: a prospective cohort study to investigate the relationship between blood metal ion levels and implant failure. J. Bone Joint Surg. Am. 96, 1091–1099.

He, T., Wu, W., Huang, Y., Zhang, X., Tang, T., Dai, K., 2013. Multiple biomarkers analysis for the early detection of prosthetic aseptic loosening of hip arthroplasty. Int. Orthop. 37, 1025–1031.

Health Canada, 2012. Metal-on-metal hip implants - information for orthopaedic surgeons regarding patient management following surgery - for health professionals. Available from: https://recalls-rappels.canada.ca/en/alert-recall/metal-metal-hip-implants-information-orthopaedic-surgeons-regarding-patient-0 (2012).

Ho, J., Mahajan, J., Taylor, M., Byers, A., Arauz, P., Kwon, Y.M., 2018. Metallosis in cemented titanium alloy total knee arthroplasty without apparent metal-on-metal articulation. Knee 25, 728–731.

Jacobs, J., Skipor, A., Black, J., Urban, R., Galante, J., 1991. Release and excretion of metal in patients who have a total hip-replacement component made of titanium-base alloy. J. Bone Joint Surg. 73-A, 1475–1486.

Jacobs, J.J., Skipor, A.K., Campbell, P.A., Hallab, N.J., Urban, R.M., Amstutz, H.C., 2004. Can metal levels be used to monitor metal-on-metal hip arthroplasties? J. Arthroplasty 19, 59–65.

Kalson, N.S., Mathews, J.A., Alvand, A., Morgan-Jones, R., Jenkins, N., Phillips, J.R.A., et al., 2020. Investigation and management of prosthetic joint infection in knee replacement: a BASK surgical practice guideline. Knee 27, 1857–1865.

Kurtz, S., Ong, K., Lau, E., Mowat, F., Halpern, M., 2007. Projections of primary and revision hip and knee arthroplasty in the United States from 2005 to 2030. J. Bone Joint Surg. Am. 89, 780–785.

Lenguerrand, E., Whitehouse, M.R., Beswick, A.D., Toms, A.D., Porter, M.L., Blom, A.W., 2017. Description of the rates, trends and surgical burden associated with revision for prosthetic joint infection following primary and revision knee replacements in England and Wales: an analysis of the National Joint Registry for England, Wales, Northern Ireland and the Isle of Man. BMJ Open 7, 1–9.

Leopold, S.S., Berger, R.A., Patterson, L., Skipor, A.K., Urban, R.M., Jacobs, J.J., 2000. Case report serum titanium level for diagnosis of a failed, metal-backed patellar component. J. Arthroplasty 15, 938–943.

Liu, T.-K., Liu, S.-H., Chang, C.-H., Yang, R.-S., 1998. Concentration of metal elements in the blood and urine in the patients with cementless total knee arthroplasty. Tohoku. J. Exp. Med. 185, 253–262.

Manisty, C., Skinner, J., Moon, J.C., 2018. Metal-on-metal hips and heart failure – can we relax? Int. J. Cardiol. 284, 65–66.

Maradit Kremers, H., Larson, D.R., Crowson, C.S., Kremers, W.K., Washington, R.E., Steiner, C.A., et al., 2015. Prevalence of total hip and knee replacement in the United States. J. Bone Joint Surg. Am. 97, 1386–1397.

Matharu, G., Mellon, S., Murray, D., Pandit, H., 2015. Follow-up of metal-on-metal hip arthroplasty patients is currently not evidence based or cost effective. J. Arthroplasty 30, 1317–1323.

McAlister, I.P., Abdel, M.P., 2016. Elevated serum titanium level as a marker for failure in a titanium modular fluted tapered stem. Orthopedics 39, e768–e770.

McNally, M., Sousa, R., Wouthuyzen-Bakker, M., Chen, A., Soriano, A., Vogely, H., et al., 2021. The EBJIS definition of periprosthetic joint infection. Bone Joint. J. 103-B, 18–25.

McPherson, E.J., Dipane, M.V., Sherif, S.M., 2014. Massive pseudotumor in a 28mm ceramic-polyethylene revision THA: a case report. Reconstr. Rev. 4, 11–17.

Megaterio, S., Galetto, F., Alossa, E., Capretto, S., 2001. Effetti a distanza del rilascio di ioni metallo in usura della testa protesica: presentazione di un caso [Systemic effects of ionic release in wear of prosthetic head: a case report]. G.I.O.T. 27, 173–175.

MHRA, 2017. All metal-on-metal (MoM) hip replacements: updated advice for follow-up of patients.

Moran, C.A., Mullick, F.G., Ishak, K.G., Johnson, F.B., Hummer, W.B., 1991. Identification of titanium in human tissues: probable role in pathologic processes. Hum. Pathol. 22, 450–454.

National Joint Registry. 16th Annual Report, 2019.

Omlor, G.W., Kretzer, J.P., Reinders, J., Streit, M.R., Bruckner, T., Gotterbarm, T., et al., 2013. In vivo serum titanium ion levels following modular neck total hip arthroplasty-10 year results in 67 patients. Acta Biomater. 9, 6278–6282.

Osmon, D.R., Berbari, E.F., Berendt, A.R., Lew, D., Zimmerli, W., Steckelberg, J.M., et al., 2013. Diagnosis and management of prosthetic joint infection: clinical practice guidelines by the infectious diseases society of America. Clin. Infect. Dis. 56, e1–e25.

Parvizi, J., Gehrke, T., Chen, A.F., 2013. Proceedings of the international consensus on periprosthetic joint infection. Bone Joint J. 95B, 1450–1452.

Parvizi, J., Tan, T.L., Goswami, K., Higuera, C., Della Valle, C., Chen, A.F., et al., 2018. The 2018 definition of periprosthetic hip and knee infection: an evidence-based and validated criteria. J. Arthroplasty 33, 1309–1314.

Parvizi, J., Zmistowski, B., Berbari, E.F., Bauer, T.W., Springer, B.D., Della Valle, C.J., et al., 2011. New definition for periprosthetic joint infection: from the workgroup of the musculoskeletal infection society. Clin. Orthop. 469, 2992–2994.

Paustenbach, D.J., Galbraith, D.A., Finley, B.L., 2014. Interpreting cobalt blood concentrations in hip implant patients. Clin. Toxicol. 52, 98–112.

Paustenbach, D.J., Tvermoes, B.E., Unice, K.M., Finley, B.L., Kerger, B.D., 2013. A review of the health hazards posed by cobalt. Crit. Rev. Toxicol. 43, 316–362.

Prentice, J.R., Clark, M.J., Hoggard, N., Morton, A.C., Tooth, C., Paley, M.N., et al., 2013. Metal-on-metal hip prostheses and systemic health: a cross-sectional association study 8 years after implantation. PLoS One 8, 1–9.

Quitmann, H., Wedemeyer, C., Von Knoch, M., Russe, K., Saxler, G., 2006. Titanium serum levels may remain elevated despite hip revision surgery for wear-through of an acetabular component. Biomed. Tech. 51, 27–29.

Romanò, C., Khawashki, H., Benzakour, T., Bozhkova, S., Del Sel, H., Hafez, M., et al., 2019. The W.A.I.O.T. definition of high-grade and low-grade peri-prosthetic joint infection. J. Clin. Med. 8, 650.

Ross, R., Deng, Y., Fang, R., Frisch, N., Jacobs, J., Sumner, D., 2018. Discovery of biomarkers to identify peri-implant osteolysis before radiographic diagnosis. J. Orthop. Res. 36, 2754–2761.

Sabah, S.A., Moon, J.C., Jenkins-Jones, S., Morgan, C.L., Currie, C.J., Wilkinson, J.M., et al., 2018. The risk of cardiac failure following metal-on-metal hip arthroplasty. Bone Joint J. 100-B, 20–27.

Sakamoto, M., Watanabe, H., Higashi, H., Kubosawa, H., 2016. Pseudotumor caused by titanium particles from a total hip prosthesis. Orthopedics 39, e162–e165.

Signore, A., et al., 2019. Consensus document for the diagnosis of prosthetic joint infections: a joint paper by the EANM, EBJIS, and ESR (with ESCMID endorsement). Eur. J. Nucl. Med. Mol. Imaging 46, 971–988.

Smith, A.J., Dieppe, P., Porter, M., Blom, A.W., 2012. Risk of cancer in first seven years after metal-on-metal hip replacement compared with other bearings and general population: linkage study between the National Joint Registry of England and Wales and hospital episode statistics. BMJ 344, e2383.

Stulberg, B.N., Merritt, K., Bauer, T.W., 1994. Metallic wear debris in metal-backed patellar failure. J. Appl. Biomater. 5, 9–16.

Sumner, D.R., Ross, R., Purdue, E., 2014. Are there biological markers for wear or corrosion? A systematic review. Clin. Orthop. Relat. Res. 472, 3728–3739.

Swiatkowska, I., 2020. Self-reported neurotoxic symptoms in hip arthroplasty patients with highly elevated blood cobalt: a case-control study. J. Patient Saf. 18, e10–e17.

Swiatkowska, I., Martin, N.G., Henckel, J., Apthorp, H., Hamshere, J., Hart, A.J., 2020. Blood and plasma titanium levels associated with well-functioning hip implants. J. Trace Elem. Med. Biol. 57, 9–17.

Tarpada, S.P., Loloi, J., Schwechter, E.M., 2020. A case of titanium pseudotumor and systemic toxicity after total hip arthroplasty polyethylene failure. Arthroplast. Today 6, 710–715.

TGA, 2017. Metal-on-metal hip replacement implants. Available from: https://www.tga.gov.au/resources/resource/guidance/metal-metal-hip-replacement-implants (2017).

Thomas, P., Bandl, W., Maier, S., Summer, B., Przybilla, B., 2006. Hypersensitivity to titanium osteosynthesis with impaired fracture healing, eczema, and T-cell hyperresponsiveness in vitro: case report and review of the literature. Contact Dermat. 55, 199–202.

Tower, S., 2010a. Arthroprosthetic cobaltism: identification of the at-risk patient. Alaska. Med. 52, 28–32.

Tower, S., 2010b. Cobalt toxicity in two hip replacement patients. State Alaska Epidemiol. Bull. 14, 1.

Tvermoes, B.E., Finley, B.L., Unice, K.M., Otani, J.M., Paustenbach, D.J., Galbraith, D.A., 2013. Cobalt whole blood concentrations in healthy adult male volunteers following two-weeks of ingesting a cobalt supplement. Food Chem. Toxicol. 53, 417–424.

Tvermoes, B.E., Unice, K.M., Paustenbach, D.J., Finley, B.L., Otani, J.M., Galbraith, D.A., 2014. Effects and blood concentrations of cobalt after ingestion of 1 mg/d by human volunteers for 90 d. Am. J. Nutr. 99, 632–646.

Unice, K.M., Kerger, B.D., Paustenbach, D.J., Finley, B.L., Tvermoes, B.E., 2014. Refined biokinetic model for humans exposed to cobalt dietary supplements and other sources of systemic cobalt exposure. Chem. Biol. Interact. 216, 53–74.

Unice, K.M., Monnot, A.D., Gaffney, S.H., Tvermoes, B.E., Thuett, K.A., Paustenbach, D.J., et al., 2012. Inorganic cobalt supplementation: prediction of cobalt levels in whole blood and urine using a biokinetic model. Food Chem. Toxicol. 50, 2456–2461.

Van Der Straeten, C., Grammatopoulos, G., Gill, H.S., Calistri, A., Campbell, P., De Smet, K. A., 2013. The 2012 Otto Aufranc Award: the interpretation of metal ion levels in unilateral and bilateral hip resurfacing. Clin. Orthop. Relat. Res. 471, 377–385.

Vendittoli, P.-A., Roy, A., Mottard, S., Girard, J., Lusignan, D., Lavigne, M., 2010. Metal ion release from bearing wear and corrosion with 28 mm and large-diameter metal-on-metal bearing articulations: a follow-up study. J. Bone Joint Surg. Br. 92-B, 12–19.

Wretenberg, P., 2008. Good function but very high concentrations of cobalt and chromium ions in blood 37 years after metal-on-metal total hip arthroplasy. Med. Devices Evid. Res. 1, 31–32.

Yao, J.J., Lewallen, E.A., Thaler, R., Dudakovic, A., Wermers, M., Day, P., et al., 2020. Challenges in the measurement and interpretation of serum titanium concentrations. Biol. Trace Elem. Res. 196, 20–26.

# Index

*Note*: Page numbers followed by "*f*" and "*t*" refer to figures and tables, respectively.

## I

## R



## S

MIX
Papier | Fördert gute Waldnutzung
FSC® C083411

Kolibri 360 GmbH
Ferdinand-Jühlke-Straße 7
99095 Erfurt, Deutschland
produktsicherheit@kolibri360.de